Almut Pitscheider

Rare-Earth Oxo- and Fluoride-Containing Borates at Extreme Conditions

Almut Pitscheider

Rare-Earth Oxo- and Fluoride-Containing Borates at Extreme Conditions

Synthesis and Investigation of Metastable High-Pressure Phases

Südwestdeutscher Verlag für Hochschulschriften

Impressum/Imprint (nur für Deutschland/only for Germany)
Bibliografische Information der Deutschen Nationalbibliothek: Die Deutsche Nationalbibliothek verzeichnet diese Publikation in der Deutschen Nationalbibliografie; detaillierte bibliografische Daten sind im Internet über http://dnb.d-nb.de abrufbar.
Alle in diesem Buch genannten Marken und Produktnamen unterliegen warenzeichen-, marken- oder patentrechtlichem Schutz bzw. sind Warenzeichen oder eingetragene Warenzeichen der jeweiligen Inhaber. Die Wiedergabe von Marken, Produktnamen, Gebrauchsnamen, Handelsnamen, Warenbezeichnungen u.s.w. in diesem Werk berechtigt auch ohne besondere Kennzeichnung nicht zu der Annahme, dass solche Namen im Sinne der Warenzeichen- und Markenschutzgesetzgebung als frei zu betrachten wären und daher von jedermann benutzt werden dürften.

Verlag: Südwestdeutscher Verlag für Hochschulschriften GmbH & Co. KG
Dudweiler Landstr. 99, 66123 Saarbrücken, Deutschland
Telefon +49 681 37 20 271-1, Telefax +49 681 37 20 271-0
Email: info@svh-verlag.de

Approved by: Insbruck, Leopold-Franzens-Universität, Dissertation, 2008

Herstellung in Deutschland:
Schaltungsdienst Lange o.H.G., Berlin
Books on Demand GmbH, Norderstedt
Reha GmbH, Saarbrücken
Amazon Distribution GmbH, Leipzig
ISBN: 978-3-8381-2885-6

Imprint (only for USA, GB)
Bibliographic information published by the Deutsche Nationalbibliothek: The Deutsche Nationalbibliothek lists this publication in the Deutsche Nationalbibliografie; detailed bibliographic data are available in the Internet at http://dnb.d-nb.de.
Any brand names and product names mentioned in this book are subject to trademark, brand or patent protection and are trademarks or registered trademarks of their respective holders. The use of brand names, product names, common names, trade names, product descriptions etc. even without a particular marking in this works is in no way to be construed to mean that such names may be regarded as unrestricted in respect of trademark and brand protection legislation and could thus be used by anyone.

Publisher: Südwestdeutscher Verlag für Hochschulschriften GmbH & Co. KG
Dudweiler Landstr. 99, 66123 Saarbrücken, Germany
Phone +49 681 37 20 271-1, Fax +49 681 37 20 271-0
Email: info@svh-verlag.de

Printed in the U.S.A.
Printed in the U.K. by (see last page)
ISBN: 978-3-8381-2885-6

Copyright © 2011 by the author and Südwestdeutscher Verlag für Hochschulschriften GmbH & Co. KG and licensors
All rights reserved. Saarbrücken 2011

Meinen Eltern

This thesis owes its existence to the help, support, and inspiration of many people. In the first place, I would like to express my sincere appreciation and gratitude to Univ.-Prof. H. Huppertz for his support and encouragement during this thesis's work and beyond. Your enthusiasm and the great working atmosphere contributed a great deal to the scientific results. I never regretted my decision to follow you to Innsbruck.

Special thanks go to Prof. W. Schnick from the LMU Munich for his continuous support of this work. During and after our movement to Innsbruck, many experiments for this thesis could be performed in his laboratories in Munich and numerous analytical measurements were possible due to his generosity.

My warm thanks go to PD Dr. O. Oeckler from the LMU Munich for his advice and support with crystallographic problems. Thank you for countless emails, measurements, and contributions for this work. Without your support, many results would still be in abeyance.

I am indebted to Dr. R. Kaindl and Dr. J. Konzett from the Institute of Mineralogy in Innsbruck, who patiently performed the IR, Raman, and microprobe measurements on my samples. Thank you for your contributions and your commitment.

Prof. R. Pöttgen, Dr. R.-D. Hoffmann, and T. Fickenscher from the University in Münster I want to thank for microprobe and magnetic measurements, as well as for crystallographic support and guidance.

My thanks go to Prof. R. Glaum and his staff from the University in Bonn for the performance of UV/Vis measurements and to Prof. J. Senker from the University in Bayreuth for his efforts concerning the solid state NMR measurements.

Special thanks go to C. Minke, T. Miller, and Dr. P. Meyer from the LMU Munich for numerous other measurements on my samples.

Our technician F. Höpperger I want to thank for the manufacture of countless assembly parts. We never had such well-fitting parts before and it made the work a great deal easier.

I want to thank my research students B. Seyr, S. Flür, S. Grutsch, and M. Mieth for a lot of laboratory work for this thesis. Special thanks go to M. Enders who came up with many results during his Master's thesis.

Many thanks go to the Huppertz group Dr. G. Heymann, S. Neumair, S. Herdlicka, and A. Siedler. Thank you for the nice working atmosphere, good cooperations, and a fun time at events like the Faschingsvorlesung!

A big thanks goes to all the scientific and non-scientific colleagues of the present and of the past for helpful hands and thoughts, fun events, and the friendly atmosphere. I fear to forget someone, so I don't mention everyone personally – I hope you know that this is for you! In this context, I would also like to thank my past colleagues and dear friends at the LMU Munich who make me feel most welcome whenever I stop by.

Special thanks go to Evelyne Manigand for helping me out with bureaucracy and for always having a sympathetic ear. You spared me a lot!

My dear colleague Gunter Heymann played a big part in making my time in Innsbruck a great one. Thank you for continuous advice, discussions, and support whenever needed, and a pleasant time as roommates. Without you and your little daughter, working in the "Kinderzimmer" would have been so much less fun!

Deep thanks go to Marion Bauer for being a wonderful friend. Thank you for delicious meals in the lunch break, special parties at your new home, and the great time spent together. I will miss you!

I owe my deepest gratitude to my husband Max Pitscheider. Thank you for everything that you are and do for me. You are my bridge over troubled water, endure all my moods and fears, and support me, whatever I do. I love you more than I can express.

My warm thanks go to my "second mom" Petra Pitscheider for her care and for making me feel at home in my new family. Thank you for giving your son to me.

Above all I want to thank my parents for their love and support throughout my life. You gave me the power and strength to be who I am and to achieve what I did. Thank you – I love you!

*The scientist is not a person who gives the right answers,
he's one who asks the right questions.*

Claude Lévi-Strauss

Contents

1 **Introduction** 1
   1.1 High Pressure in the Laboratory 1
   1.2 Materials Under Pressure 5
   1.3 High-Pressure Chemistry of Borates 10

2 **Experimental Methods** 19
   2.1 Glove Box 19
   2.2 High-Pressure Synthesis 20
   2.3 The Walker-Type Module 21
   2.4 The Multianvil Device 23
   2.5 Sample Preparation 28
   2.6 Calibration of the Multianvil Device 36
      2.6.1 Pressure Calibration 36
      2.6.2 Temperature Calibration 40
   2.7 Recovery of the Sample 41
   2.8 Experimental Danger 42

3 **Analytical Methods** 43
   3.1 X-Ray Diffraction Methods 43
      3.1.1 Basic Principles of X-Ray Diffraction 43
      3.1.2 X-Ray Powder Diffraction 46
      3.1.3 Temperature-Programmed X-Ray Powder Diffraction 46
      3.1.4 Single Crystal X-Ray Diffraction 47
      3.1.5 Computer Programs for X-Ray Diffraction Data 48
   3.2 Spectroscopic Methods 48
      3.2.1 Infrared Spectroscopy 49
      3.2.2 Raman Spectroscopy 50
      3.2.3 UV/Vis Spectroscopy 51
   3.3 Elemental Analysis 51
   3.4 Theoretical Calculations 52
      3.4.1 Lattice Energy Calculations According to the MAPLE Concept 52
      3.4.2 Bond-Length Bond-Strength Calculations 53
      3.4.3 Charge Distribution Calculations with the CHARDI Concept 54

4 **Experimental Part** 57

- 4.1 Rare-Earth Borates 57
  - 4.1.1 Introduction 57
  - 4.1.2 $Ho_{31}O_{27}(BO_3)_3(BO_4)_6$ 59
  - 4.1.3 λ-$PrBO_3$ 71
  - 4.1.4 π-$ErBO_3$ 79
- 4.2 Rare-Earth Fluoride and Fluorido Borates 85
  - 4.2.1 $Yb_5(BO_3)_2F_9$ 89
  - 4.2.2 $Er_5(BO_3)_2F_9$ 99
  - 4.2.3 $Tm_5(BO_3)_2F_9$ 107
  - 4.2.4 $Gd_4B_4O_{11}F_2$ 115
  - 4.2.5 $RE_4B_4O_{11}F_2$ (RE = Eu, Dy) 125
  - 4.2.6 $La_4B_4O_{11}F_2$ 133
  - 4.2.7 $Pr_4B_3O_{10}F$ 145
  - 4.2.8 "$RE_5(BO_{3.66}F_{0.34})_3F$" ($RE$ = Gd, Yb) 153
  - 4.2.9 $RE_3(BO_3)_2F_3$ (RE = Gd, Dy) 163
  - 4.2.10 $RE_{12}B_{11}O_{31}F_7$ (RE = La, Pr, Nd, Sm) 181
- 4.3 Rare-Earth Oxides 187

5 Prospects 193
6 Summary 197
- 6.1 High-Pressure / High-Temperature Synthesis 197
- 6.2 Rare-Earth Oxoborates 197
- 6.3 Rare-Earth Fluoride and Fluorido Borates 200
- 6.4 Rare-Earth Oxides 205

7 Appendix 207
- 7.1 List of Abbreviations and Special Characters 207
- 7.2 Awards 208
- 7.3 Publications 208
  - 7.3.1 Conference Contributions 208
  - 7.3.2 Papers 209
- 7.4 CSD Numbers 211
- 7.5 References 213

1 Introduction

1.1 High Pressure in the Laboratory

Classical solid-state reactions are carried out by heating a mixture of two or more solid components in order to surpass the usually high activation energy. After a distortion or destruction of the crystal structure of the educts, the formation of a new component becomes energetically more favorable and a new product can be obtained. Besides the mere increase of temperature, an additional variation of the parameter pressure can be used in order to affect a solid-state reaction. While we are generally working in laboratories at a constant ambient pressure of about 1 atm, natural chemical processes in the Earth's interior or in outer space take place under extreme pressure conditions, as depicted in Figure 1.1-1.

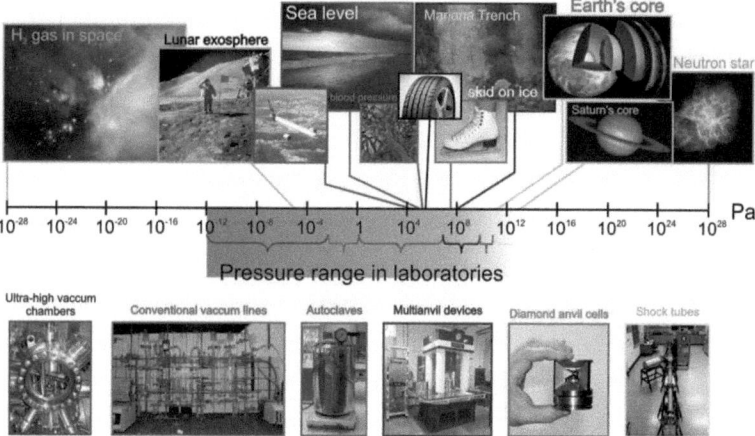

Figure 1.1-1
Pressure scale depicting the pressure range applicable in modern laboratories

Using vacuum pumps or pressure devices, it is possible to greatly extend the pressure range under which experiments can be carried out in the laboratory (Figure 1.1-1, bottom). Since about 90 % of the matter in the universe is exposed to pressures exceeding 10 GPa, scientific research under extreme pressure conditions is crucial to understand chemical and physical processes taking place in the earth's interior or in outer space. In contrast to nature, the realization of extreme pressures in the laboratory is very challenging, technically and physically.

Autoclaves are the most common laboratory devices to apply higher-than-ambient pressures. Functioning like a pressure cooker, pressures up to 2 GPa are generated for sterilization purposes, solvothermal syntheses, vulcanization processes, supercritical fluids research, organic reactions and other applications [1-3].

Higher pressures can be reached with piston cylinder apparatuses, consisting of a piston forcing into a cylinder and thus compressing the material placed into the furnace assembly. Developed by Boyd and England, piston cylinders reach pressures and temperatures up to 5 GPa and 1700 °C [4].

An increase of the sample size and pressure was realized with Hall's development of the belt module (Figure 1.1-2) [5]. Here, two conical carboloy pistons push from both sides into a specially shaped carboloy chamber. Both, chamber and pistons, receive lateral support from stressed binding rings. So far, the belt-module is one of the most used devices in preparative high-pressure research up to a pres-sure and temperature range of 10 GPa and 2000 °C [6, 7]. The first flat anvil apparatus was developed by Bridgman and compressed thin disks of solid materials with two anvils [8]. Even though this device could evoke pressures up to 10 GPa, the sample obtained was very small. Like the piston cylinder and belt

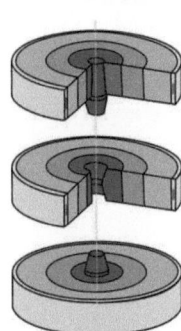

Figure 1.1-2
Belt module

apparatuses, the Bridgman anvil apparatus applies pressure only along one axis.

To obtain constant, steady pressure throughout the sample, it is desirable to expose the sample to more isotropic pressure. This is done by an increase of the number of anvils compressing the reaction chamber. Based on the Bridgman anvil apparatus, Hall developed the first multianvil device, making use of four tetrahedrally arranged anvils (Figure 1.1-3) [9]. The anvils were driven together by hydraulic rams along lines intersecting at tetrahedral angles in the center of a regular tetrahedral volume, enclosed by the anvils themselves. A pyrophyllite tetrahedron served as pressure medium, realizing pressures of 10 GPa and temperatures of 3000 °C.

Figure 1.1-3
Tetrahedral multi-anvil device

Further developments of the multianvil techniques include hexahedral devices [10, 11] and subsequently moved on to the currently used octahedral assemblies. The first octahedral-anvil device was developed by Kawai *et al.*, who used anvils segmented from a sphere, which surrounded an octahedral cavity [12]. In a double staging variant, two sets of anvils were used inside the sphere (Figure 1.1-4) [5]. Since the embedding of the sphere in oil was rather unhandy, the sphere surrounding the inner anvils was removed and replaced by rigid driving units. Those were made up of two identical components, set in opposition. As the two components approach each other, the inner anvils converge and on their part compress a sample placed in the center. In 1990, Walker *et al.* developed a multianvil module (called the "Walker-module") [13]. In Figure 1.1-5, this modern arrangement is depicted as it is

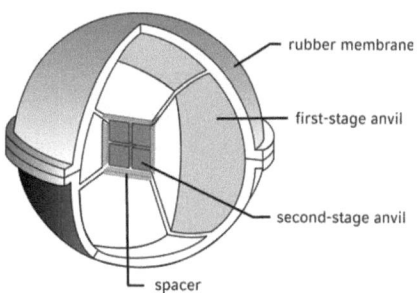

Figure 1.1-4
Double-staged spherical multi-anvil device

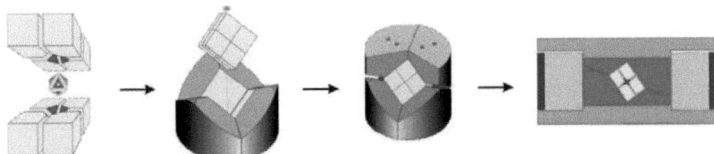

Figure 1.1-5
Principle of modern multianvil devices

used in the multianvil device at Innsbruck University. Further details about the assembly preparation are given in Sections 2.3 - 2.5. Octahedral multianvil devices are suitable for synthesis in a pressure range up to 25 GPa and temperatures up to 3000 K. More information can be found in a review from Onodera, who summarized various versions of octahedral-anvil and related devices for generating high pressures, with details of their design, construction, fabrication, operation, and calibration [14]. By substitution of the regularly used tungsten carbide anvils against smaller sintered diamond anvils, Ito *et al.* realized pressures higher than 35 GPa [15].

Much higher pressures (in special cases up to 550 GPa) can be reached using a diamond anvil cell (DAC) (Figure 1.1-6) [16, 17]. Here, the sample is compressed between two diamonds, which allows only sample amounts too small for practical applications. Nevertheless, the great advantage of this cell is its built-in optical window, which allows the real-time visual observation and a wide variety of *in situ* measurements (Raman and Brillouin scattering, infrared absorption, X-ray structure determination) through the diamonds. With use of high-power lasers, temperatures of almost 6700 K were reached [17].

Even higher pressures can be reached with dynamic pressure techniques (Figure 1.1-7). Here, one or mul-

Figure 1.1-6
Diamond anvil cell

tiple shock waves are sent through the sample by either detonating an explosive charge or allowing a hyper-velocity projectile to slam into a plate, on which the sample is mounted. The exact pressure obtainable *via* this method is difficult to measure. In 1996, Weir *et al.* performed shock wave experiments on liquid hydrogen and oxygen, generating pressures and temperatures in the range of 90-180 GPa and 2200-2400 K, respectively [18], postulating the formation of metallic hydrogen and oxygen at very low temperatures for pressures above 300 GPa. Even though the requirements

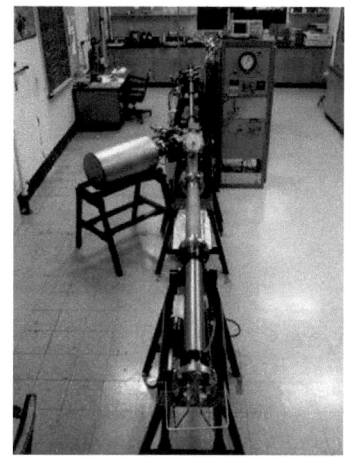

Figure 1.1-7
Shock tube at Dayton University

of space, equipment, and financial resources are enormous, the shock wave technique is the only way to reach pressures significantly above 500 GPa in combination with very high temperatures and large sample volumes.

1.2   Materials Under Pressure

The application of high pressures on chemical substances has an impact on their structural features. Chemical bonds inside of the material can either be changed, leading to structural transformations, or newly created between different precursors, resulting in the formation of new materials.

Pressure-induced structural transformations in materials occur due to compression and densification of the material. An increase of pressure leads to a shortening of the interatomic distances and results in an electrostatic repulsion between similarly charged ions. The current structure of the material becomes thus instable and unfavorable. A structural transformation takes place, often accompanied by an increase of the coordination numbers of atoms, leading to a lengthening of the chemical bonds. Thus, high-pressure phases often show higher coordinated atoms and larger bond lengths in denser crystal structures, than the corresponding ambient-pressure phases. These findings are summarized in the so-called pressure-coordination-rule [19].

The reaction of two different substances involves the formation of new chemical bonds between them. Under high-pressure conditions, the thermodynamically stable product can be different from the one that would be formed under ambient pressure conditions. This is also an effect of the densification, *e.g.* preferring the formation of solid phases. In agreement with "Le Chatelier's rule", denser structures are formed. Even though the thermodynamically favorable structures often differ between ambient and high-pressure conditions, it is still possible to handle high-pressure phases under ambient pressure conditions after the synthesis. Due to kinetic effects, most high-pressure phases are metastable. This means that they can be kept in their current structure below a specific retransformation temperature. Diamond, for example, is synthesized at 6 GPa and 1500 °C from graphite with a catalyst (Figure 1.2-1). A retransformation only occurs when diamond is heated, so that enough energy for a structural transformation is put into the system. Above 700 °C, a noticeable decrease of hardness occurs and graphitization starts, depending on the size of the diamonds. Recently, boron-doped diamond (containing about 3 % boron) moved into the focus. Due to their stability and their very wide potential window in aqueous solution, boron-doped diamond electrodes are excellent for anodic oxidation and are suitable for wastewater treatment [20]. Superconductivity for boron-doped diamond below 4 K was observed by Ekimov *et al.* [21].

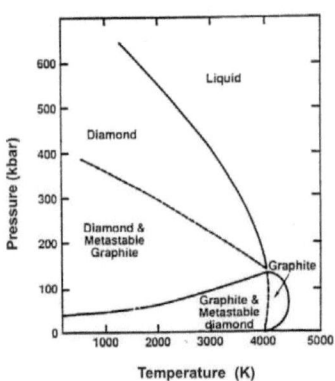

Figure 1.2-1
Phase diagram of diamond

1.2 Materials Under Pressure

The synthesis of superhard diamond from graphite is one of the reasons why high-pressure / high-temperature research is done. While diamond is still the hardest and one of the most important abrasives in industry, the graphitization process mentioned above is quite cumbersome. Therefore, superhard materials with better thermal stability or even harder materials than diamond are in the focus of high-pressure research. The most important technical abrasives are corundum, silicon carbide, boron nitride, and diamond. While corundum is an ambient-pressure material, the very hard silicon carbide types HPSiC (hot pressed SiC) and LPSSiC (liquid phase sintered SiC) are densified under pressures up to 30 MPa [22]. Of boron nitride, only the cubic high-pressure phase c-BN can be used as an abrasive. Its structure is commensurate with that of cubic diamond (Figure 1.2-2) and it is the second-hardest material known [23]. Hexagonal BN crystallizes in layers as isotypic graphite and is thus a soft material. In contrast to diamond, c-BN is thermally stable even above 1000 °C.

In 2001, Solozhenko *et al.* reported the synthesis of a pseudo-diamond-structured phase with the composition BC_2N in a multianvil press at 18 GPa

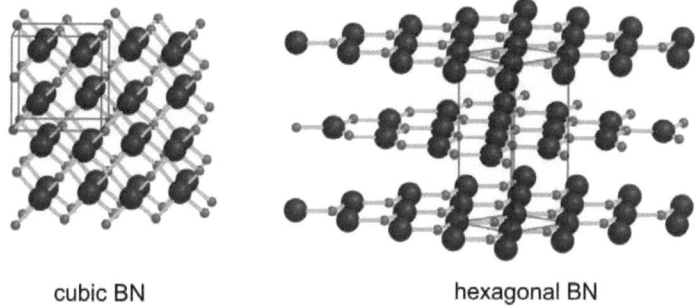

cubic BN   hexagonal BN

Figure 1.2-2
Crystal structures of cubic and hexagonal boron nitride

and 2200 K [24]. Several measurements and theoretical studies concerning the hardness of this substance were undertaken, several of which indicate that cubic BC_2N is even harder than cubic boron nitride and thus the world's second-hardest material [25]. Nevertheless, there is still a heavy discussion in progress, concerning the hardness of c-BC$_2$N and whether it truly is a new compound or simply a mixture of diamond and c-BN.

The third-hardest material, according to current knowledge, is the high-pressure modification of Si$_3$N$_4$, termed γ-Si$_3$N$_4$, synthesized at 15 GPa and 2000 K. The ambient pressure phases α- and β-Si$_3$N$_4$ are already hard ceramics used in high-endurance and high-temperature applications [26]. While these polymorphs are built up from SiN$_4$-tetrahedra, γ-Si$_3$N$_4$ crystallizes in a spinel structure with SiN$_4$- and SiN$_6$-polyhedra (Figure 1.2-3) and is thus another example for the pres-sure-induced densification accompanied by an increase of coordination numbers. A post-spinel phase, δ-Si$_3$N$_4$, is predicted to form at pressures around 160 GPa, built up solely from SiN$_6$-polyhedra [27].

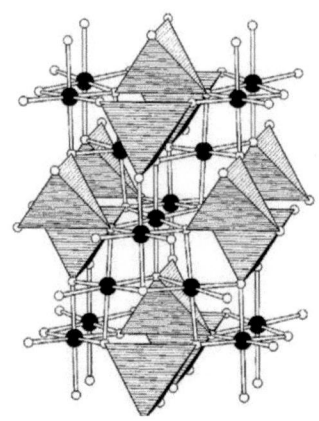

Figure 1.2-3
Crystal structure of γ-Si$_3$N$_4$ (black spheres: N, hollow spheres: Si) [26]

Similar interest is directed towards the corresponding carbon compound C$_3$N$_4$. Since Liu and Cohen predicted in 1985 that this material would be harder than diamond [28], the existence and the structure of β-C$_3$N$_4$ were lengthily discussed. Due to their low thermodynamic stability with respect to the elements (C and N$_2$), syntheses of this material failed until in 1993, thin films of C-N products were obtained [29]. Later on, several successful syntheses of C$_3$N$_4$ were reported, e.g. Nguyen and Jeanloz [30] described the first example of a dense carbon nitride formed in the bulk, synthesized at high pressure and high temperature, and Win et al. reported the synthesis of β-C$_3$N$_4$ nano-sized crystals [31]. Nevertheless, serious doubts that these potentially superhard materials have actually been made will remain until large crystals of nitride are synthesized, precisely characterized, and their mechanical properties tested.

1.2 Materials Under Pressure

Not only the achievement and stabilization of high atomic coordination number, but also the stabilization of high oxidation states for transition metals under high pressure is possible. Therefore, high oxygen pressures are applied to the samples (either externally [32] or generated *in situ* by the thermal decomposition of unstable oxides like CrO_3 and $KClO_3$ under high-pressure [33, 34]). With this technique, *e.g.* oxides with Fe^{5+}, Co^{4+}, Ni^{3+}, and Cu^{3+} cations were obtained.

1.3 High-Pressure Chemistry of Borates

Boron is an element with several outstanding properties. It is the only metalloid in the 13th group of the periodic table and differs remarkably from the other elements of this group, also due to its small atomic radius. It is able to form multicenter bonds and thus many complex modifications containing B_{12}-icosahedra are known. Even amorphous boron contains these regular icosahedra, which are bound randomly to each other without long-range order [35]. Binary and multinary boron compounds also show a variety of complex structures, *e.g.* in borides and boranes. Being isoelectronic to carbon, several boron nitrides show structural analogy to organic carbon compounds.

Chemically, boron is closer to silicon than to aluminum, as can be expected from the diagonal relationship principle in the periodic table. Boron and silicon are both semiconductors, form halides, that are hydrolyzed in water, and have acidic oxides. The oxygen compounds of boron are closely related to those of silicon. Silicates are comprised of SiO_4-tetrahedra, which can be connected *via* common corners to form rings, chains, sheets, and networks. In Figure 1.3-1, this is depicted using the example of β-quartz. With boron being an element of the 13th group of the periodic table, borates form tetrahedral BO_4-groups as well as planar or aplanar BO_3-groups. The structural diversity

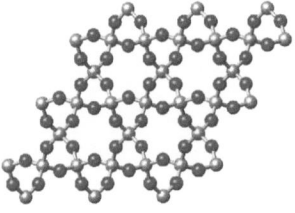

Figure 1.3-1

Crystal structure of β-quartz from SiO_4-tetrahedra

Table 1.3-1

Boron minerals

Borax	$Na_2[B_4O_5(OH)_4] \cdot 8\ H_2O$
Kernit	$Na_2[B_4O_5(OH)_4] \cdot 2\ H_2O$
Ulexit	$NaCa[B_5O_6(OH)_6] \cdot 5\ H_2O$
Colemanit	$Ca[B_3O_4(OH)_3] \cdot H_2O$
Sassolite	H_3BO_3

1.3 High-Pressure Chemistry of Borates

that can be obtained by the interconnection of these building blocks exceeds those of the silicates by far.

In nature, boron is mainly present in the form of several borate minerals (Table 1.3-1), of which borax (Figure 1.3-2a) is the commercially most important. Main borax deposits are in Turkey and California. Kernite (Figure 1.3-2b) was discovered in 1926 in Kern County, California and later named thereafter. It was the only known source of kernite for many years but is now mined in Argentina, Spain, and Turkey. Ulexite and Colemanite (Figure 1.3-2c and d) are found in California and Nevada, in Chile, and in Kazakhstan. All these minerals were naturally formed during the Cenozoic Era as a result of evaporation of geothermal springs. Sassolite (Figure 1.3-2e) is the mineral form of boron acid and occurs in volcanic fumaroles and hot springs. The mineral may be found in lagoons throughout Tuscany and Sasso in Italy.

Turkey and the United States are the leading producers of borates with 35 % and 29 % of the world supply, based on B_2O_3 content (Table 1.3-2). Other countries, that are significant exporters, include Argentina and Chile. Europe and Japan rely on imports for their boron min-eral and chemical supplies.

Figure 1.3-2
The borate minerals borax (a), kernite (b), ulexite (c), colemanite (d), and sassolite (e)

Table 1.3-2

World production and reserves (in 10^3 t) of boron materials in 2005 [36]

	Production	Reserves
Turkey	1,700	60,000
United States	1,230	40,000
Chile	600	?
Argentina	550	2,000
Russia	500	40,000
China	140	25,000
Bolivia	100	?
Kazakhstan	30	?
Peru	10	4,000
Iran	3	1,000
World total (rounded)	4,860	170,000

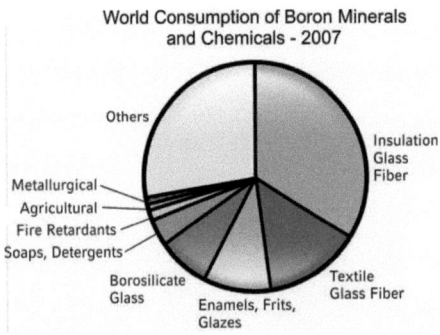

Figure 1.3-3

Application areas of boron in industry [37]

The leading boron producing companies are Rio Tinto Plc with boron mines in Argentina and the United States and Eti Holding A.S. with mines in Turkey [37].

The decision to use mineral versus boron chemicals is determined by the full cost associated with either alternative. The most important boron compounds in terms of volume are borax, boric acid, boron oxide, boron carbide, boron hydroxide, boron nitride, elemental boron, and ferroboron.

Not only due their structural diversity, but also due to their properties, borates are of great interest for scientific and economic applications. Boron compounds are used in a wide variety of products and manufacturing processes (Figure 1.3-3). Borax, for example, is a component of many detergents, cosmetics, and enamel glazes and also applied for many laboratory purposes, e.g. in biochemical buffer solutions, as a flux in metallurgy,

1.3 High-Pressure Chemistry of Borates

and as a precursor for other boron compounds.

The largest application areas for borates are fiberglasses, ceramics, and borosilicate glass (Figure 1.3-3). The latter is made from 70–80 % SiO_2 and 7-13 % B_2O_3. It further contains 4-8 % alkaline metal oxides, 2-7 % Al_2O_3 and up to 5 % alkaline earth metal oxides. The resulting glass is highly resistant against chemicals and is thus used for laboratory glassware and housekeeping purposes. Due to its low thermal expansion coefficient, borosilicate glass is heat resistant and unaffected by sudden temperature fluctuations.

In the detergent industry, sodium peroxoborates are used as bleaching agents. Since boron is an essential ultratrace element and plant nutrient, borates are important components of fertilizer. A growing and important use of zinc borate, ammonium pentaborate, and boric oxide is as fire retardants in the plastics industry. Zinc borates with the composition $xZnO.yB_2O_3.zH_2O$ (e.g. Firebrake®) is of commercial importance because it is water insoluble and does not release water until heated to 290 °C.

Besides the industrial application of natural borate minerals, synthetic borates become more and more important. In the Inorganic Crystal Structure Database (ICSD), more than 2100 B-O-containing compounds are listed today [38]. The primary field of interest is on the optic properties of borates. For example, SrB_4O_7:Eu is a phosphor used for UV-emitting medical lamps, (Y, Gd)BO_3:Eu is the red-emitting component in plasma display panels, and $GdMgB_5O_{10}$:Ce,Tb is a green luminescent phosphor in fluorescent tubes. Many studies concerning borate phosphors for white-emitting LED lamps are currently undertaken.

Many borates also show non-linear optical (NLO) properties. In these materials, the dielectric polarization responds nonlinearly to the electric field of the light. One of the thereby created optical phenomena is frequency doubling or second-harmonic generation. With this technique, the 1064-nm output from Nd:YAG lasers or the 800-nm output from Ti:sapphire lasers can be converted to visible light, with wavelengths of 532 nm (green) or 400 nm (violet), respectively. For NLO applications, β-BaB_2O_4 (BBO) [39-41], LiB_3O_5 (LBO) [42-45], α-BiB_3O_6 (BIBO) [46-48], and $CsLiB_6O_{10}$ (CLBO) [49-52] are widely used.

Generally, most research activities concerning borates are conducted under ambient pressure conditions. Limited high-pressure studies are carried out by geologist concerning borate minerals. The Huppertz group was the first to study the field of synthetic borates systematically under high-pressure conditions. It appeared that this kind of synthesis had several advantages. As mentioned above, borates form tetrahedral BO_4-groups and trigonal BO_3-groups. In accordance with the pressure-coordination-rule [19], BO_3-groups tend to transform into BO_4-tetrahedra under increasing pressure. Another typical high-pressure effect is the densification of the structures, leading to a different connection of the BO_3- and BO_4-groups, *e.g. via* common edges instead of common corners or by an increase of the coordination number of the bridging atoms. The pressure-homology-rule [19] specifies that the structure of a high-pressure phase may be similar to that of the analog compound of a related heavier element at ambient conditions. Last but not least, a pressure induced crystallization effect is observed. Borates, being glass formers in general, show an increased willingness to grow crystals under elevated pressure by changing the Gibbs free energy. Thus it is possible, to greatly extend the structural diversity of this substance class by pressure regulation. These general findings are shown in detail on the following examples.

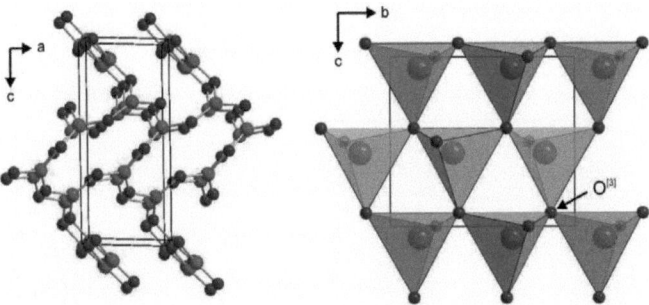

Figure 1.3-4
Crystal structures of B_2O_3-I (left) and B_2O_3-II (right). Red spheres: B^{3+}, blue spheres: O^{2-}.

1.3 High-Pressure Chemistry of Borates

Boron oxide can be found in two modifications: the ambient-pressure phase B_2O_3-I [53] and the high-pressure phase B_2O_3-II [54]. B_2O_3-I comprises of BO_3-groups (Figure 1.3-4 left), which are all connected *via* common corners. B_2O_3-II exhibits only BO_4-tetrahedra (Figure 1.3-4 right), which are also connected *via* common corners. In contrast to B_2O_3-I, the bridging oxygen atoms of the tetrahedra are threefold coordinated by boron, which leads to a much denser structure.

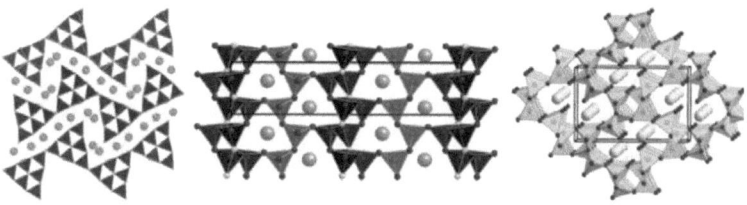

Figure 1.3-5
Crystal structures of β-$RE(BO_2)_3$ (RE = Nd, Sm, Gd-Lu) (left), γ-$RE(BO_2)_3$ (RE = La-Nd) (middle), and δ-$RE(BO_2)_3$ (RE = La, Ce) (right).

Among the *meta*-borates $RE(BO_2)_3$, a monoclinic ambient pressure phase α-$RE(BO_2)_3$ (RE = Y, La - Nd, Sm - Lu) is known. These compounds exhibit both BO_3- and BO_4-groups. The orthorhombic borates β-$RE(BO_2)_3$ (RE = Tb [55], Nd, Sm, Gd-Lu [56,57]) are built up from corrugated sheets of corner-sharing tetrahedra (Figure 1.3-5 left) and are obtainable both at ambient and at higher pressure. Via high-pressure / high-temperature syntheses, two other *meta*-borates comprising only of BO_4-tetrahedra can be synthesized (Figure 1.3-5 middle and right). The orthorhombic compounds γ-$RE(BO_2)_3$ (RE = La – Nd) [58] and the monoclinic phases δ-$RE(BO_2)_3$ (RE = La ,Ce) [59, 60] form a BO_4-network.

For a long time it was believed that the connection of BO_4-tetrahedra exclusively occurs *via* common corners. In 2001, Huppertz *et al.* obtained the borate $Dy_4B_6O_{15}$ (Figure 1.3-6, top left), the first to form edge-sharing BO_4-tetrahedra under high-pressure conditions [61]. Over the following years, this revolutionary structural feature could be realized in several other borates, namely $Ho_4B_6O_{15}$ [62] and α-$RE_2B_4O_9$ (RE = Sm - Ho) [63, 64] (Figure 1.3-6, top right). Interestingly, the corresponding water-containing phases $RE_4B_6O_{14}(OH)_2$ (RE = Dy, Ho) [65] showed only the regular common-corner connection. With β-/HP-MB_2O_4, (M = Fe, Co Ni) [66-68], the first compound consisting solely of edge-sharing BO_4-tetrahedra was synthesized (Figure 1.3-6, bottom left). To our astonishment, the structural feature of edge-sharing BO_4 tetrahedra was recently found in the compound $KZnB_3O_6$ [69, 70], which was synthesized under ambient-pressure conditions (Figure 1.3-6, bottom right). This result means that the structural motif of edge-sharing BO_4-tetrahedra is no longer a domain of high-pressure chemistry, but still favored under these conditions, as there exist eleven different compounds with three structure types prepared under high-pressure and only one compound synthesized under ambient-pressure conditions.

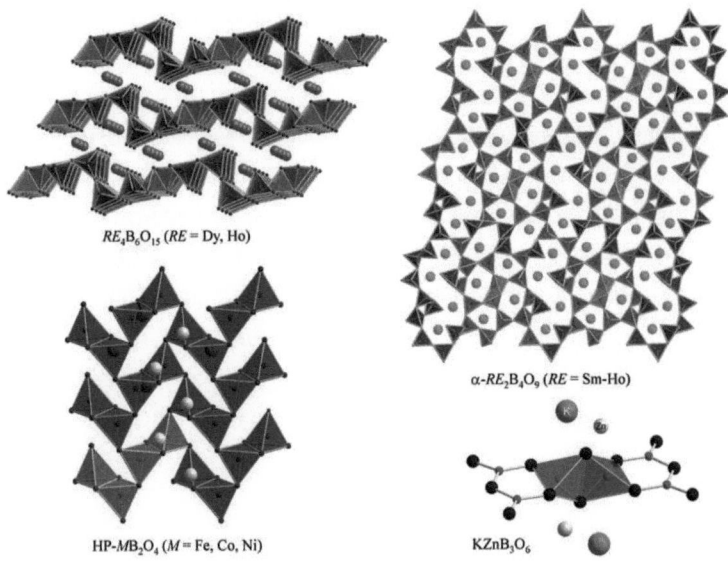

Figure 1.3-6

Crystal structures of $RE_4B_6O_{15}$ (RE = Dy, Ho), α-$RE_2B_4O_9$ (RE = Sm – Ho), β-/HP-MB_2O_4, (M = Fe, Co Ni), and $KZnB_3O_6$, with edge-sharing BO_4-tetrahedra (red).

1.3 High-Pressure Chemistry of Borates

According to the pressure-homology-rule [19], several high-pressure borates are known, which crystallize in a structure similar or closely related to the structure of a heavier element at ambient conditions. CdB_2O_4, for example, represents a basic structure for the related compounds $BaGa_2O_4$ and $KAlSiO_4$. The rare-earth borates $RE_3B_5O_{12}$ (RE = Tm – Lu) [71] also exhibit a silicate-analog structure, homeotype to that of semenovite.

In the ternary systems Zr-B-O and Hf-B-O, only glasses were known, until Knyrim *et al.* synthesized the phases $\beta\text{-}MB_2O_5$ (M = Zr [72], Hf [73]), comprised of BO_4-tetrahedra (Figure 1.3-7). These compounds are not only examples for pressure-induced crystallization, they also show structures related to the silicate minerals gadolinite-(Y) and datolite, which exhibit a similar arrangement of SiO_4-tetrahedra.

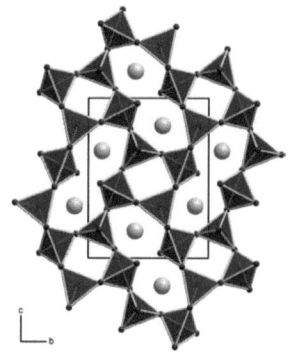

Figure 1.3-7
Crystal structure of $\beta\text{-}MB_2O_5$ (*M* = Zr, Hf)

Tin-based amorphous composite oxide (TCO) glasses are investigated as electrode materials for lithium-ion rechargeable batteries [74]. In this regard, many studies concerning the structure of tin borate glasses were undertaken. High-pressure / high-temperature synthesis finally resulted in the first crystalline compound $\beta\text{-}SnB_4O_7$ in the system Sn-B-O, which fully comprised of BO_4-tetrahedra [75]. This new approximant of the glassy state made a correlation of the oxidation state of tin atoms coordinated to

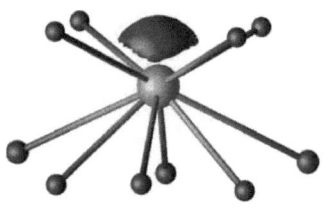

Figure 1.3-8
Lone pair at the Sn^{2+} ion in $\beta\text{-}SnB_4O_7$, calculated with the electron localization function (ELF).

oxygen atoms possible and revealed the existence of an electron lone pair at the tin atom (Figure 1.3-8).

High-pressure investigations of materials have been shown to be a valuable tool concerning the synthesis of new compounds with fascinating structures and outstanding properties. As mentioned above, the high-pressure / high-temperature synthesis of transition-metal and rare-earth borates have been performed by the Huppertz group for several years with great success. During this thesis, further work on rare-earth oxoborates was performed. After a detailed description of the experimental methods used during this work, these results are presented in Section 4.1. We additionally decided to extend the field of high-pressure rare-earth borates concerning the syntheses of rare-earth fluoride and fluorido borates. This thesis presents the first attempts of high-pressure research on fluorine-containing borates. In Section 4.2, an insight in the chemistry of fluoride and fluorido borates is given, followed by a summary of the published results.

2 Experimental Methods

2.1 Glove Box

All the syntheses carried out during this thesis were solid state reactions, starting from a mixture of solid educts. To have a quick and thorough reaction, the mixture should be as homogeneous as possible, so the educts are ground together in an agate pestle. Depending on the educts, the oxygen and moisture sensitivity of the compounds has to be taken into account, especially when increasing their reactive surface during pestling. For example, B_2O_3 and La_2O_3, which were often used during this thesis, quickly transform into their hydroxides $B(OH)_3$ and $La(OH)_3$ when exposed to air humidity. Rare-earth fluorides form hydroxides by evaporating gaseous HF. It is therefore inevitable to handle these chemicals in a glove box under a dry nitrogen atmosphere, as depicted in Figure 2.1-1. Inside, all reaction mixtures for the syntheses reported in Section 4 were prepared and exposed to air minutes before the high-pressure experiment was started.

Figure 2.1-1
Glove box with nitrogen atmosphere in the inner compartment

2.2 High-Pressure Synthesis

All the high pressure / high-temperature syntheses presented in this thesis were performed in a 1000 t multianvil press with modified Walker-type module (both devices from Voggenreiter, Mainleus, Germany). Figure 2.2-1 displays the press, including the module, and it's control devices. In the following, the layout of the high-pressure devices (Sections 2.3 and 2.4), the sample preparation (Section 2.5), and the calibration (Section 2.6) are discussed.

Figure 2.2-1
High-pressure device with the main hydraulic unit, the worm gear screw jack driven by a servomotor, the press including the Walker-type module, and the control unit (from left to right).

2.3 The Walker-Type Module

A multianvil device is the most economic way to combine sufficiently high, quasi-hydrostatic pressure with acceptable sample volumes for analytical purposes (see Section 1.1). The original multianvil module developed by Walker [13]

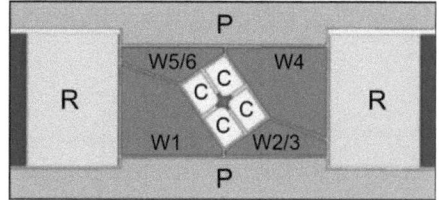

Figure 2.3-1
Cross section of the Walker-type module

was combined with several improvements of Frost [76] and scaled up to a maximum load of 1000 t by the company Voggenreiter. A schematic drawing of the module is depicted in Figure 2.3-1. The pressure medium (red octahedron) is placed in the center of eight tungsten carbide cubes (C), which are surrounded by six tool steel wedges (W_{1-6}). The whole assembly is positioned in a containment ring (R). The loading of the module is accomplished through pressure-distribution plates (P). The strain of the module under high loads, is mainly absorbed by the containment ring (HSM, 1.2343, $R_c = 52$), made from tool steel with an inner distance of 17.8 cm, an outer distance of 33.8 cm, and a total length of 19.4 cm. For safety reasons, the containment ring is surrounded by an additional safety ring (Höver, 1.4541) with 33.8 cm (inner distance) × 37.8 cm (outer distance) and the identical length. The total height of the rings is 19.4 cm. Figure 2.3-2 shows a photograph of the massive containment ring surrounded by the safety ring, incorporating three wedges with wiring ports.

The wedges are also made from tool steel (HSM, 1.3343), hardened to $R_c = 62$. The square face side dimensions are 6.00 cm × 6.00 cm (for 32 mm tungsten car-

Figure 2.3-2
Containment ring (matt) and surrounding safety ring (shiny)

Figure 2.3-3
Steel tool wedge for the Walker-type module

bide cubes) with an angle of 35°26' to the axis of the module. The lowest corner of the square face has a distance of 2.00 cm from the basis and the total height is 98 mm. Figure 2.3-3 gives a view of one of the wedges with the square face in front and wiring channels on both sides. Three wedges can be fitted on the bottom of the module with a gap of approximately 1 mm between them.

On the top and the bottom of the module, pressure-distribution plates are placed. The main part of these plates (37.8 cm in diameter, 3.9 cm thick) is from an Al alloy (Alimex, AMP 8000), while the raised part of the plate, which directly lays on the wedges, is made by tool steel (Höver, 1.4548.4). With this arrangement, deformations of the Al plates are prevented, while at the same time the plates are as lightweight as possible. The latter is extremely important, since the top pressure-distribution plate is used as a lid, which has to be moved manually. Figure 2.3-5 displays the top and the bottom pressure-distribution plates. Between the two parts of the pressure-distribution plates, cooling water can flow through a spiral (Figure 2.3-4). Both, the pressure-distribution plates and the wedges are equipped with several tunnels and shafts for wiring. The wiring is important for calibrations or experiments with thermocouples, which is discussed in Section 2.6.

Figure 2.3-5
Top and the bottom pressure-distribution plates

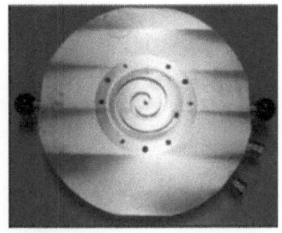

Figure 2.3-4
Cooling spiral inside the plates

2.4 The Multianvil Device

The high-pressure laboratory of the Institute of General and Inorganic Chemistry at Innsbruck University is equipped with a 1000 t hydraulic downstoking press, which was constructed and set up by the company Voggenreiter (Mainleus, Germany). A schematic drawing is shown in Figure 2.4-1. This device can establish sample pressures up to 25 GPa at a maximum temperature of 2700 K. It is constructed from large metal frames inside of which the table

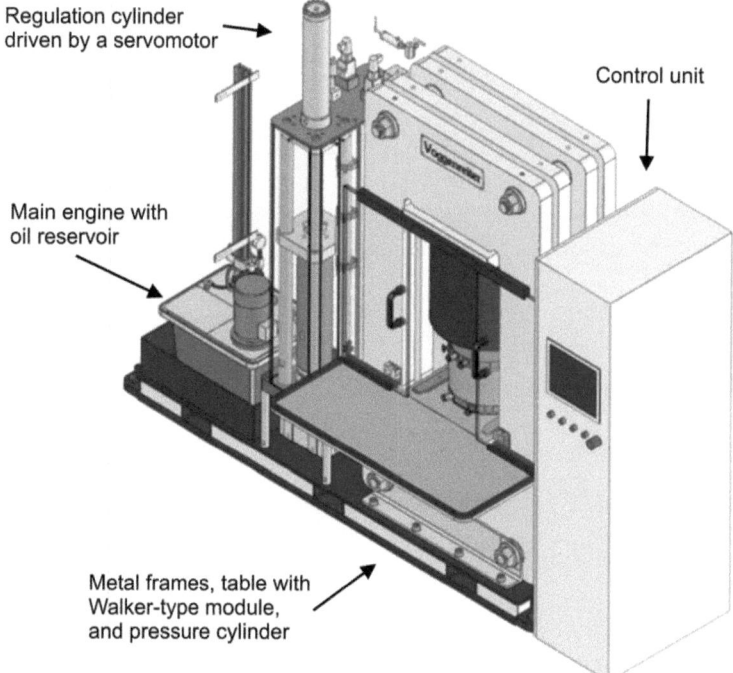

Figure 2.4-1
High-pressure frame press from Voggenreiter

with the Walker-type module and the pressure cylinder are mounted.

The pressure is built up *via* a hydraulic system. The hydraulic oil circuit can be divided into three parts:

a) the main pressure cylinder inside the frames
b) the regulation cylinder with a worm gear screw jack driven by a servomotor
c) the main engine with the oil reservoir.

Figure 2.4-2 shows photographs of these compounds, while Figure 2.4-3

Figure 2.4-2
Main pressure cylinder (left), regulation cylinder with a worm gear screw jack (middle) and the main engine with the oil reservoir (right).

gives a diagram of the oil circuit divided in the three mentioned parts (red line).

The heating principle of the multianvil device is based on resistance heating. Therefore, graphite furnaces surround the sample in the center of the assembly (Section 2.5). *Via* power connections at the top and the bottom pres-

2.4 The Multianvil Device

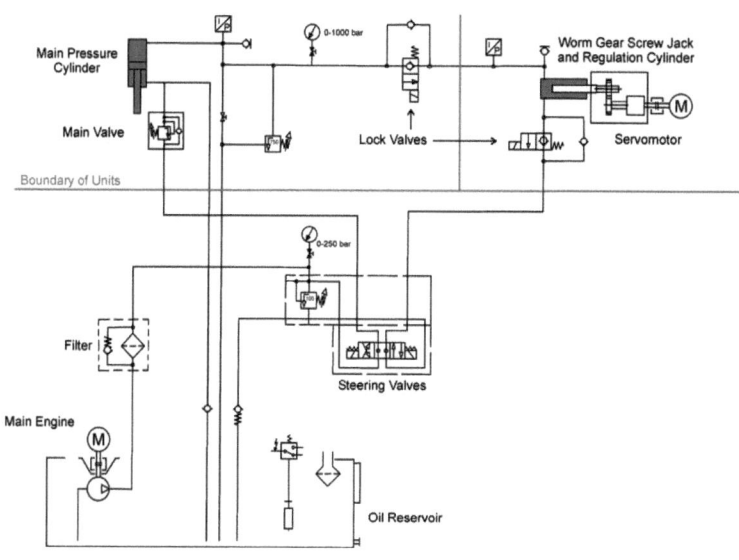

Figure 2.4-3
Diagram of the oil circuit of the 1000 t-press

sure-distribution plates of the module, a current can flow through the assembly and heat up the furnaces. For heating purposes, a special heating- and temperature control-unit was designed. It consists of a main transformer regulated *via* an Eurotherm 2404 temperature controller (Limburg a. d. Lahn, Germany), which can either be programmed or directly connected to a thermocouple inside the assembly. The topical values of voltage, current, and power are shown on three displays next to the temperature controller. An additional security circuit is built up outside the heating unit for surveillance of the water cooling. Any failure in the cooling circuit leads to a direct disconnection of the main current.

Monitoring voltage, current, power, and temperature are essential in the beginning of the heating phase. Deviations from known experimental values clearly indicate malfunctions like short circuits (high current, low voltage) or misalignments (low current, high voltage).

In order to start a high-pressure experiment, the sample is placed inside of the Walker-type module (for sample preparation, see Section 2.5) and the

Figure 2.4-4
Induction swith for the movement control of the pressure ram.

module is centered beneath the main pressure cylinder. The main engine now closes the 1 cm gap between the ram and the top pressure distribution plate of the Walker-type module with high velocity. To avoid a hard put down, induction switches stop the driving of the ram just before it touches the pressure distribution plate. Figure 2.4-4 shows the induction switch on the right side of the distance piece regulating the fast movement of the ram. In the next step, the oil pressure in the main cy-linder is regulated by the servomotor. By moving the worm gear screw jack into the outer regulation cylinder, very precise compression and decompression is possible. Up to an oil pres-sure of 10 bar (equaling a load of 14.3 t), the compression of the mod-ule takes place in a relatively fast time of approximately 20 min, closing all gaps between the wedges, cubes, and the octahedral cavity. Further compression is predetermined by the ramp given for the experiment. The compression rate has a maximum value of 72 bar per hour, whereas the decompression time requires the threefold time.

The main engine, the servomotor, and the valves are controlled by a Programmable Logic Controller (PLC, Simatic S7-300) equipped with a serial RS232C interface. *Via* a second control unit (Windows PC), the essential informations concerning the experimental profile (*e.g.* the pressure and temperature ramps) are entered. A special program named PRESSCONTROL [77] is used for communication, calibration, and surveillance, transferring the pressure ramps to the PLC, reading out the current system pressure, and controlling all

2.4 The Multianvil Device

actions of the PLC. The transfer of data from PRESSCONTROL to the temperature controller is performed *via* a RS232C interface.

Figure 2.4-5 shows the graphical interface of the program. All experimental parameters have to be entered, *e.g.* type and size of the octahedron, type of furnace, and the pressure / temperature ramps. For heating, the program distinguishes between temperature-controlled heating *via* a thermocouple or the use of predetermined power / temperature curves for the corresponding assembly without thermocouple. These curves are obtained by calibration experiments, which are discussed in Section 2.6. The oil pressure is used as input parameter for the intended load, whereby 700 bar correspond to the maximum load of 1000 t. All entered parameters can be saved, so a convenient recall of the experimental conditions for a repetition is possible. After entering the experimental details, the pressure and the temperature ramp are shown as black lines in two diagrams. During the high-pressure experiment, the actual oil pressure and temperature given by the thermocouple are continuously read out and displayed on the screen.

An additional security feature is installed for the case of a so-called blowout - a sudden decrease of the inner pressure of the assembly, indicating severe instabilities. Automatically, slow pressure reduction is started and a blow-

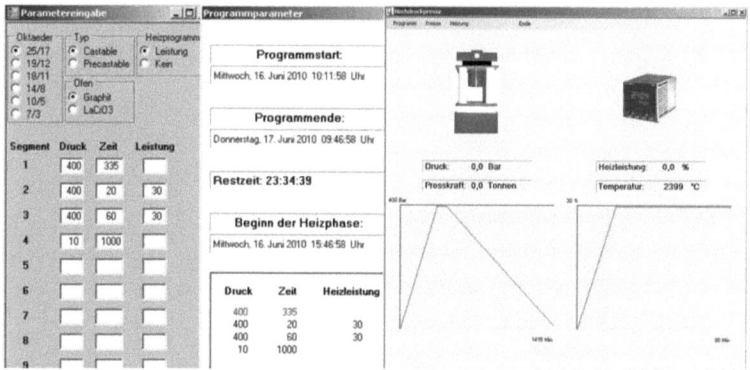

Figure 2.4-5

Graphical interface for parameter input and surveillance of the pressure / temperature program.

out warning is displayed on the graphical interface.

2.5 Sample Preparation

The pressure distribution inside of the multianvil module is based on the six wedges compressing a cubic arrangement of eight inner anvils (Section 2.3). These inner anvils are tungsten carbide cubes with an edge length of 32 mm and truncated corners to form a triangular face. Eight cubes with eight truncated corners, separated by gaskets, form an octahedral cavity (Figure 2.5-1). Inside this cavity, the octahedral pressure-transmitting medium is placed.

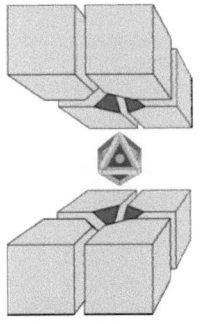

Figure 2.5-1
Octahedral cavity formed by the inner anvils

There exist several different assemblies, which are clearly defined by their octahedral edge length (OEL) and truncation edge length (TEL) of the corresponding tungsten carbide cube. For instance, an 18/11 assembly describes an octahedron with an edge length of 18 mm, including eight tungsten carbide cubes exhibiting truncated triangular faces with an edge length of 11 mm. The choice of the assembly type depends on the desired sample pressure and its corresponding sample volume. Typical assemblies for the 6-8 type multianvil apparatus are displayed in Table 2.5-1 [78]. During this thesis, only 18/11 and 14/8 assemblies were used.

The quality, technical specifications and suitability of the tungsten carbide used for the inner anvils, produced by different manufactures, are highly variable [79]. The average lifetime of a set of cubes differs from approximately five to twenty runs of high-pressure / high-temperature experiments, depending on the experimental conditions. Cubes applied in this work were Hawedia HA-7%Co (Marklkofen, Germany), Ceratizit "TSM20" (Reutte, Austria) and Kennametal O2F and THM-U (Friedrichsdorf, Deutschland).

2.5 Sample Preparation

Table 2.5-1
Typical multianvil assembly sizes (OEL = octahedral edge length, TEL = truncation edge length).

	25 mm	18 mm	14 mm	10 mm
TEL	17 mm	11 mm	8 mm	4 mm
Pressure (max.)	4 GPa	10 GPa	13 GPa	16 GPa
Sample volume	~ 80 mm^3	~ 35 mm^3	~ 9 mm^3	~ 4 mm^3
amount	~ 100 - 60 mg	~ 50 - 30 mg	~ 30 - 20 mg	~ 15 - 10 mg

The pressure transmitting medium is a prefabricated, sintered magnesium oxide octahedron doped with 5 % magnesium chromite (Ceramic Substrates & Components LTD, Isle of Wight). In order to place the sample inside of this octahedron, a hole is drilled between two adjacent of its faces. This hole is then filled with several assembly parts, which are crucial for the subsequent heating of the sample. A general survey about these parts is given in Figure 2.5-2. The sample is filled in a crucible and closed with a lid (a), both made from hexagonal boron nitride (HeBoSint® S100 or P100, Henze, Kempten, Germany). The crucible is then placed in a small graphite tube (b), which is then centered in a larger graphite cylinder (c) with the help of two MgO spacers (d) (Magnorite MN399CX, Saint-Gobain Industrial Ceramics,

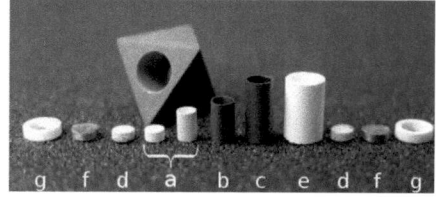

Figure 2.5-2
Assembly parts for the sample preparation

Worcester, MA, USA). The graphite (RW403, SGL Carbon, Bonn Germany) functions as a resistance heater while the use of two furnaces with a stepped wall thickness reduces the thermal gradient along the sample [80]. The crucible with the surround-ing furnaces is then placed into an additional zirconia sleeve (e) (cesima ceramics, Wust, Germany) for thermal isolation of the furnace against the MgO octahedron. The furnaces are contacted *via* molybdenum plates (f) (Mo007905, Goodfellow, Huntingdon, England) at the top and bottom, which are hold in place by two MgO rings (g) that are glued to the zirconia sleeve. This whole arrangement is now place inside the octahedron. A cross section through a completed octahedron is shown in Figure 2.5-3.

Hexagonal boron nitride is the crucible material of choice due to its chemical inertness under high-pressure / high-temperature conditions. As it is a quite soft and easy machinable material, the manufacturing of different capsule sizes starting from a h-BN bar is quite easy and cheap. Nevertheless, there are samples, that react with the crucible material under the extreme conditions applied. Then a crucible inlay has to be made from metal foils of copper, molybdenum, platinum, or gold for the experiments. Another weak spot can be found with the graphite furnaces. They can be used successfully as a heater material up to pressures of 10 GPa and temperatures of 1500 °C. Above these limits, the material starts to convert to diamond and its performance collapses. As an alternative, LaCrO$_3$ (Cherry-O, Amagasaki-City, Japan) can be used as a heater material to generate temperatures up to 3000 K. The different assembly sizes displayed in Table 2.5-1 naturally result in differing dimensions for the assembly parts. For the 18/11 and 14/8 assemblies used in this thesis, they are displayed in Figure 2.5-4 and Figure 2.5-5.

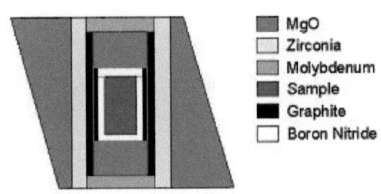

Figure 2.5-3
Cross section of the prepared octahedron

2.5 Sample Preparation

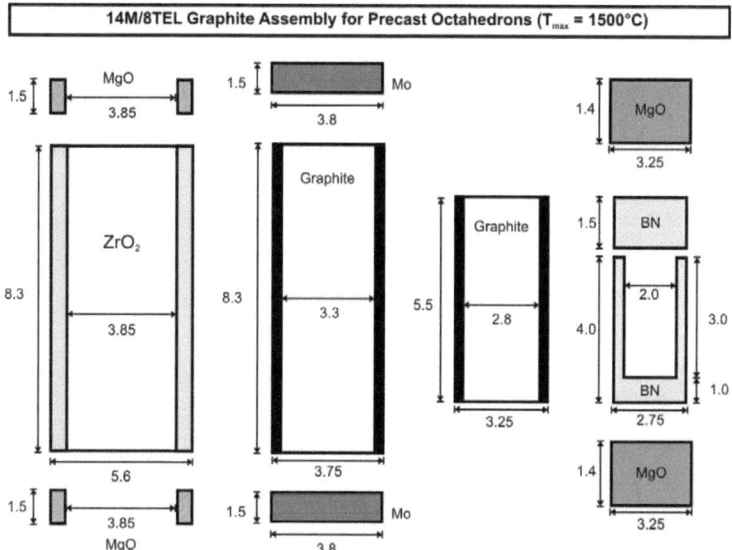

Figure 2.5-4

Dimensions (mm) of the pieces for the inner octahedral part of a **14/8** assembly

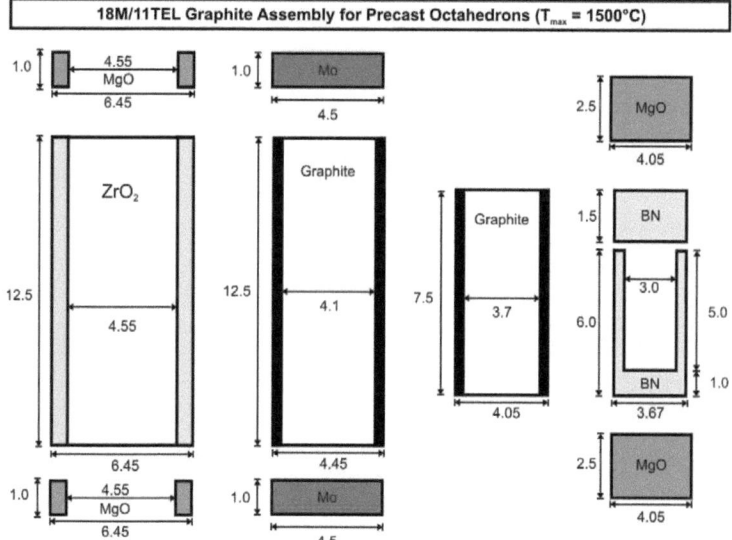

Figure 2.5-5

Dimensions (mm) of the pieces for the inner octahedral part of a **18/11** assembly

After setting the inner part into the drilled octahedron, the eight tungsten carbide cubes are arranged around it, as seen schematically in Figure 2.5-1. In order to compress the octahedron, the inner anvil cubes may not touch each other. This is granted by the application of pyrophyllite gaskets, which are directly cut from mineral blocks in our workshop. The triangular faces of four of the truncated cubes are surrounded by affixed gaskets of different lengths (Figure 2.5-6). When assembled, the gaskets mate and thus a sealing and stabilization effect is achieved. For fixation, an adhesion is used, which does not contain dissolver (Multi-talent Liquid Glue, Tombow Pencil Co., Tokyo, Japan). This is important to keep the gaskets in place even under elevated temperatures.

Figure 2.5-6
Arrangement of pyrophyllite gaskets on the tungsten carbide cubes

On the same four cubes, cardboard (imitation bristol, 369 g/cm^3) is applied on all three faces surrounding the triangular faces, while the other four cubes are laminated with PTFE tape (SK-05-AD, Fiberflon, Konstanz, Germany), as shown in Figure 2.5-7. In the final arrangement of the cubes, each cardboard site is neighboured to a PTFE-taped site (Figure 2.5-8 left). This combination was found to be very effective at slowing and

Figure 2.5-7
Tungsten carbide cubes with PTFE foil and cardboard

2.5 Sample Preparation

stabilizing the extruding of the gaskets under high pressures. When assembling the cubes around the octahedron, they are glued to pads of fiber glass (Type 2372.4, Menzel & Seyfried, Gröbenzell, Germany) with instant adhesive (Toolcraft Rapid 150, Conrad Electronic, Hirschau, Germany) for stabilization reasons (Figure 2.5-8). Furthermore, the pads with a thickness of 0.8 mm relieve any stress-causing irregularities on the interfaces between inner anvil and wedge. This is extremely important for convergence during loading. Figure 2.5-8 right shows a completely assembled cube hold together by six pads. The top and the bottom pads have an incision in each of which a copper stripe (about 1.5 cm × 3.0 cm, thickness: 0.1 mm) is inserted. *Via* these copper stripes, the electric current flows from the wedges through the pads into the two opposite cubes, whose truncated faces lie directly on the molybdenum

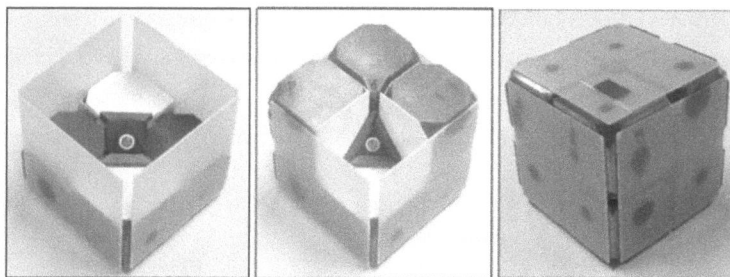

Figure 2.5-8
Different stages of the assembling of the inner anvils

plates.

In Figure 2.5-9, the procedure of positioning the assembled cubes into the Walker-type module is demonstrates schematically. Three wedges form a nest (Figure 2.5-9a), wherein the cubic arrangement of the inner anvils with the octahedron inside is positioned along its threefold axis. Afterwards, the top wedges take their positions on top of the cube, leaving a gap between top and bottom wedges (Figure 2.5-9b). The complete assemblage takes place in the containment ring (Figure 2.5-9c). For reasons of clarity, the containment ring is not drawn in the previous pictures.

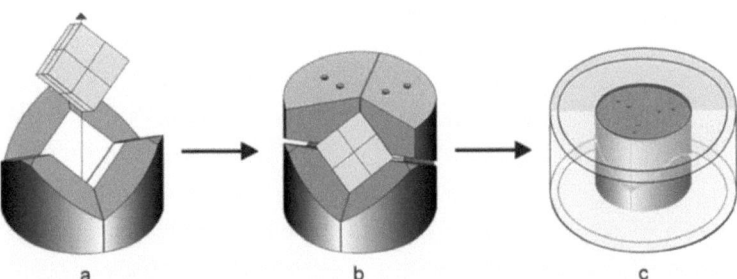

Figure 2.5-9
Schematic drawing of the assemblage of the Walker-type module [5]

Photographs of the placement procedure are displayed in Figure 2.5-10. The small gap between the wedges and the bore of the containment ring is filled with two sheets of PET-foil (OAN/IA, D-K Kunststoff-Folien GmbH, Dessau, Germany). Depending on the applied pressure during the experiment, two foils with different thickness (0.50 mm for low pressures and 0.75 mm for higher pressures) can be chosen. The outer sheet covers the complete inner containment ring and juts out 1.5 cm above it. The inner sheet of foil covers

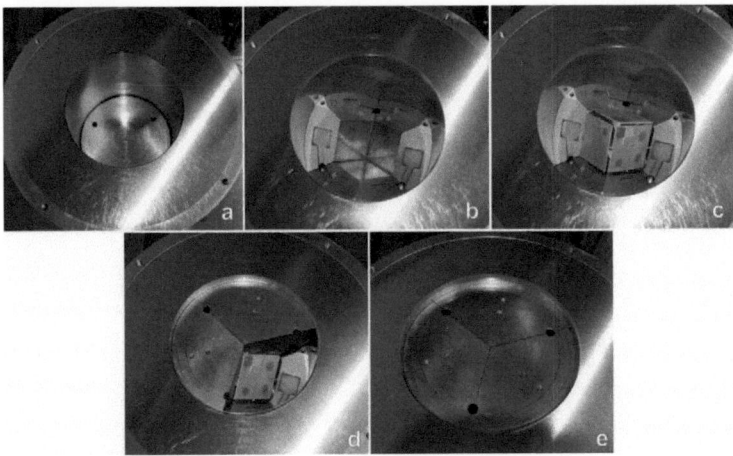

Figure 2.5-10
Assembling of the Walker-type module:
a) empty Walker-type module   b) nest formed by three steel wedges
c) placement of the cube inside the nest   d) insertion of the top wedges
e) loaded Walker-type module

2.5 Sample Preparation

the outside of each wedge. Both foils are fixed to the steel parts by a thin film of polytetraflouroethylen (PTFE) spray. The polyester sheets fulfill several functions: The lubricated interfaces between containment ring, wedges, and polyester sheets provide a slip surface to accommodate the motion of the wedges, while the polyester sheets provide electrical insulation of the containment ring from the wedges. After assembling the Walker-type module as in Figure 2.5-10, the pressure distribution plate is lifted on top of the last three wedges with care, so that the projecting polyester sheet is not damaged.

Figure 2.5-11
Module under load

Figure 2.5-11 shows the module under load. On the front side of the pressure distribution plates, the cooling hoses are attached (Figure 2.5-12 left). On the rear side of the bottom plate, there is a small metal extension, which moves directly into a small shoe connecting the plate with one pole of the electrical circuit for heating. The second pole is fixed at the ram and contacts the pressure distribution plate, when the press is closed (Figure 2.5-12 right). The bottom pressure distribution plate and the ram are isolated through massive glass fiber plates from the module.

Figure 2.5-12
Left: Cooling hoses attached to the front side of the Walker-type module
Right: Power connection on the rear side of the Walker-type module

2.6 Calibration of the Multianvil Device

2.6.1 Pressure Calibration

The calculation of the effective pressure inside a multianvil device in dependence of the applied force is nearly impossible, since too many factors have to be taken into account. For example, the gasket and side-support arrangements absorb a sizable fraction of the force applied to the sample, making it impossible to determine the exact pressure present in the sample capsule. Therefore, calibration procedures are applied, that calibrate pressures as a function of hydraulic oil pressure and rely on known phase transitions. Data points concerning the transition pressure for certain materials were collected during experiments in diamond anvil cells, where *in situ* pressure measurements are carried out *via* the ruby luminescence. Common materials used for calibrating the assemblies up to a pressure of 13 GPa are bismuth (I-II, II-III, and III-V transitions at 2.55, 3.15, and 7.70 GPa, respectively), representing one of the most studied elements at high-pressure [81-92], and ZnTe, showing a semi-conductor to metal transitions (6 GPa –anomaly–, LPP-HPP1 at 9.6 GPa, and HPP1-HPP2 at 12.1 GPa) [93-97].

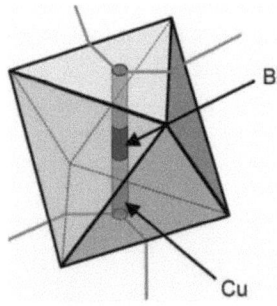

Figure 2.6-1
Schematic view of the octahedron filled with bismuth, copper electrodes, and four cable connections

For the calibration experiments with bismuth, a 3.8 mm-long cylinder of Bi metal is centered in a hole of 1.7 mm diameter between opposite faces of the octahedron (Figure 2.6-1). On both sides of the Bi rod, a copper electrode seals the remaining hole. Cable connections from copper wire, inserted between the Bi and the Cu pieces, allow electrical communication between the bismuth and the resistance measuring circuit outside the press. In contrast to Walker [13], the opposing tungsten carbide cubes contacting the copper

2.6 Calibration of the Multianvil Device

electrodes, were not part of the circuit in the here presented Bi calibrations. The bismuth resistance was measured as a voltage drop across the circuit, using a constant current of 100 mA recorded with the help of the program PRESSCONTROL [77]. Figure 2.6-2 shows a typical chart for the relative resistance of Bi in a 14/8 precast octahedron in dependence of the ram oil pressure, recorded at the LMU Munich. The diagram clearly exhibits three sharp resistance changes due to the phase transformations I-II, II-III, and III-V. On a structural level, rhombohedral Bi-I (isotypic to As) transforms into Bi-II with a monoclinic structure, and further into Bi-III, being composed of a tetragonal host structure and an interpenetrating guest component, which is incommensurate with the host [81]. Further increase in pressure results in a phase transformation to the body-centered cubic phase Bi-V above 7.7 GPa.

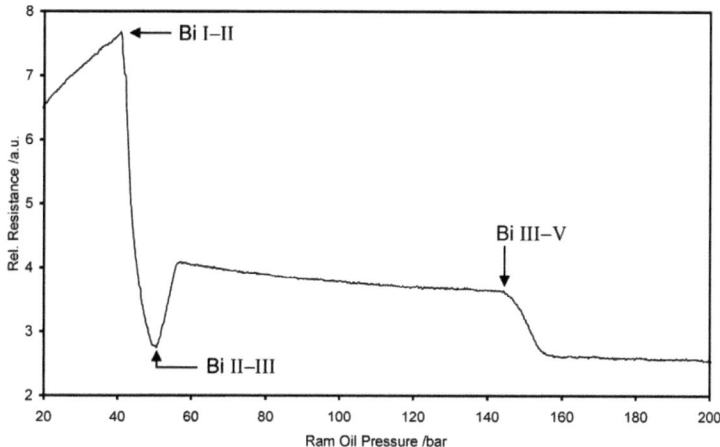

Figure 2.6-2
Relative resistance of Bi in dependence of the ram oil pressure

For calibrations with ZnTe, a powdered sample (99.998 %, Alfa Aesar) was filled into a 1.6 mm hole inside the octahedron and compressed to a thickness of 1.9 mm between the Cu rods. Due to the much larger resistance of ZnTe (semi-conductor) compared to tungsten carbide, the cubes were part of the circuit. Figure 2.6-3 shows the resistance changes in ZnTe in dependence on the hydraulic oil pressure (recorded at the LMU Munich), exhibiting three anomalies (around 6, 9.6 and 12 GPa). In the compression process, the anomaly at 6 GPa is explained by a change of the band gap in the zinc blende type ZnTe, which is stable up to ~9.5 GPa. With increasing pressure, zinc telluride transforms to a semiconducting cinnabar-type phase (LPP-HPP1) [95, 96], followed by a transition to a metallic orthorhombic (*Cmcm*) phase (HPP1-HPP2) [97], which can be regarded as a distorted rocksalt lattice.

Figure 2.6-4 represents the pressure calibration curves obtained from various measurements for the 18/11 and 14/8 assemblies at Innsbruck University.

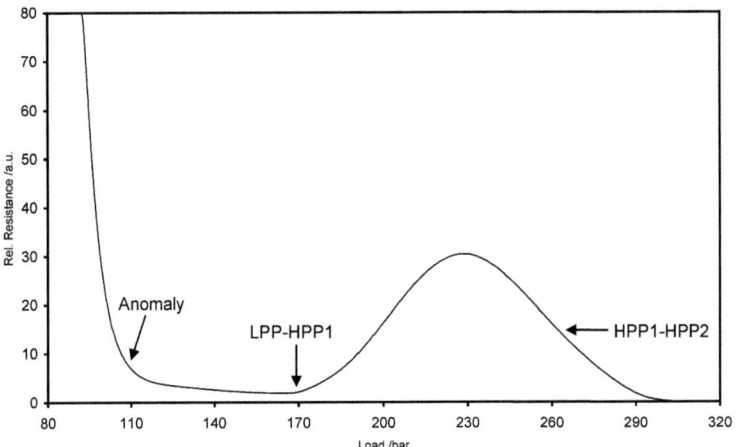

Figure 2.6-3

Relative resistance of ZnTe in dependence of the ram oil pressure

2.6 Calibration of the Multianvil Device

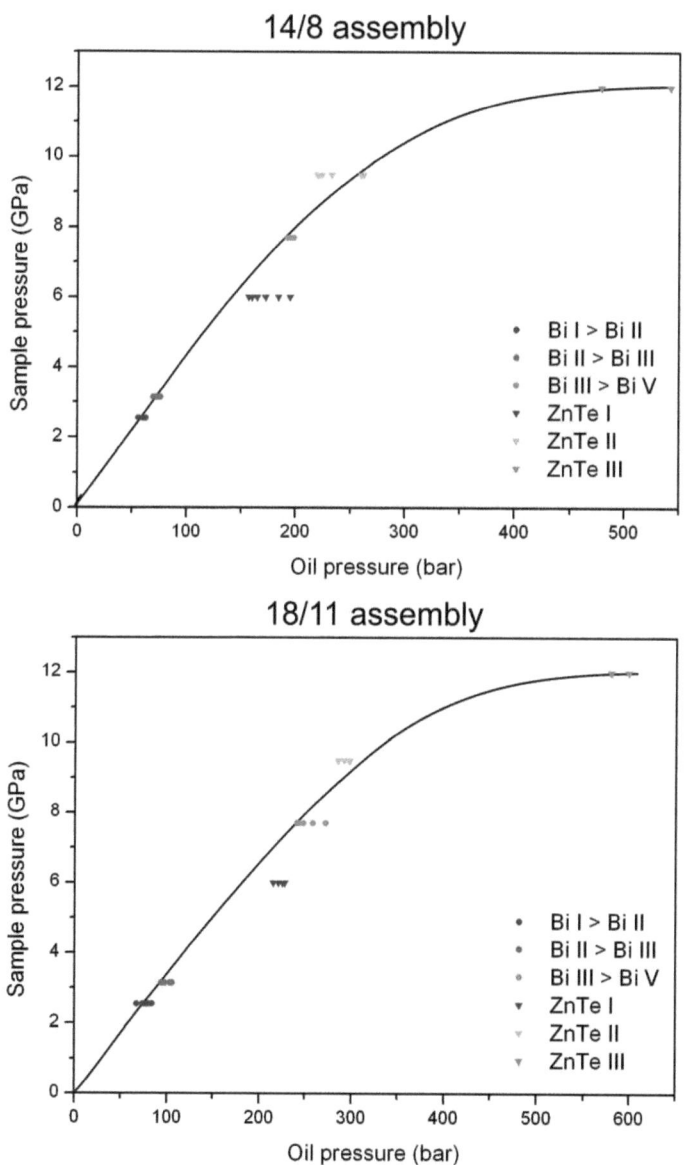

Figure 2.6-4
Calibration curves for the assemblies, obtained from the pressure phase transformation measurements

2.6.2 Temperature Calibration

Temperature calibration is the second part of the calibration procedure. The temperature is measured using a thermocouple, which can be inserted along the axis or perpendicular to the heater axis. The latter design may results in disturbance of the temperature field because of a local increase of the electrical resistance at the holes drilled through the furnaces, but it runs much more stable and was thus used for our calibrations. As thermocouple types, Pt-$Pt_{87}Rh_{13}$ (R K1, R PT13RH K1, Ögussa, Vienna, Austria) was used for temperatures up to 1500 °C, and $W_{97}Re_3$-$W_{75}Re_{25}$ above 1500 °C (Typ D, Johnson-Matthey, London, UK). Recorded temperatures are affected by several effects. Besides a thermal gradient inside the sample, which requires very exact placing of the thermocouple, one of the main uncertainties arises from the pressure effect on the thermocouple electromotive force (emf) [98]. Experimental difficulties prevented the clear measurement of this effect. Curves of the measured temperatures in dependence of the heating power were recorded for every assembly, to be able to perform synthesis in a roughly estimated temperature range without fitting a thermocouple into the assembly. An example for such curves is given in Figure 2.6-5.

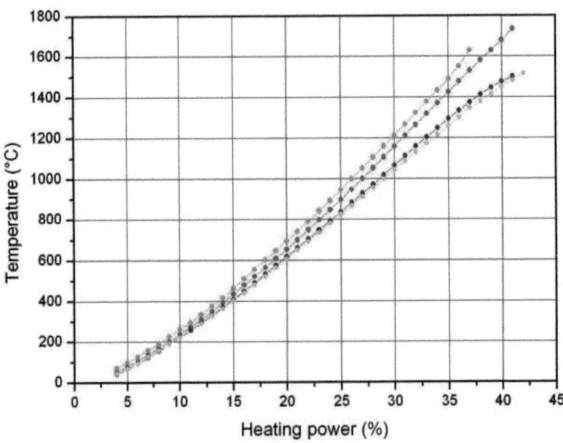

Figure 2.6-5
Temperature calibration curves for a 14/8 assembly at 500 bar oil pressure

2.7 Recovery of the Sample

After decompression of the Walker-type module, the top pressure distribution plate is lifted, the three top wedges are removed, and the eight inner cubic anvils with the octahedron inside carefully raised out of the nest.

Figure 2.7-1 gives a view of the eight anvils separated after the experiment. The gaskets are mashed, prevented from extruding by the cardboards and the PTFE-tape. The recovered octahedron is cracked open with a center punch and broken apart (Figure 2.7-2). The graphite cylinder including the capsule can be easily isolated from the octahedron, and the sample is carefully separated from the boron nitride capsule. During this process, possible reactions of the crucible material are revealed. Different stages of this process are shown in Figure 2.7-3.

Figure 2.7-1
Cubic anvils with octahedron after the experiment

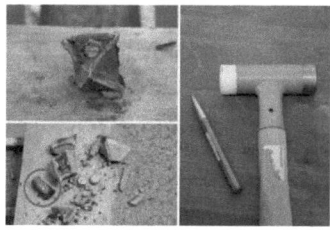

Figure 2.7-2
The recovered octahedron (top left) is broken apart and the graphite capsule (red circle, bottom left) is obtained.

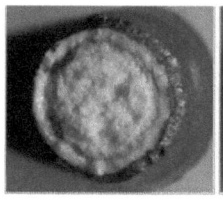 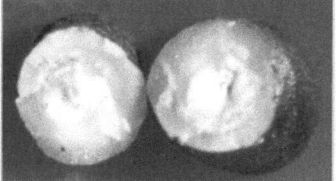 

Figure 2.7-3
Left: staring reaction.   Middle advanced reaction.   Right: Complete reaction

2.8 Experimental Danger

During the high-pressure experiments, the Walker-type module and its inner parts are put under extreme strain. Due to the robustness of the module, dangers caused by ruptures of the rings are excluded. Nevertheless, safety planes of thick acrylic glass shield the ram and the module during experiments. The most common accidental event in the laboratory is the so-called blowout - a sudden decrease of the inner pressure of the assembly, indicating severe instabilities. This may either occur due to an expansion or explosion of the sample or the failure of one of the gaskets. This usually occurs during the heating period and can result in a very loud bang. Therefore, it is required to wear ear protectors during the heating phase of an experiment. Automatic safety shutdown of the heating and a complete pressure reduction is programmed for the case of a blowout.

The wedges and lid of the Walker-type module are massive steel parts and thus very heavy. Since they are moved manually to and from the module, safety shoes with steel toe cap have to be worn while operating the module parts.

Figure 2.8-1
Broken cube with sharp-edged fragments

Special care is needed, when handling the eight tungsten carbide cubic anvils after the experiment. Especially when the cubes were used at high-temperatures and high-pressure, they start to built up tensions, which can lead to explosive destructions under normal pressure conditions. Figure 2.8-1 shows cubes, which form sharp-edged fragments when broken. Therefore, it is necessary to wear goggles while working with used cubes, and to keep the cubes under a protective shield, if not in use.

3 Analytical Methods

3.1 X-Ray Diffraction Methods

X-ray scattering techniques reveal information about the crystallographic structure, chemical composition, and physical properties of materials. These techniques are based on observing the scattered intensity of an X-ray beam hitting a sample as a function of incident and scattered angle, polarization, and wavelength or energy. From the angles and intensities of these diffracted beams, a three-dimensional picture of the density of electrons within the crystal can be derived. From this electron density, the mean positions of the atoms in the crystal can be determined, as well as their chemical bonds, their disorder and various other information. X-ray diffraction methods are the most common analytical procedures in solid-state chemistry. Powder X-ray diffraction (Section 3.1.2) allows a quick identification of the sample and delivers data for powder structure determination. If a single crystal is available, single crystal X-ray diffraction (Section 3.1.4) is the basis for structure determination and refinement.

3.1.1 Basic Principles of X-Ray Diffraction

X-ray diffraction analysis of crystals is based on the fact that the dimensions of their lattice spacing (0.03 – 0.5 nm) [99] is comparable to the X-ray radiation wavelength. Common radiation sources are copper or molybdenum anodes, delivering wavelengths of CuK_α = 154.18 pm and MoK_α = 71.073 pm, respectively. The periodically arranged atoms or ions in a crystal build up a three-dimensional grating on which X-rays can be dispersed. For a crystalline solid, the waves are scattered from lattice planes separated by the interplanar distance d_{hkl}. Where the scattered waves interfere constructively, they remain in phase since the path length of each wave is equal to an integer multiple of the wavelength λ. This is described in the Bragg equation (3.1), which correlates the interplanar spacing d_{hkl} with the diffraction angle θ_{hkl}:

$$2 d_{hkl} \sin \theta_{hkl} = n \lambda \tag{3.1}$$

A graphical display of the Bragg equation is given in Figure 3.1-1.

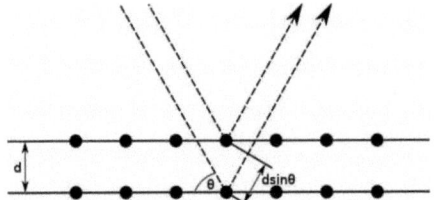

Figure 3.1-1
Reflection of X-rays on parallel lattice planes.

Due to the three-dimensional periodicity of crystals, the dispersed waves show angle-dependent intensity maxima. This is described by the three Laue equations (3.2), which correlate the lattice periods *a*, *b*, and *c* (corresponding to the x-, y-, and z-directions) and the angle of incidence μ and the diffraction angle ν of the X-ray beam with the diffraction order *n*. When all of these three equations are fulfilled, sharp reflections occur.

$$\begin{aligned} a \cos \mu_a + a \cos \nu_a &= n_1 \lambda \\ b \cos \mu_b + b \cos \nu_b &= n_2 \lambda \\ c \cos \mu_c + c \cos \nu_c &= n_3 \lambda \end{aligned} \tag{3.2}$$

When X-rays are scattered on a crystal structure, they interact with the electrons of the crystal atoms. The intensity of a reflection is dependent on the number of electrons and therefore dependent on the chemical element. The scattered total intensities I_{hkl} are proportional to the square of the structure amplitude $|F_{hkl}|^2$ (Equation (3.3)), considering several correction terms during data reduction.

3.1 X-Ray Diffraction Methods

$$I_{hkl} \propto L \cdot P \cdot |F_{hkl}|^2 \qquad (3.3)$$

(L = Lorentz factor, P = polarisation factor)

The structure factor F_{hkl} describes the way in which an incident beam is scattered by the atoms of a crystal unit cell, taking into account both the atom positions (x_j, y_j, z_j) and its scattering factor f_j. This allows a summation over all atoms j included in the cell (Equation (3.4)).

$$F_{hkl} = \sum_j f_j\, e^{2\pi i\,(hx_j + ky_j + lz_j)} \qquad (3.4)$$

Since the X-ray beam interacts with the electrons of the crystal atoms, information about the electron density in the crystal can be gathered. Following from Equation (3.3), the structure factor F_{hkl} is directly proportional to the total intensities I_{hkl}. If the structure factors are known in its amplitude and phase, the atomic positions can be uniquely determined. The electron density function ρ_{xyz} can be calculated by a Fourier Transformation of the structure factor F_{hkl} (Equation (3.5)) and an interpretation of the electron density finally yields the crystal structure.

$$\rho_{xyz} = \frac{1}{V} \sum_{hkl} F_{hkl}\, e^{-2\pi i\,(hx + ky + lz)} \qquad (3.5)$$

Since the atoms are spatially distributed in the unit cell, there will be a difference in phase when considering the scattered amplitude from two atoms. When recording intensities, only the structure amplitude $|F_{hkl}|^2$ (Equation (3.3)) is accounted for, so that the phases, which are the moduli of the structure factor, are lost. This is referred to as the "crystallographic phase problem", which has to be solved *via* mathematical operations like the Direct or the Patterson methods [100, 101], resulting in a crude structural model of the crystal. Structural refinement is done by optimization on full-matrix least-squares on F^2. Additional atoms are assigned to residual peaks of the electron density by difference Fourier Synthesis with associated refinement, until no significant residual peaks remain.

3.1.2 X-Ray Powder Diffraction

Relative to other methods of analysis, X-ray powder diffraction allows for rapid, non-destructive analysis of without the need for extensive sample preparation [102]. This gives the ability to quickly analyze unknown materials and perform materials characterization of multi-component mixtures. Identification is performed by comparison of the diffraction pattern to a known standard or to a database such as the Cambridge Structural Database (CSD). Also, the crystal structures of known materials can be refined from powder data, using the Rietveld method, a so-called full pattern analysis technique.

X-ray powder diffraction was performed on a STOE Stadi P diffractometer (STOE & Cie, Darmstadt, Germany) with Ge(111)-monochromatized MoK$_{\alpha1}$ radiation (λ = 70.93 pm). Measurements were performed in transmission geometry, placing a powder sample between two acetate films. The intensities were collected by a position sensitive detector (PSD) with an opening angle of $2\theta = 5°$

3.1.3 Temperature-Programmed X-Ray Powder Diffraction

When a sample is heated, especially in the case of a metastable high-pressure phase, phase transitions may occur. During temperature-programmed *in situ* X-ray powder diffraction, the sample is heated gradually and a powder diffraction pattern is recorded after each heating step. If the material melts to an isotropic liquid, all sharp lines will disappear and be replaced by a broad amorphous pattern. If the transition produces another crystalline phase, one set of lines will suddenly be replaced by another set. In some cases, lines will split or coalesce, *e.g.* if the material undergoes a continuous, second order phase transition.

3.1 X-Ray Diffraction Methods

Temperature-programmed X-ray powder diffraction was performed at the Ludwig-Maximilians-University (LMU) in Munich on a STOE Stadi P diffractometer (MoK$_{\alpha 1}$ radiation, λ= 70.93 pm) with a computer-controlled STOE furnace. All measurements were carried out in Debeye-Scherrer geometry (silica capillary, Fa. Hilgenberger, Malsfeld, Germany, $\varnothing$ext 0.1-0.5 mm). An electrically heated graphite tube fixed the capillary vertical in respect to the scattering plane. Borings in the tube permitted unobstructed pathways for the primary beam and the scattered radiation. The temperature, measured by a thermocouple inside the graphite tube, was kept constant within 0.2 °. The samples were heated from room temperature to 500 °C in 100 °C steps, and from 500 °C to 1100 °C in 50 °C steps. The heating rate was set to 40 °C/min. Afterwards, the samples was cooled down to 500 °C in 50 °C steps, and from 500 °C to room temperature in 100 °C steps (heating rate: 50 °C/min). After each heating step, a diffraction pattern was recorded over the desired angular range.

3.1.4 Single Crystal X-Ray Diffraction

Via single crystal X-ray diffraction, data for the crystal structure solution and refinement was collected. Small single crystals were isolated under a polarizing stereo microscope (M125, Leica, Bensheim, Germany) with a planapochromate objective. In a thin film of paraffin oil, the mechanical fragmentation of the crystal from a crystalline aggregate of the sample is done. The crystals are then moved from the oil and mounted on thin glass fibers with colorless nail polish. After a short quality check, the single crystal intensity data were measured on a KappaCCD 4-circle diffractometer (BRUKER AXS/Enraf-Nonius, Karlsruhe, Germany), equipped with graphite-monochromatized Mo-K$_\alpha$ radiation (71.073 pm), a Micracol Fiber Optics collimator, and a Nonius FR590 generator. Usually, a raw data reduction was carried out with instrumental specific software, typically accounting for Lorentz-, polarization-, and isotropic extinction corrections.

3.1.5 Computer Programs for X-Ray Diffraction Data

Powder diffractograms were recorded and handled with the STOE program package WinXPow [103]: Indexing of the diffraction patterns was performed with the algorithms of TREOR [104-106], ITO [107], and DICVOL [108], while the simulation of powder patterns based on single crystal data was done with THEO [109]. The integrated search routine „SEARCH-MATCH" [110], which refers to the JCPDS - database [111] was used for phase analysis.

Data reduction and absorption correction were performed with the programs SCALEPACK [112], X-RED [113], X-SHAPE [114], and HABITUS [115]. Data analyzation, space group determination, and semi-empirical absorption correction was done with X-PREP [116]. Crystal structures were solved by Direct or Patterson methods with SHELXS-97 [117,118] and refined with SHELXL-97 (full-matrix least-squares against F^2) [119]. Both programs are combined under the graphical user interface WinGX [120]. Structure evaluation and verification was done with PLATON [121].

Crystal structure visualization was accomplished with the graphic program DIAMOND [122].

3.2 Spectroscopic Methods

Vibrational spectroscopy is used for structure characterization of molecular arrangements, since vibrations of groups in molecules affect the electromagnetic radiation. In infrared (IR) spectroscopy, light of all frequencies is passed through a sample and the intensity of the transmitted light is measured at each frequency. The molecule absorbs light of a given frequency, thus the light is attenuated when it passes through the sample. In Raman scattering, the vibrating group in the molecule interacts with monochromatic light, and the frequency of the light changes. The shift in energy gives information about the phonon modes in the system. Ultraviolet-visible (UV-Vis) spectroscopy uses light in the visible and adjacent (near-UV and near-infrared (NIR)) ranges. The absorption in the visible range directly affects the perceived color of the chem-

3.2 Spectroscopic Methods

icals involved. In this region of the electromagnetic spectrum, molecules undergo electronic transitions. When a vibrational spectrum is obtained, peaks can be assigned to typical vibrations in databases. For this thesis, IR and Raman spectroscopy were mainly used to distinguish between BO_3- and BO_4-groups, while UV/Vis spectroscopy was used to determine the electronic states of the samples. Data interpretation and handling were performed with the program ORIGIN [123].

3.2.1 Infrared Spectroscopy

The infrared spectrum of a sample is recorded by passing a beam of infrared light through the sample. This can be done with a monochromatic beam, which changes in wavelength over time, or by using a Fourier Transform (FT) instrument to measure all wavelengths at once. When the frequency of the IR light is the same as the vibrational frequency of a molecule, absorption occurs. Examination of the transmitted light reveals how much energy was absorbed at each wavelength. Analysis of these absorption characteristics reveals details about the molecular structure of the sample. In order for a vibrational mode in a molecule to be IR active, it must be associated with changes in the permanent dipole.

IR measurements on bulk samples were recorded in reflection on a Nicolet 5700 FTIR spectrometer (Thermo Fisher Scientific Inc., Dreieich, Germany), scanning a range from 400 to 4000 cm^{-1}. Before measuring, the sample was thoroughly dried under high vacuum for several days.

Single crystals of the sample were either measured in transmission as an FTIR spectrum or in attenuated total reflection (ATR). FTIR absorbance spectra were recorded on crystals on a BaF_2 plate using a VERTEX 70 spectrometer attached to a HYPERION 3000 microscope, a MIR light source, and a LN-MCT detector (Bruker Optics GmbH, Ettlingen, Germany). For FTIR-ATR spectra, a frustrum shaped germanium ATR-crystal with a tip diameter of 100 µm was pressed on the surface of the crystals with a power of 5 N. Minimum-maximum normalization for FTIR and an enhanced ATR-correction [124] for FTIR-ATR spectra were done with the OPUS 6.5 software (Bruker Optics GmbH,

Ettlingen, Germany). Background correction and peak fitting followed *via* polynomial and folded Gauss-Lorentz functions.

3.2.2 Raman Spectroscopy

This spectroscopic method relies on the inelastic Raman scattering, of monochromatic light. The sample is illuminated with a laser beam in the visible, near infrared, or near ultraviolet range. The laser light interacts with phonons or other excitations in the system, resulting in the energy of the laser photons being shifted up or down. Wavelengths close to the laser line, due to elastic Rayleigh scattering, are filtered out while the rest of the collected light is dispersed onto a detector.

Confocal Raman spectra of single crystals were measured on a Lab Ram-HR 800 Raman micro-spectrometer (HORIBA Jobin Yvon, Unterhaching Germany) under an 100× objective (numerical aperture N.A. = 0.9, Olympus, Hamburg, Germany). The crystals were either excited by the 532 nm emission line of a 30 mW Nd-YAG-laser (green), by the 632 nm emission line of a He-Ne-laser (red), or by a 785 nm diode laser (NIR). The size of the laser spot on the crystal surface was approximately 1 µm. The scattered light was dispersed by a grating with 1800 lines/mm and collected by a 1024 × 256 open electrode CCD detector. Third order polynomial background subtraction, normalization, and band fitting by Gauss-Lorentz functions were done by the LABSPEC 5 software (HORIBA Jobin Yvon, Unterhaching Germany).

When rare-earth elements and compounds are excited with lasers, luminescence effects can be observed. To make sure that the observed bands in the Raman spectra are no luminescence bands, several measurements with different laser wavelengths were performed on each crystal. While real Raman bands are wavelength independent, luminescence bands shift when the excitation wavelength is changed.

3.2.3 UV/Vis Spectroscopy

Depending on the size and the optical quality of the crystals, UV/Vis spectra can be measured on single crystals or on polycrystalline agglomerates. Single crystals have the advantage of providing polarization information as well [125]. UV/Vis/NIR electronic absorption spectra were measured using a strongly modified CARY 17 microcrystal spectrophotometer (Spectra Services, ANU Canberra, Australia) [126].

3.3 Elemental Analysis

The investigation of the morphology and composition of single crystals or the complete sample is of great importance. Elemental analysis can detect impurities, that are not visible in the powder diffraction pattern and verify the sum formula derived from crystal stucture refinement.

During this thesis, energy dispersive X-ray spectroscopy (EDX) and electronprobe microanalysis (EPMA) were used to determine elemental compositions of the samples. Both analytical techniques are carried out in a scanning electron microscope (SEM), which images the sample surface by scanning it with a high-energy beam of electrons in a raster scan pattern. The electrons interact with the atoms that make up the sample producing signals, that contain information about the sample's surface topography, composition and other properties such as electrical conductivity. When the atoms of a sample are hit by an electron beam, it may excite an electron in an inner shell of the atom, ejecting it from the shell. An electron from an outer, higher-energy shell then fills the hole, and the difference in energy between the higher-energy shell and the lower energy shell may be released in the form of an X-ray. The number and energy of the X-rays emitted from a specimen can be measured by an energy dispersive spectrometer. Since each element has a unique atomic structure, the emitted X-rays are characteristic and allow the unique identification of the elements. For EDX analyses, an energy dispersive X-ray detector is used, while EPMA devices use a wave dispersive X-ray detector. The latter are much more sensitive to low elemental concentrations than are EDX detectors.

Concentrations in the range of 500-1000 ppm can generally be measured, and for some elements within some types of materials, the detection limit can be near 20 ppm. Nevertheless, elements lighter than oxygen cannot be measured without reservations, which is a drawback for the measurement of borates.

EDX measurements were conducted by a scanning electron microscope (JEOL Ltd., JSM-6500F, Tokyo, Japan) with field emission source (maximum resolution: 1.5 nm) and an EDX detector (model 7418, Oxford Instruments, Oxfordshire, UK). Single crystals or crystalline aggregates were fixed on a brass sample carrier with self-adhesive carbon foils (Plano, Wetzlar, Germany).

EPMA measurements were performed on an electron supermicropobe (JEOL Ltd., JXA-8100, Tokyo, Japan) equipped with an energy-dispersive system (Thermo Noran Inc., Madison, Wisconsin, USA) and five wavelength spectrometers (14 crystals, optimization for low Z-elements). The samples were prepared by embedding bulk material in epoxy resin and polishing with diamond polishing paste.

3.4 Theoretical Calculations

3.4.1 Lattice Energy Calculations According to the MAPLE Concept

The MAPLE concept (Madelung Part of Lattice Energy) [127-129] is a valuable tool for the verification of derived crystal structures. It sums up the coordinative contribution of ligands to its central atoms and thus considers the electrostatic interactions in an ionic crystal in dependence of atomic distance, charge, and coordination. For every atom, partial MAPLE values are calculated and summed up to a total value for the compound. A calculation of the lattice energy is performed *via* the Born-Landé equation (3.10), regarding the attractive and repulsive forces between the ions.

3.4 Theoretical Calculations

$$E = -\frac{N_A M z^+ z^- e^2}{4\pi\varepsilon_0 r_0}\left(1 - \frac{1}{n}\right) \quad (3.6)$$

N_A: Avogadro constant
M: Madelung constant
z: charge number of ions
e: elementary charge
ε_0: permittivity of free space
r_0: closest ionic distance
n: Born exponent

The Madelung constant M_i is used in determining the electrostatic potential of a single ion i, accounting for long-range interactions with neighboring ions j:

$$M_i = \sum_j \frac{z_j}{r_{ij}/r_0} \quad (3.7)$$

z: charge number of ions
r_0: closest ionic distance

MAPLE values are additive with high accuracy, so that the sum of the total MAPLE values of the starting materials are comparable to the MAPLE value of the product.

3.4.2 Bond-Length Bond-Strength Calculations

In solid state compounds, the bond-length bond-strength concept allows the interpretation and evaluation of chemical bond distances. Historically founded on Pauling's defined bond-grade [130], the concept was mainly applied on metals and intermetallic phases. Later on, the concept was extended to a multitude of compounds by Brown [131], as well as by Breese and O'Keefe [132]. The correlation of bond lengths and bond valences allows a prediction of bond distances in solid state compounds with known valences. Contrary, it is possible to calculate valence sums v_{ij} from bond distances d_{ij} between two atoms i and j (Equation (3.8)). Experimental bond distances derived from crystal structure determinations can be used to check crystal structures on their plausibility.

$$v_{ij} = exp\left[\frac{(R_{ij} - d_{ij})}{b}\right] \qquad (3.8)$$

The constant b in Equation (3.8) was determined to 37 pm by Brown and Altermatt [131]. The bond valence parameter R_{ij} is characteristic for each elemental combination and was determined from known compounds [132, 133]. The total valence sum V_i sums up the bond valences v_{ij} of all bonds starting from atom i:

$$V_i = \sum_j v_{ij} \qquad (3.9)$$

All bond-length bond-strength calculations in this thesis were carried out with the program VALIST [134].

3.4.3 Charge Distribution Calculations with the CHARDI Concept

The CHARDI concept (<u>Cha</u>rge <u>Di</u>stribution in Solids) [135,136] combines Pauling's bond-grade [130] and the effective coordination number (ECoN). In contrast to the MAPLE concept, CHARDI considers anion-anion and cation-cation interactions as well. The ECoN contribution $\Delta E(ij \rightarrow k)$ is based on the average distance $d(ij \rightarrow k)$ between the cations K_{ij} (crystallographic site j) and anions A_k. The summation of these contributions provides a partial effective coordination number Δ(ECoN) for every anion A_k as ligand of cation K_{ij} j. In consideration of the number of anion A_k surrounding K_{ij}, a part of the charge distribution $\Delta q(ij \rightarrow k)$ is obtained. According to Equations (3.10) and (3.11) the charge Q of the cations K_{ij} and anions A_k are calculated, respectively.

3.4 Theoretical Calculations

$$Q_{cation} = -\sum_i \sum_j \Delta q(ij \rightarrow k)_{cation} \tag{3.10}$$

$$Q_{anion} = -\sum_k \Delta q(ij \rightarrow k)_{anion} \tag{3.11}$$

4 Experimental Part

4.1 Rare-Earth Borates

4.1.1 Introduction

As mentioned in the general introduction on high-pressure borates (Section 1.3), a lot of research has been performed on the chemistry of transition metal and rare-earth borates. Concerning trivalent rare-earth oxoborates, at the beginning of our research in 1999 the composition $REBO_3$ [137-168], REB_3O_6 [55-60, 163-166, 169-185], "RE_3BO_6" [153, 186-188], $RE_{26}O_{27}(BO_3)_8$ [189, 190], $RE_{17.33}(BO_3)_4(B_2O_5)_2O_{16}$ [191, 192], and $RE_{8.66}(BO_3)_2(B_2O_5)O_8$ [193] were known. Partly, several polymorphs of specific compositions could be obtained at higher temperatures and pressures. Afterwards, the compositions REB_5O_9 [194, 195] and $RE_4B_{14}O_{27}$ [196] were discovered under normal pressure conditions by Li et al. and Nikelski et al., respectively. Applying high-pressure conditions, the Huppertz group was able to add four new compositions to the system, namely $RE_4B_6O_{15}$ [61, 62], α-/β-$RE_2B_4O_9$ [63, 64, 197-199], $RE_3B_5O_{12}$ [71, 200], and $RE_4B_{10}O_{21}$ [201, 202]. In the case of $RE_4B_6O_{15}$, the new structural motive of edge-sharing BO_4 tetrahedra was observed for the first time. Table 4.1-1 gives a survey of all known compositions of ternary rare-earth oxoborates with trivalent rare-earth cations, ordered by the ratio $RE_2O_3 : B_2O_3$.

During this thesis, the high-pressure investigations on the field of rare-earth borates were carried on. The new holmium borate $Ho_{31}O_{27}(BO_3)_3(BO_4)_6$ could be obtained by S. Hering and the first single crystal structures of λ-$PrBO_3$ and π-$ErBO_3$ were derived. The latter is a very important contribution to the ongoing discussion concerning the crystal structure of these orthoborates. In the following, the publications on these compounds are presented.

Table 4.1-1
Known phases in the system RE_2O_3/B_2O_3.

Composition	$RE_2O_3 : B_2O_3$	RE	Comments
α-REB_5O_9	1 : 5	Sm-Er	Pentaoxoborates [194]
β-REB_5O_9	1 : 5	La, Ce	Pentaoxoborates [195]
$RE_4B_{14}O_{27}$	1 : 3.5	La	[196]
α-REB_3O_6	1 : 3	La-Nd, Sm-Tb	[165, 166, 169-179, 181, 182, 184]
β-REB_3O_6	1 : 3	Nd, Sm, Gd-Lu	[55-57, 165]
γ-REB_3O_6	1 : 3	La-Nd	[58, 183]
δ-REB_3O_6	1 : 3	La, Ce	[59, 60]
$RE_4B_{10}O_{21}$	1 : 2.5	Pr	[201, 202]
α-$RE_2B_4O_9$	1 : 2	Sm-Ho	[63, 64, 197]
β-$RE_2B_4O_9$	1 : 2	Gd, Dy	[198, 199]
$RE_3B_5O_{12}$	3 : 5	Sc, Er-Lu	[71, 200]
$RE_4B_6O_{15}$	2 : 3	Dy, Ho	[61, 62]
π-$REBO_3$	1 : 1	Y, Nd, Sm-Lu	LT Pseudo hex. phases [137-139, 146-148, 153]
μ-$REBO_3$	1 : 1	Y, Sm-Gd, Dy-Lu	HT Calcite related structure [140, 142, 148, 153]
λ-$REBO_3$	1 : 1	La-Nd, Sm, Eu	Aragonite structure [137, 150, 161, 162, 166]
β-$REBO_3$	1 : 1	Sc, Er-Lu	Calcite structure [137, 138, 140, 151, 154-156]
ν-$REBO_3$	1 : 1	Ce-Nd, Sm-Dy	Tric. (H-$NdBO_3$) [149, 152, 157, 159, 163, 164]
χ-$REBO_3$	1 : 1	Dy-Er	Triclinic phases [167, 168]
H-$REBO_3$	1 : 1	La, Ce, Nd	Monocl. (H-$LaBO_3$) [158-160]
$RE_{8.66}O_8(BO_3)_2(B_2O_5)$	8.66 : 4	Ho	[193]
$RE_{17.33}O_{16}(BO_3)_4(B_2O_5)_2$	~8.7 : 4	Y, Gd	[191, 192]
RE_3BO_6	3 : 1	Y, La, Pr-Lu	($(REO)_3BO_3$) [153, 186-188]
$RE_{26}O_{27}(BO_3)_8$	13 : 4	La, Nd	(8 $RE_3BO_6 \cdot RE_2O_3$) [189, 190]

4.1 Rare-Earth Borates

4.1.2 $Ho_{31}O_{27}(BO_3)_3(BO_4)_6$

In collaboration with S. Hering (LMU Munich), a crystalline sample containing $Ho_{31}O_{27}(BO_3)_3(BO_4)_6$ was synthesized under high-pressure / high-temperature conditions of 7.5-11.5 GPa and 1200 °C (Figure 4.1-1). As seen in Table 4.1-1, the rare-earth oxoborates can be classified according to their $RE_2O_3 : B_2O_3$ ratio. The new compound $Ho_{31}O_{27}(BO_3)_3(BO_4)_6$ shows the lowest percentage of boron in a trivalent rare-earth oxoborate, synthesized so far. This becomes clear when looking at its crystal structure: isolated BO_4-tetrahedra and BO_3-units in form are embedded in a complex holmium oxide network. Therefore, this compound could also be termed as a holmium oxide borate.

In the past, all oxoborate compounds synthesized at pressures exceeding 8 GPa revealed exclusively BO_4-tetrahedra. The fact that $Ho_{31}O_{27}(BO_3)_3(BO_4)_6$ exhibits trigonal planar coordinated boron atoms is exceptional for a material synthesized under these conditions. All the more it is astonishing that the BO_3-group is stable in the structure of $Ho_{31}O_{27}(BO_3)_3(BO_4)_6$ in the plethora of oxygen atoms.

The crystal structure and properties of this oxoborate are described in the following publication. S. Hering synthesized the compound and solved the crystal structure. The preparation of single crystals and the IR and Raman measurements on the crystals were done during this thesis.

Figure 4.1-1
Crystalline sample containing $Ho_{31}O_{27}(BO_3)_3(BO_4)_6$

Solid State Sciences 12 (2010) 1993–2002

Contents lists available at ScienceDirect

Solid State Sciences

journal homepage: www.elsevier.com/locate/ssscie

High-pressure synthesis and crystal structure of the new holmium oxoborate $Ho_{31}O_{27}(BO_3)_3(BO_4)_6$

Stefanie A. Hering [a], Almut Haberer [b], Reinhard Kaindl [c], Hubert Huppertz [b,*]

[a] Department Chemie, Ludwig-Maximilians-Universität München, Butenandtstrasse 5-13 (Haus D), D-81377 München, Germany
[b] Institut für Allgemeine, Anorganische und Theoretische Chemie, Leopold-Franzens-Universität Innsbruck, Innrain 52a, A-6020 Innsbruck, Austria
[c] Institut für Mineralogie und Petrographie, Leopold-Franzens-Universität Innsbruck, Innrain 52, A-6020 Innsbruck, Austria

ARTICLE INFO

Article history:
Received 20 May 2010
Received in revised form
6 July 2010
Accepted 24 August 2010
Available online 17 September 2010

Keywords:
Holmium oxoborate
Borates
High-pressure
Multianvil
Crystal structure
Rare-earth oxoborate

ABSTRACT

The new rare-earth oxoborate $Ho_{31}O_{27}(BO_3)_3(BO_4)_6$ was synthesized in a Walker-type module under high-pressure/high-temperature conditions of 7.5–11.5 GPa and 1200 °C. The compound crystallizes in the space group $R\bar{3}$ (no. 148) with the lattice parameters $a = 2657.9(4)$ pm and $c = 1146.9(2)$ pm ($Z = 6$). The structure exhibits 11 different Ho cations next to three and six independent BO_3- and BO_4-groups, respectively. With an elemental ratio of Ho:B = 3.44 (31:9), $Ho_{31}O_{27}(BO_3)_3(BO_4)_6$ represents the rare-earth richest rare-earth oxoborate known to this day.

© 2010 Elsevier Masson SAS. All rights reserved.

1. Introduction

In the last decade, the substance class of oxoborates [1–5] was in the focus of our scientific research with respect to their structural behaviour and new synthetic possibilities under high-pressure/high-temperature conditions. The multianvil technique [6] allowed the access to several new metastable polymorphs and compositions, unattainable under normal pressure conditions. Adjacent to a multitude of new main group [7–10] and transition metal oxoborates [11–22], special attention was given to the ternary rare-earth (RE) oxoborates with trivalent rare-earth cations [23]. At the beginning of our research in 1999, trivalent rare-earth oxoborates of the composition $REBO_3$ [24–55], REB_3O_6 [50–53,56–78], "RE_3BO_6" [23,40,79,80], $RE_{26}O_{27}(BO_3)_8$ [81,82], $RE_{17.33}(BO_3)_4(B_2O_5)_2O_{16}$ [83,84], and $RE_{8.66}(BO_3)_2(B_2O_5)_2O_8$ [85] were known. Partly, several polymorphs of specific compositions could be obtained at higher temperatures and pressures. Afterwards, the compositions REB_5O_9 [86,87] and $RE_4B_{14}O_{27}$ [88] were discovered under normal pressure conditions by Li et al. and Nikelski et al., respectively. Applying high-pressure conditions, we were able to add four new compositions to the system, namely $RE_4B_6O_{15}$ [89,90],

$RE_2B_4O_9$ [91–95], $RE_3B_5O_{12}$ [96], and $RE_4B_{10}O_{21}$ [97,98]. In the case of $RE_4B_6O_{15}$, the new structural motive of edge-sharing BO_4-tetrahedra was observed for the first time. Table 1 gives a survey of all known compositions (to the best of our knowledge) of ternary rare-earth oxoborates with trivalent rare-earth cations, ordered by the ratio $RE_2O_3:B_2O_3$. So, at the top of Table 1, the rare-earth oxoborates α-/β-REB_5O_9 show the highest fraction of boron, while at the bottom of Table 1 the compounds with the lowest boron contents are positioned. As the last one, the new compound $RE_{31}O_{27}(BO_3)_3(BO_4)_6$ ($RE = Ho$) can be found, exhibiting the lowest percentage of boron in a trivalent rare-earth oxoborate, synthesized so far. In the following, we report about the high-pressure synthesis, structure, and properties of the new holmium oxoborate $Ho_{31}O_{27}(BO_3)_3(BO_4)_6$.

2. Experimental section

The new oxoborate $Ho_{31}O_{27}(BO_3)_3(BO_4)_6$ was synthesized under high-pressure/high-temperature conditions in a modified Walker-type module in combination with a 1000 t press (both devices from the company Voggenreiter, Mainleus, Germany). As pressure medium, precastable MgO-octahedra (Ceramic Substrates & Components, Isle of Wight, UK) with edge lengths of 14 or 18 mm (14/8 or 18/11-assembly) were applied. Eight tungsten carbide

* Corresponding author. Fax: +435125072934.
E-mail address: hubert.huppertz@uibk.ac.at (H. Huppertz).

1293-2558/$ – see front matter © 2010 Elsevier Masson SAS. All rights reserved.
doi:10.1016/j.solidstatesciences.2010.08.016

4.1 Rare-Earth Borates

Table 1
Known phases in the system RE_2O_3/B_2O_3.

Composition	RE_2O_3: B_2O_3	RE:B	RE	Comments
α-REB_5O_9	1:5	0.2	Sm–Er	Pentaoxoborates [86]
β-REB_5O_9	1:5	0.2	La, Ce	Pentaoxoborates [87]
$RE_4B_{14}O_{27}$	1:3.5	0.285	La	[88]
α-REB_3O_6	1:3	0.33	La–Nd, Sm–Tb	[52,53,56,58–66,68,69,74]
β-REB_3O_6	1:3	0.33	Nd, Sm, Gd–Lu	[52,75–78]
γ-REB_3O_6	1:3	0.33	La–Nd	[70,71]
δ-REB_3O_6	1:3	0.33	La, Ce	[72,73]
$RE_4B_{10}O_{21}$	1:2.5	0.4	Pr	[97,98]
α-$RE_2B_4O_9$	1:2	0.5	Sm–Ho	[91–93]
β-$RE_2B_4O_9$	1:2	0.5	Gd, Dy	[94,95]
$RE_3B_5O_{12}$	3:5	0.6	Er–Lu	[96]
$RE_4B_6O_{15}$	2:3	0.66	Dy, Ho	[89,90]
π-$REBO_3$	1:1	1	Y, Nd, Sm–Lu	LT Pseudo hex. phases [24–26,33–35,40]
μ-$REBO_3$	1:1	1	Y, Sm–Gd, Dy–Lu	HT Calcite related structure [27,29,35,40]
λ-$REBO_3$	1:1	1	La–Nd, Sm, Eu	Aragonite structure [24,37,48,49,53]
β-$REBO_3$	1:1	1	Sc, Er–Lu	Calcite structure [24,25,27,38,41–43]
ν-$REBO_3$	1:1	1	Ce–Nd, Sm–Dy	Tric. (H–$NdBO_3$) [36,39,44,46,50,51]
χ-$REBO_3$	1:1	1	Dy–Er	Triclinic phases [54,55]
H-$REBO_3$	1:1	1	La, Ce, Nd	Monocl. (H–$LaBO_3$) [45–47]
$RE_{4.66}O_6(BO_3)_2(B_2O_5)$	8.66:4	2.165	Ho	[85]
$RE_{17.33}O_{16}(BO_3)_4(B_2O_5)_2$	~8.7:4	2.175	Y, Gd	[83,84]
RE_3BO_6	3:1	3	Y, La, Pr–Lu	$((REO)_3BO_3)$ [23,40,79,80]
$RE_{26}O_{27}(BO_3)_8$	13:4	3.25	La, Nd	(8 RE_3BO_6·RE_2O_3) [81,82]
$RE_{31}O_{27}(BO_3)_3(BO_4)_6$	31:9	3.44	Ho	[This work]

Abbreviations: LT low-temperature; HT high-temperature.

cubes (TSM 10, Ceratizit, Austria) with the truncation edge lengths of 8 or 11 mm compressed the octahedra. The samples were filled into a cylindrical boron nitride crucible (BNP GmbH, HeBoSint® S10, Germany) (volume: ~9 mm³ (14/8) up to ~35 mm³ (18/11)) and sealed with a fitting boron nitride plate. Further information on the construction of the assemblies are given in references [6,99,100].

According to Eq. (1), fine powders of C-Ho_2O_3 (99.995%, Alfa Aesar, Emmerich, Germany) and B_2O_3 (99.9%, Strem Chemicals, Newburyport, USA) were mixed, ground together, and filled into the boron nitride crucibles.

$$15.5\ Ho_2O_3 + 4.5\ B_2O_3 \xrightarrow[1200\ ^\circ C]{7.5-11.5\ GPa} Ho_{31}O_{27}(BO_3)_3(BO_4)_6 \quad (1)$$

The assemblies were compressed to 11.5 GPa in 3.5 h, followed by a heating period of 15 min, in which the samples reached a temperature of 1200 °C. The temperature was held for 10 min, with a following segment of tempering (10 min) at 450 °C to quench the sample finally to room temperature within 1 min. After decompression (10.5 h), the recovered MgO-octahedra were broken apart to isolate the samples for further analytical investigations. The surrounding hexagonal boron nitride crucibles were carefully separated from the samples to obtain the air- and humidity-resistant compound $Ho_{31}O_{27}(BO_3)_3(BO_4)_6$ as a crystalline dark grey to pink (neon lamps) or yellow-orange (naturally light sources) solid according to the lighting conditions (alexandrite effect [101,102]).

To estimate the formation region of the compound with respect to the parameters pressure and temperature, we performed several different high-pressure/high-temperature experiments with varying pressures and temperatures at a constant composition of the starting materials. These experiments led to the finding that pressure could be reduced down to 7.5 GPa at the same synthesis temperature of 1200 °C, always leading to the compound $Ho_{31}O_{27}(BO_3)_3(BO_4)_6$ in good quality.

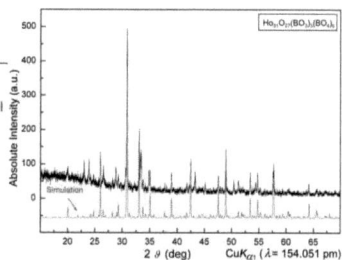

Fig. 1. Powder diffraction pattern of the metastable compound $Ho_{31}O_{27}(BO_3)_3(BO_4)_6$ with the corresponding simulation from the single crystal data. Residual reflections belong to a not yet identified side phase.

Table 2
Crystal data and structure refinement of $Ho_{31}O_{27}(BO_3)_3(BO_4)_6$.

Empirical formula	$Ho_{31}O_{27}(BO_3)_3(BO_4)_6$
Molar mass, g mol⁻¹	6170.12
Crystal system	rhombohedral
Space group	$R\bar{3}$ (no.148)
Powder diffractometer	Stoe Stadi P
Radiation	$CuK_{\alpha1}$ ($\lambda = 154.051$ pm)
Powder data	
a, pm	2658.2(6)
c, pm	1145.5(3)
V, nm³	7.010(3)
Single crystal data	
Single crystal diffractometer	Enraf-Nonius Kappa-CCD
Radiation	MoK_α ($\lambda = 71.073$ pm) (graphite monochromator)
a, pm	2657.9(4)
c, pm	1146.9(2)
V, nm³	7.017(2)
Formula units per cell	$Z = 6$
Calculated density, g cm⁻³	8.76
Crystal size, mm³	0.01 × 0.01 × 0.01
Temperature, K	293(2)
Detector distance, mm	35.0
Exposure time, min	25.0
Absorption coefficient, mm⁻¹	51.9
$F(000)$, e	15612
θ range, degree (°)	3.7–31
Range in hkl	±38; ±38; ±16
Reflections total/independent	28543/4960
R_{int}	0.1009
Reflections with $I \geq 2\sigma(I)$	3660
R_σ	0.0683
Data/ref. parameters	4960/297
Absorption correction	Multi-scan (SCALEPACK [104])
Final $R1/wR2$ [$I \geq 2\sigma(I)$]	0.0339/0.0536
Final $R1/wR2$ (all data)	0.0631/0.0580
Goodness-of-fit on F_o^2	1.036
Largest diff. peak and hole, e Å⁻³	6.95/−3.08

3. Crystal structure analysis

The samples were characterized by X-ray powder diffraction, using a Stoe Stadi P diffractometer with monochromatized Cu$K_{\alpha 1}$ ($\lambda = 154.051$ pm) radiation (Stoe & Cie, Darmstadt, Germany). Fig. 1 shows a typical powder pattern, exhibiting $Ho_{31}O_{27}(BO_3)_3(BO_4)_6$, as well as reflections of a not yet identified side phase. The experimental powder pattern tallies well with the theoretical powder pattern (Fig. 1, bottom), simulated from the single crystal data.

All powder diffractograms were measured and handled with the Stoe program package WINXPOW [103]. The powder diffraction pattern was indexed on the basis of a trigonal cell with the lattice parameters $a = 2658.2(6)$ and $c = 1145.5(3)$ pm and a volume of $7.010(3)$ nm^3.

Small single crystals of $Ho_{31}O_{27}(BO_3)_3(BO_4)_6$ were mechanically isolated and tested for suitability on a precession camera in Laue mode (Buerger Precession camera 205, Huber Diffraktionstechnik GmbH, Rimsting, Germany), combined with an image plate system (Fujifilm BAS-2500 Bio Imaging Analyser, Fuji Photo Film corporation, Japan). Single crystal intensity data were collected at room temperature from a small, regular crystal with the size $0.01 \times 0.01 \times 0.01$ mm^3, using an Enraf-Nonius Kappa-CCD single crystal diffractometer (Enraf-Nonius, Delft, Netherlands) with graphite monochromatized MoK_α ($\lambda = 71.073$ pm) radiation. A multi-scan absorption correction (SCALEPACK [104]) was applied to the intensity data.

According to the systematic extinctions, the space groups $R\bar{3}$ (no. 148) and $R3$ (no. 146) were derived. Structure solution and parameter refinement (full-matrix least squares against F^2) were successfully achieved in the space group $R\bar{3}$ (no. 148) by the SHELX-97 software set [105,106]. With the exception of the boron atom B2, all atoms could be refined with anisotropic displacement parameters. The high standard deviations of the B–O bond-lengths were

Table 3
Atomic coordinates and isotropic equivalent displacement parameters (U_{eq}/Å^2) for $Ho_{31}O_{27}(BO_3)_3(BO_4)_6$ (space group: $R\bar{3}$, No. 148). U_{eq} is defined as one third of the trace of the orthogonalized U_{ij} tensor.

Atom	Wyckoff site	x	y	z	U_{eq}
Ho1	18f	0.04045(2)	0.13786(2)	0.24508(4)	0.0074(2)
Ho2	18f	0.09366(2)	0.29530(2)	0.25530(4)	0.0074(2)
Ho3	18f	0.00696(2)	0.33401(2)	0.09479(4)	0.0113(2)
Ho4	18f	0.09249(2)	0.20633(2)	0.92080(4)	0.0082(2)
Ho5	18f	0.95947(2)	0.19489(2)	0.09030(4)	0.0074(2)
Ho6	18f	0.95431(2)	0.19082(2)	0.44075(4)	0.0084(2)
Ho7	18f	0.12113(2)	0.08890(2)	0.42691(4)	0.0099(2)
Ho8	18f	0.90851(2)	0.03105(2)	0.07486(4)	0.0086(2)
Ho9	18f	0.23591(2)	0.35261(2)	0.09477(4)	0.0084(2)
Ho10	18f	0.12334(2)	0.44307(2)	0.38490(4)	0.0102(2)
Ho11	6c	0	0	0.25271(0)	0.0170(2)
O1	18f	0.0857(3)	0.3476(3)	0.4003(5)	0.006(2)
O2	18f	0.0580(3)	0.2083(2)	0.8364(5)	0.007(2)
O3	18f	0.1526(3)	0.2744(3)	0.0280(5)	0.007(2)
O4	18f	0.9482(3)	0.1263(3)	0.2446(5)	0.010(2)
O5	18f	0.9970(3)	0.2626(3)	0.2484(5)	0.011(2)
O6	18f	0.8271(3)	0.9442(3)	0.0306(5)	0.007(2)
O7	18f	0.0700(3)	0.9950(3)	0.3978(5)	0.008(2)
O8	18f	0.1760(3)	0.3888(3)	0.8967(5)	0.010(2)
O9	18f	0.0504(3)	0.2053(3)	0.3891(6)	0.010(2)
O10	18f	0.0926(3)	0.3359(3)	0.0883(5)	0.007(2)
O11	18f	0.2186(3)	0.1384(3)	0.3678(5)	0.009(2)
O12	18f	0.9429(3)	0.1253(3)	0.5538(6)	0.013(2)
O13	18f	0.0023(3)	0.2570(3)	0.5430(6)	0.011(2)
O14	18f	0.0023(3)	0.0945(3)	0.2549(5)	0.012(2)
O15	18f	0.1453(3)	0.2832(3)	0.4314(5)	0.014(2)
O16	18f	0.2029(3)	0.4406(3)	0.4521(5)	0.009(2)
O17	18f	0.9928(3)	0.2640(3)	0.9481(6)	0.013(2)
O18	18f	0.2704(3)	0.3571(3)	0.9185(6)	0.010(2)
O19	18f	0.0044(3)	0.0692(3)	0.1042(6)	0.010(2)
O20	18f	0.0511(3)	0.2073(3)	0.0902(6)	0.015(2)
B1	18f	0.2211(5)	0.1167(5)	0.252(2)	0.009(2)
B2	18f	0.2203(5)	0.4457(5)	0.576(2)	0.007(2)
B3	18f	0.1070(5)	0.2232(5)	0.421(2)	0.014(3)

Table 4
Anisotropic displacement parameters (U_{ij}/Å^2) for $Ho_{31}O_{27}(BO_3)_3(BO_4)_6$ (space group $R\bar{3}$, No. 148).

	U_{11}	U_{22}	U_{33}	U_{23}	U_{13}	U_{12}
Ho1	0.0072(2)	0.0068(2)	0.0084(2)	0.0002(2)	0.0010(2)	0.0036(2)
Ho2	0.0071(2)	0.0081(2)	0.0070(2)	0.0007(2)	0.0001(2)	0.0038(2)
Ho3	0.0076(2)	0.0072(2)	0.0191(3)	−0.0023(2)	−0.0026(2)	0.0036(2)
Ho4	0.0085(2)	0.0083(2)	0.0082(2)	−0.0007(2)	−0.0010(2)	0.0043(2)
Ho5	0.0066(2)	0.0075(2)	0.0072(2)	−0.0003(2)	−0.0006(2)	0.0029(2)
Ho6	0.0091(2)	0.0082(2)	0.0088(2)	−0.0010(2)	−0.0007(2)	0.0050(2)
Ho7	0.0122(2)	0.0095(2)	0.0083(2)	0.0008(2)	0.0008(2)	0.0057(2)
Ho8	0.0087(3)	0.0073(2)	0.0081(2)	0.0005(2)	0.0003(2)	0.0027(2)
Ho9	0.0070(2)	0.0098(2)	0.0070(2)	−0.0003(2)	0.0001(2)	0.0031(2)
Ho10	0.0091(2)	0.0083(2)	0.0134(3)	−0.0007(2)	−0.0018(2)	0.0047(2)
Ho11	0.0086(3)	U_{11}	0.0338(6)	0	0	0.0043(2)
O1	0.008(3)	0.004(3)	0.004(3)	−0.001(3)	0.003(3)	0.002(3)
O2	0.009(3)	0.006(3)	0.004(3)	0.000(3)	0.005(3)	0.003(3)
O3	0.008(3)	0.009(3)	0.009(4)	0.003(3)	0.001(3)	0.005(3)
O4	0.002(3)	0.010(4)	0.014(4)	−0.005(3)	−0.003(3)	0.002(3)
O5	0.007(3)	0.016(4)	0.007(4)	−0.002(3)	−0.002(3)	0.004(3)
O6	0.004(3)	0.007(3)	0.008(3)	0.001(3)	0.005(3)	0.001(3)
O7	0.012(4)	0.013(4)	0.005(4)	0.002(3)	0.004(3)	0.010(3)
O8	0.007(3)	0.006(3)	0.010(4)	−0.001(3)	0.000(3)	−0.002(3)
O9	0.007(3)	0.016(4)	0.010(4)	−0.004(3)	−0.002(3)	0.006(3)
O10	0.007(3)	0.008(3)	0.005(3)	0.001(3)	0.003(3)	0.003(3)
O11	0.008(3)	0.017(4)	0.002(3)	−0.002(3)	−0.001(3)	0.007(3)
O12	0.010(4)	0.017(4)	0.011(4)	0.002(3)	0.001(3)	0.006(3)
O13	0.006(3)	0.010(4)	0.010(4)	0.003(3)	0.000(3)	0.003(3)
O14	0.021(4)	0.017(4)	0.001(3)	0.000(3)	0.001(3)	0.011(3)
O15	0.016(4)	0.014(4)	0.006(4)	−0.004(3)	0.004(3)	0.003(3)
O16	0.008(3)	0.009(3)	0.008(4)	−0.001(3)	0.000(3)	0.003(3)
O17	0.013(4)	0.008(4)	0.011(4)	−0.001(3)	0.004(3)	0.001(3)
O18	0.006(3)	0.013(4)	0.015(4)	−0.006(3)	−0.002(3)	0.007(3)
O19	0.008(4)	0.010(4)	0.014(4)	0.001(3)	0.001(3)	0.005(3)
O20	0.013(4)	0.019(4)	0.013(4)	0.005(3)	0.008(3)	0.007(3)
B1	0.015(6)	0.015(6)	0.003(6)	0.002(5)	0.002(5)	0.011(5)
B3	0.020(7)	0.024(7)	0.004(6)	0.004(5)	0.004(5)	0.016(6)

Table 5
Interatomic distances (pm) and angles (O−B−O) (°), derived from single crystal data of $Ho_{31}O_{27}(BO_3)_3(BO_4)_6$.

Ho1−O (C.N. = 7)	∅ = 232.4	Ho2−O (C.N. = 8)	∅ = 244.7	Ho3−O (C.N. = 8)	∅ = 244.3
Ho4−O (C.N. = 8)	∅ = 239.6	Ho5−O (C.N. = 8)	∅ = 241.8	Ho6−O (C.N. = 9)	∅ = 247.8
Ho7−O (C.N. = 9)	∅ = 254.6	Ho8−O (C.N. = 9)	∅ = 248.7	Ho9−O (C.N. = 7)	∅ = 235.9
Ho10−O (C.N. = 8)	∅ = 248.2	Ho11−O (C.N. = 9)	∅ = 249.9		

B1−O11	146(2)	B2−O3	148(2)	B3−O9	138(2)
B1−O8	148(2)	B2−O16	148(2)	B3−O15	140(2)
B1−O2	148(2)	B2−O6	150(2)	B3−O13	142(2)
B1−O4	152(2)	B2−O20	150(2)		∅ = 140
	∅ = 149		∅ = 149		
O11−B1−O8	111.9(8)	O3−B2−O16	109.9(8)	O9−B3−O15	117.3(9)
O11−B1−O2	108.5(8)	O3−B2−O6	106.4(8)	O9−B3−O12	121.3(10)
O8−B1−O2	111.7(8)	O16−B2−O6	111.5(8)	O15−B3−O12	121.4(9)
O11−B1−O4	109.8(8)	O3−B2−O20	113.8(8)		
O8−B1−O4	103.4(7)	O16−B2−O20	106.0(8)		
O2−B1−O4	111.5(8)	O6−B2−O20	109.3(7)		
	∅ = 109.5		∅ = 109.5		∅ = 120.0

4.1 Rare-Earth Borates

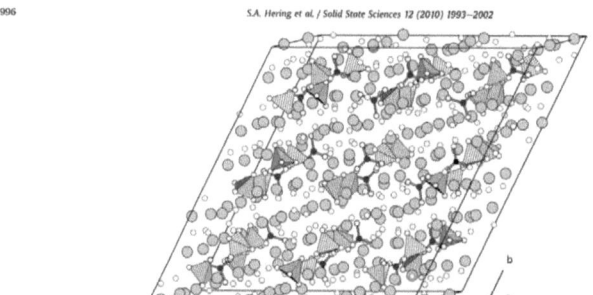

Fig. 2. Crystal structure of $Ho_{31}O_{27}(BO_3)_3(BO_4)_6$, showing BO_3- and BO_4-units, embedded in a complex holmium oxide network (grey spheres: holmium; open spheres: oxygen; black spheres: boron).

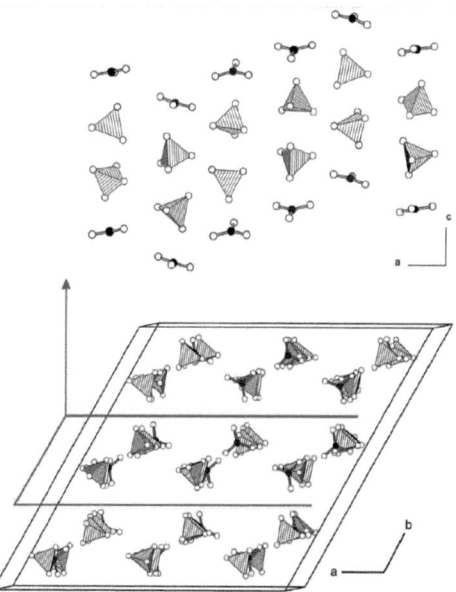

Fig. 3. Arrangement of the isolated trigonal BO_3-units and BO_4-tetrahedra in the crystal structure of $Ho_{31}O_{27}(BO_3)_3(BO_4)_6$. One section of the unit cell (indicated by the arrow) is shown in detail from a different viewing angle (top).

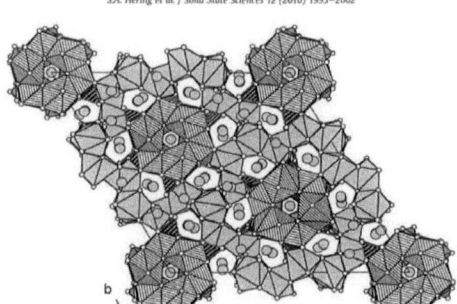

Fig. 4. Crystal structure of $Ho_{31}O_{27}(BO_3)_3(BO_4)_6$ (view along $[00\bar{1}]$) with the coordination polyhedra of Ho1, Ho2, Ho3, Ho7, and Ho8. The BO_4-tetrahedra and BO_3-groups are shown as black tetrahedra and ball-stick model, respectively (grey spheres: holmium; open spheres: oxygen; black spheres: boron).

traced back to the fact that the refinement of the light boron and oxygen atoms next to the heavy holmium atoms is always problematic (prolate ellipsoids).

All relevant information and details of the single crystal data collection are listed in Table 2. Furthermore, the positional parameters (Table 3), anisotropic displacement parameters (Table 4), and interatomic distances and angles (Table 5) are listed.

Further details of the crystal structure investigation are available from the Fachinformationszentrum (FIZ) Karlsruhe, D-76344 Eggenstein-Leopoldshafen, Germany (fax: (+49) 7247 808 666; e-mail: crysdata@fiz-karlsruhe.de), on quoting the depository number CSD-421761.

4. Results and discussion

4.1. Crystal structure of $Ho_{31}O_{27}(BO_3)_3(BO_4)_6$

Fig. 2 shows the structure of the holmium oxoborate $Ho_{31}O_{27}(BO_3)_3(BO_4)_6$, displaying isolated BO_4-tetrahedra as dark hatched polyhedra and BO_3-units in form of ball-stick models, embedded in a complex holmium oxide network. Three crystallographically different boron atoms occur in the structure, from which B1 and B2 are tetrahedrally coordinated, while B3 is coordinated trigonally planar by three oxygen atoms. Eleven crystallographically independent Ho^{3+} cations can be distinguished, which

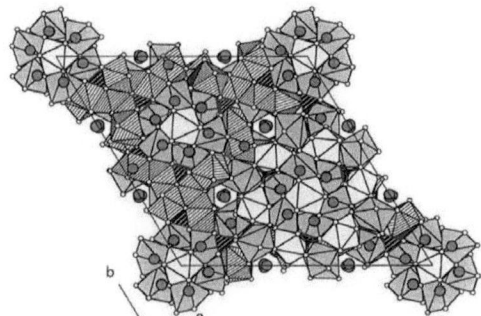

Fig. 5. Crystal structure of $Ho_{31}O_{27}(BO_3)_3(BO_4)_6$ (view along $[00\bar{1}]$) with the coordination polyhedra of Ho1, Ho4, Ho5, Ho6, Ho9, Ho10, and Ho11 (grey spheres: holmium; open spheres: oxygen; black spheres: boron).

4.1 Rare-Earth Borates

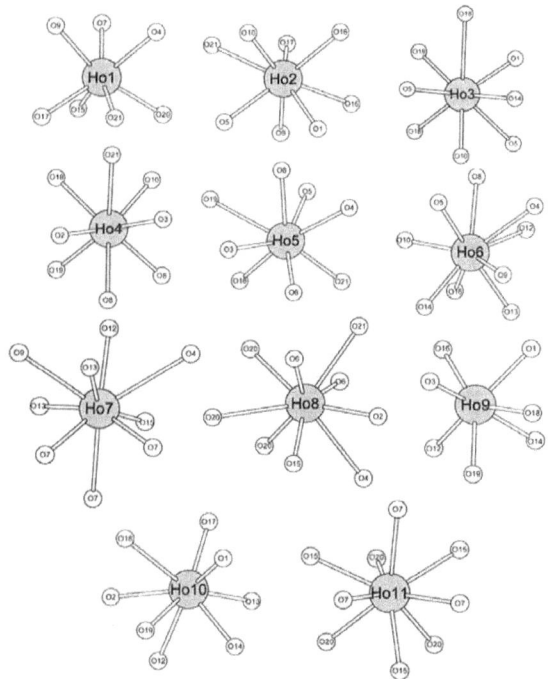

Fig. 6. Coordination spheres of the Ho–O-polyhedra in the crystal structure of $Ho_{31}O_{27}(BO_3)_3(BO_4)_6$.

Table 6
Charge distribution in $Ho_{31}O_{27}(BO_3)_3(BO_4)_6$, calculated with the bond-length/bond-strength-concept (ΣV) [109–112] and the Chardi concept (ΣQ) [113,114].

	Ho1	Ho2	Ho3	Ho4	Ho5	Ho6	Ho7	Ho8	Ho9	Ho10	Ho11
ΣV	3.18	2.86	2.72	3.33	2.98	2.94	3.04	3.23	3.03	2.74	2.49
ΣQ	3.09	2.90	2.93	2.94	3.40	3.00	2.92	2.97	2.82	2.79	2.98

	B1	B2	B3
ΣV	2.93	2.93	2.75
ΣQ	3.15	3.07	3.03

	O1	O2	O3	O4	O5	O6	O7	O8	O9	O10
ΣV	−2.23	−2.18	−2.19	−1.92	−1.67	−2.03	−1.97	−1.85	−2.06	−2.43
ΣQ	−2.42	−2.26	−2.25	−1.76	−1.70	−2.07	−2.06	−1.77	−2.09	−2.46

	O11	O12	O13	O14	O15	O16	O17	O18	O19	O20
ΣV	−1.88	−1.92	−1.97	−2.19	−1.94	−1.87	−1.64	−2.26	−1.83	−2.12
ΣQ	−1.90	−1.90	−2.11	−2.28	−1.95	−1.95	−1.42	−2.25	−1.78	−1.63

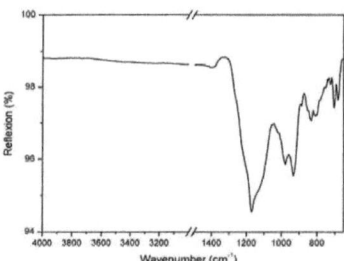

Fig. 7. FTIR spectrum of $Ho_{31}O_{27}(BO_3)_3(BO_4)_6$ bulk material.

are seven- (Ho1, Ho9), eight- (Ho2–Ho5, Ho10), and nine-fold coordinated (Ho6–Ho8, Ho11) by oxygen atoms. For a better understanding of the arrangement of the BO_3- and BO_4-units, Fig. 3 displays only these groups without the holmium ions and those oxygen atoms, which are not bound to boron. In the bottom of Fig. 3, the unit cell with a view along [001] is shown. In the top area of Fig. 3, the central area is picked out with a view along [0$\bar{1}$0]. Along the c-direction, a sequence of two BO_4-tetrahedra and one trigonal BO_3-unit can be found, while a corrugated sequence of BO_3-units and BO_4-tetrahedra can be observed along the a-axis. It is noteworthy that all BO_3- and BO_4-groups in the structure of $Ho_{31}O_{27}(BO_3)_3(BO_4)_6$ are isolated without any linkage to each other. Screening the literature, several oxoborates with isolated BO_3-groups (e.g. ludwigite ($Mg_2FeO_2(BO_3)$) [107]) and a few compounds with isolated BO_4-tetrahedra (e.g. sinhalite ($AlMgBO_4$) [108]) are known. As far as we know, $Ho_{31}O_{27}(BO_3)_3(BO_4)_6$ is the first oxoborate, in which isolated BO_3-groups and isolated BO_4-tetrahedra occur in the same structure next to each other.

The whole structure of $Ho_{31}O_{27}(BO_3)_3(BO_4)_6$ is built up from the above mentioned Ho–O-polyhedra, which are linked among themselves and additionally via the trigonal BO_3- and tetrahedral BO_4-units. The linkages of the Ho–O-polyhedra include corner-, edge-, and face-sharing of the units. Fig. 4 displays a part of the structure, viewing along the c-axis, showing the polyhedra of Ho1,

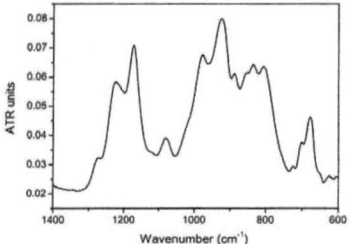

Fig. 8. FTIR-ATR spectrum of a $Ho_{31}O_{27}(BO_3)_3(BO_4)_6$ single crystal.

Ho2, Ho3, Ho7, Ho8 and the linkage to the corresponding BO_3-units and BO_4-tetrahedra. Fig. 5 displays the structure with the coordination polyhedra of Ho1, Ho4, Ho5, Ho6 Ho10, Ho11, and their linkage to the BO_3-units and BO_4-tetrahedra.

The average bond distances B–O inside the BO_4-tetrahedra show a typical mean value of 149 pm (Table 5). This value is in agreement with the typical bond length of 147.6 pm for solitary BO_4-tetrahedra [1]. The BO_3-units exhibit an average B3–O bond length of 140 pm (Table 5), which is slightly longer compared to the distances in e.g. β-YbBO$_3$ (137.8(4) pm) [43]. The average O–B–O angles show ideal values of 109.5° (B1 and B2) for the tetrahedrally coordinated boron atoms and 120.0° in the BO_3-triangles, due to their trigonal planar structure (Table 5).

The Ho–O-distances vary between 214.7 and 302.1 pm (Table 5), which is in good agreement to the bond-lengths found in $Ho_4B_6O_{15}$ (222.4(4)–262.4(4) pm) [90] or $Ho_{6.66}(BO_3)_3O_5$ (221.6(6)–286.9(13) pm) [85]. In Fig. 6, the coordination spheres of the individual Ho^{3+}-cations in $Ho_{31}O_{27}(BO_3)_3(BO_4)_6$ are shown, ranging between seven and nine coordinating oxygen anions.

The calculation of the charge distribution of the atoms in $Ho_{31}O_{27}(BO_3)_3(BO_4)_6$ via bond-valence sums by the bond-length/bond-strength-concept (ΣV) [109–112] and with the Chardi concept (Charge Distribution in Solids) (ΣQ) [113,114] confirmed the formal valence states in the oxoborate. For the calculation, bond-valence parameters of $R_{ij} = 137.1$ for the B–O bonds and $R_{ij} = 202.3$ for Ho–O bonds were applied [111]. Table 6 shows that the values for holmium and boron are in good accordance with the expected values. The calculations of the bond-valence sums for the oxygen atoms are in the expected range, except the values for O5, O10, and O17. Those values differ from the expected contribution of −2. Similar deviations for a part of the oxygen anions can be found in B-Ho$_2$O$_3$ [115], e.g. the six-fold coordinated oxygen anion in the crystal structure of B-Ho$_2$O$_3$ shows values of $\Sigma V = -1.47$ and $\Sigma Q = -1.22$, while the similarly coordinated anion O17 in $Ho_{31}O_{27}(BO_3)_3(BO_4)_6$ possesses values of $\Sigma V = -1.64$ and $\Sigma Q = -1.42$.

The calculation of the Madelung part of lattice energy (Maple) [116–118] for $Ho_{31}O_{27}(BO_3)_3(BO_4)_6$ allowed to compare the results with the corresponding Maple calculations, starting from the binary educts C-Ho$_2$O$_3$ (cubic) [119] and the high-pressure modification of boron oxide (B$_2$O$_3$-II) [120]. The additive potential of the Maple values allows to calculate hypothetical values for the compound $Ho_{31}O_{27}(BO_3)_3(BO_4)_6$, starting from the end members holmium oxide and boron oxide.

The calculations for $Ho_{31}O_{27}(BO_3)_3(BO_4)_6$ resulted in a value of 333,313 kJ/mol compared to 335,495 kJ/mol [15.5 × C-Ho$_2$O$_3$ (15275.6 kJ/mol) + 4.5 × B$_2$O$_3$-II (21,938 kJ/mol)] (deviation 0.65%), derived from the ambient pressure phase C–Ho$_2$O$_3$ and the high-pressure modification of boron oxide. Taking into account the recently published high-pressure modification of holmium oxide [115] (monoclinic B-Ho$_2$O$_3$), the Maple value comes to 333,291 kJ/mol [15.5 × B-Ho$_2$O$_3$ (15133.4 kJ/mol) + 4.5 × B$_2$O$_3$-II (21938 kJ/mol)] (deviation 0.7%).

4.2. Placement of $Ho_{31}O_{27}(BO_3)_3(BO_4)_6$ in the crystal chemistry of oxoborates

Up to now, $Ho_{31}O_{27}(BO_3)_3(BO_4)_6$ represents the rare-earth richest rare-earth oxoborate exhibiting exclusively isolated BO_3- and BO_4-groups next to each other. On the strengths of our past experience, we would have expected the exclusive occurrence of BO_4-tetrahedra for a compound, which was synthesized under extreme pressure conditions of 11.5 GPa. But $Ho_{31}O_{27}(BO_3)_3(BO_4)_6$ exhibits trigonal planar coordinated boron atoms, which are exceptional for a material synthesized under these conditions. In the past, all oxoborate compounds synthesized at pressures

4.1 Rare-Earth Borates

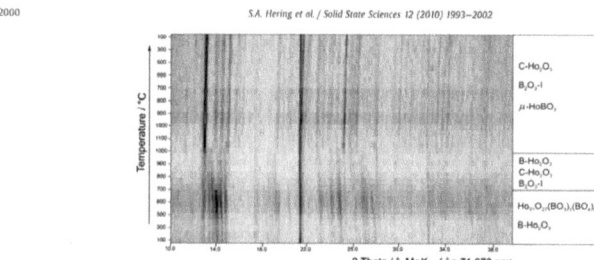

Fig. 9. *In-situ* temperature-programmed X-ray powder diffraction patterns, showing the decomposition of $Ho_{31}O_{27}(BO_3)_3(BO_4)_6$ at 700 °C.

exceeding 8 GPa revealed exclusively BO_4-tetrahedra. This behaviour is expected for oxoborates in accordance with the pressure coordination rule [121]. In a few cases, the formation of edge-sharing BO_4-tetrahedra was observed as a new structural motif for oxoborates under high-pressure conditions (e.g. HP-NiB_2O_4 [18], β-FeB_2O_4 [19], $RE_4B_6O_{15}$ (RE = Dy, Ho) [35,90], and α-$RE_2B_4O_9$ (RE = Eu, Gd, Tb, Dy) [91].) All the more it is astonishing that the BO_3-group is stable in the structure of $Ho_{31}O_{27}(BO_3)_3(BO_4)_6$, especially in the plethora of oxygen atoms.

4.3. IR spectroscopy

An FTIR spectrum of the dried $Ho_{31}O_{27}(BO_3)_3(BO_4)_6$ bulk material was recorded on a Spektrum BX II FTIR-spectrometer (Perkin Elmer, USA), equipped with a Dura sampler diamante-ATR, scanning in the range from 650 to 4500 cm^{-1}. The measurement was done at room temperature. In Fig. 7, the spectral region in the range from 4000 to 3000 cm^{-1} and from 1500 to 650 cm^{-1} is shown.

Additionally, an FTIR-ATR (Attenuated Total Reflection) spectrum of a $Ho_{31}O_{27}(BO_3)_3(BO_4)_6$ single crystal was recorded with a Bruker Vertex 70 FTIR-spectrometer (resolution ~ 0.5 cm^{-1}), attached to a Hyperion 3000 microscope in a spectral range from 600 to 4000 cm^{-1}. A frustrum-shaped germanium ATR-crystal with a tip diameter of 100 μm was pressed onto the surface of the borate crystal with a power of 5 N, which crushed it into pieces of μm-size. 64 scans for the sample and the background were acquired. Beside spectra correction for atmospheric influences, an enhanced ATR-correction [122], using the OPUS 6.5 software, was performed. Background correction and peak fitting followed via polynomial and folded Gaussian–Lorentzian functions. The FTIR-ATR spectrum in the range from 1400 to 600 cm^{-1} is given in Fig. 8.

The IR spectra of both measurements are quite similar, showing the most intense absorption areas between 1250 and 1150 cm^{-1}, 1000 and 900 cm^{-1}, 850 and 750 cm^{-1}, and around 700 cm^{-1}.

Absorptions between 1100 and 800 cm^{-1} are characteristic for tetrahedral BO_4-groups, compared to borates like β-$RE(BO_2)_3$ (RE = Dy–Lu) [78] or π-$REBO_3$ (RE = Y, Gd) [35,123], which possess exclusively BO_4-tetrahedra. Typical absorptions for triangular BO_3-groups include bending modes below 750 cm^{-1}, symmetric stretching modes around 800 cm^{-1}, and asymmetric stretching vibrations above 1200 cm^{-1}, as found in $LaBO_3$ [124,125] or χ-$REBO_3$ (RE = Dy, Er) [54,55]. Weak absorption bands were detected during single crystal measurements in the area between 3300 and 3400 cm^{-1}, indicating the presence of hydroxyl groups. Due to the small amount, these bands are most likely evoked by a surface humidity of the crystals. Corresponding absorptions for the dried bulk material were not detected (Fig. 7).

4.4. Raman spectroscopy

In order to complete the spectroscopic characterization, Raman measurements were performed on $Ho_{31}O_{27}(BO_3)_3(BO_4)_6$ single crystals. The compound showed strong luminescence, which is typical for rare-earth-containing phases. Unfortunately, a laser-excited luminescence spectrum showed that the areas with the strongest luminescence were around 550, between 630 and 680, and around 760 nm. Because these wavelengths correspond to the wavelengths of our lasers used for Raman spectroscopy, it was not possible to distinguish between luminescence bands and real absorption bands. Therefore, a characterization of the compound via Raman spectroscopy was not possible.

4.5. Thermal behaviour

The temperature stability of the oxoborate $Ho_{31}O_{27}(BO_3)_3(BO_4)_6$ was examined via an *in-situ* temperature-programmed X-ray powder diffraction experiment in air, carried out by a Stoe Stadi P powder diffractometer (Mo$K_{\alpha 1}$; λ = 70.93 pm) with a computer controlled Stoe furnace. The sample was enclosed in a silica glass capillary (Hilgenberg, Germany, $\emptyset_{ext.}$ 0.3 mm) and heated up from room temperature to 1100 °C in steps of 50 °C. Then the sample was cooled down to 500 °C in steps of 50 °C and further on in steps of 100 °C down to room temperature. Fig. 9 shows the investigated sample, consisting of $Ho_{31}O_{27}(BO_3)_3(BO_4)_6$ and a small fraction of the high-pressure modification of holmium oxide B-Ho_2O_3 (monoclinic). The latter derives from the transformation of the cubic starting material C-Ho_2O_3 under the applied conditions of 11.5 GPa. Interestingly, the synthesis at 7.5 GPa, as described in the Experimental section, did not produce the high-pressure modification B-Ho_2O_3. A detailed structural characterization of B-Ho_2O_3 can be found in Ref. [115].

The high-temperature powder patterns revealed a temperature stability of the oxoborate $Ho_{31}O_{27}(BO_3)_3(BO_4)_6$ up to a temperature of 700 °C (Fig. 9). The rising in temperature up to 950 °C led to a complete decomposition of $Ho_{31}O_{27}(BO_3)_3(BO_4)_6$; the reflections of monoclinic B-Ho_2O_3 disappeared, too. The latter ones transformed via an amorphous transition state to cubic C-Ho_2O_3, which crystallized at 1000 °C. Therewith, the high-pressure modification monoclinic B-Ho$_2$O$_3$ was re-transformed into the cubic normal pressure modification via heating, which is in accordance to the results obtained from Hoekstra et al. [126]. At 1050 °C, additional reflections of μ-HoBO$_3$ (high-temperature modification of HoBO$_3$) emerged, which result from a reaction of C-Ho$_2$O$_3$ with the boron-containing decomposition products of Ho$_{31}$O$_{27}$(BO$_3$)$_3$(BO$_4$)$_6$. The reflection at 21.3° refers to the Stoe furnace.

5. Conclusion

The new compound Ho$_{31}$O$_{27}$(BO$_3$)$_3$(BO$_4$)$_6$ shows a remarkable composition with a ratio of rare-earth:boron = 31:9. This is the highest known ratio for a rare-earth borate so far. The crystal structure resembles a holmium oxide matrix, in which a small amount of boron is incorporated. Therefore, this compound could also be termed as a holmium oxide borate. For the future, both, an increase and a further reduction of the boron content is planned via high-pressure/high-temperature chemistry in order to explore the phase diagrams of the rare-earth borates in more detail. Especially in the boron oxide-rich part of the phase diagrams, the parameter pressure seems to be crucial for the synthesis of new crystalline compounds, repressing the perpetual tendency of glass formation in borates.

Acknowledgments

The authors gratefully acknowledge the continuous support of this work by Prof. Dr. W. Schnick, Department Chemie of the University of Munich (LMU). Special thanks go to Dr. P. Mayer and T. Miller for collecting the single crystal data. Furthermore, we would like to thank A. Sattler (University of Munich (LMU)) for measuring the bulk IR data. This work was financially supported by the Deutsche Forschungsgemeinschaft (HU966/2-3) and the Fonds der Chemischen Industrie.

References

[1] F.C. Hawthorne, P.C. Burns, J.D. Grice, Reviews in Mineralogy (Chapter 2), in: Boron: Mineralogy, Petrology, and Geochemistry, vol. 33. Mineralogical Society of America, Washington, 1996.
[2] P. Becker, Adv. Mater. 10 (1998) 979.
[3] T. Sasaki, Y. Mori, M. Yoshimura, Y.K. Yap, T. Kamimura, Mater. Sci. Eng. 30 (2000) 1.
[4] D.A. Keszler, Curr. Opin. Solid State Mater. Sci. 1 (1996) 204.
[5] D.A. Keszler, Curr. Opin. Solid State Mater. Sci. 4 (1999) 155.
[6] H. Huppertz, Z. Kristallogr. 219 (2004) 330.
[7] H. Huppertz, Z. Naturforsch. 58b (2003) 257.
[8] J.S. Knyrim, F.M. Schappacher, R. Pöttgen, J. Schmedt auf der Günne, D. Johrendt, H. Huppertz, Chem. Mater. 19 (2007) 254.
[9] J.S. Knyrim, P. Becker, D. Johrendt, H. Huppertz, Angew. Chem. 118 (2006) 8419;
Angew. Chem. Int. Ed. 45 (2006) 8239.
[10] J.S. Knyrim, S.R. Römer, W. Schnick, H. Huppertz, Solid State Sci. 11 (2009) 343.
[11] H. Huppertz, G. Heymann, Solid State Sci. 5 (2003) 281.
[12] H. Emme, M. Weil, H. Huppertz, Z. Naturforsch. 60b (2005) 815.
[13] J.S. Knyrim, H. Huppertz, J. Solid State Chem. 180 (2007) 742.
[14] J.S. Knyrim, J. Friedrichs, S. Neumair, F. Roeßner, Y. Floredo, S. Jakob, D. Johrendt, R. Glaum, H. Huppertz, Solid State Sci. 10 (2008) 168.
[15] J.S. Knyrim, H. Huppertz, Z. Naturforsch. 63b (2008) 707.
[16] A. Haberer, H. Huppertz, Z. Naturforsch. 63b (2008) 713.
[17] J.S. Knyrim, H. Emme, M. Döblinger, O. Oeckler, M. Weil, H. Huppertz, Chem. Eur. J. 19 (2008) 6149.
[18] J.S. Knyrim, F. Roeßner, S. Jakob, D. Johrendt, I. Kinski, R. Glaum, H. Huppertz, Angew. Chem. 119 (2007) 9256;
Angew. Chem. Int. Ed. 46 (2007) 9097.
[19] S.C. Neumair, R. Glaum, H. Huppertz, Z. Naturforsch. 64b (2009) 883.
[20] J.S. Knyrim, H. Huppertz, J. Solid State Chem. 181 (2008) 2092.
[21] A. Haberer, H. Huppertz, J. Solid State Chem. 182 (2009) 484.
[22] S.C. Neumair, J.S. Knyrim, R. Glaum, H. Huppertz, Z. Anorg. Allg. Chem. 635 (2009) 2002.
[23] Gmelin Handbook of Inorganic and Organometallic Chemistry C11b, eight ed. Springer Verlag, Berlin, 1991.
[24] E.M. Levin, R.S. Roth, J.B. Martin, Am. Mineral 46 (1961) 1030.
[25] S. Hosokawa, Y. Tanaka, S. Iwamoto, M. Inoue, J. Mater. Sci. 43 (2008) 2276.
[26] J.H. Lin, S. Sheptyakov, Y. Wang, P. Allenspach, Chem. Mater. 16 (2004) 2418.
[27] T.A. Bither, H.S. Young, J. Solid State Chem. 6 (1973) 502.
[28] F. Bartram, E.J. Felten, Rare Earth Res. Conf. (1961) 329.
[29] W.F. Bradley, D.L. Graf, R.S. Roth, Acta Crystallogr. 20 (1966) 283.
[30] J.-Y. Henry, Mat. Res. Bull. 11 (1976) 577.
[31] R.E. Newnham, M.J. Redman, R.P. Santoro, J. Am. Ceram. Soc. 46 (1963) 253.
[32] R.S. Roth, J.L. Waring, E.M. Levin, Proc. 3rd Conf. Rare Earth Res., Clearwater, 1963, Fla., pp. 153.
[33] P.E.D. Morgan, P.J. Carroll, F.F. Lange, Mat. Res. Bull. 12 (1977) 251.
[34] G. Chadeyron, M. El-Ghozzi, R. Mahiou, A. Arbus, J.C. Cousseins, J. Solid State Chem. 128 (1997) 261.
[35] M. Ren, J.H. Lin, Y. Dong, L.Q. Yang, M.Z. Su, L.P. You, Chem. Mater. 11 (1999) 1576.
[36] K.K. Palkina, V.G. Kuznetsov, L.A. Butman, B.F. Dzhurinskii, Acad. Sci. USSR 2 (1976) 286.
[37] H.J. Meyer, Naturwissenschaften 56 (1969) 458.
[38] H.J. Meyer, A. Skokan, Naturwissenschaften 58 (1971) 566.
[39] H.J. Meyer, Naturwissenschaften 59 (1972) 215.
[40] M.Th. Cohen-Adad, O. Aloui-Lebbou, C. Goutaudier, G. Panczer, C. Dujardin, C. Pedrini, P. Florian, D. Massiot, F. Gerard, Ch. Kappenstein, J. Solid State Chem. 154 (2000) 204.
[41] D.A. Keszler, H. Sun, Acta Crystallogr. C44 (1988) 1505.
[42] S.C. Abrahams, J.L. Bernstein, E.T. Keve, J. Appl. Crystallogr. 4 (1971) 284.
[43] H. Huppertz, Z. Naturforsch. 56b (2001) 697.
[44] G. Corbel, M. Leblanc, E. Antic-Fidancev, M. Lemaître-Blaise, J.C. Krupa, J. Alloys Compd. 287 (1999) 71.
[45] R. Böhlhoff, H.U. Bambauer, W. Hoffmann, Z. Kristallogr. Krist. 133 (1971) 386.
[46] R. Böhlhoff, H.U. Bambauer, W. Hoffmann, Naturwissenschaften 57 (1970) 129.
[47] S. Lemanceau, G. Bertrand-Chadeyron, R. Mahiou, M. El-Ghozzi, J.C. Cousseins, P. Conflant, R.N. Vannier, J. Solid State Chem. 148 (1999) 229.
[48] J. Weidelt, H.U. Bambauer, Naturwissenschaften 55 (1968) 342.
[49] G.K. Abdullaev, Kh.S. Mamedov, G.G. Dzhafarov, Azerbaidzhanskii Khimicheskii Zhurnal (1976) 117.
[50] S. Noirault, O. Joubert, M.T. Caldes, Y. Piffard, Acta Crystallogr. E 62 (2006) i228.
[51] H. Emme, H. Huppertz, Acta Crystallogr. C 60 (2004) i117.
[52] F. Goubin, Y. Montardi, P. Deniard, X. Rocquefelte, R. Brec, S. Jobic, J. Solid State Chem. 177 (2004) 89.
[53] H. Müller-Bunz, T. Nikelski, Th. Schleid, Z. Naturforsch. 58b (2003) 375.
[54] H. Huppertz, B. von der Eltz, R.-D. Hoffmann, H. Piotrowski, J. Solid State Chem. 166 (2002) 203.
[55] R.-D. Hoffmann, H. Huppertz, unpublished results.
[56] H.U. Bambauer, J. Weidelt, J.-St. Ysker, Z. Kristallogr. 130 (1969) 207.
[57] H.U. Bambauer, J. Weidelt, J.-St. Ysker, Naturwissenschaften 55 (1968) 81.
[58] J.-St. Ysker, W. Hoffmann, Naturwissenschaften 57 (1970) 129.
[59] G.K. Abdullaev, Kh.S. Mamedov, G.G. Dzhafarov, Sov. Phys. Crystallogr. 26 (1981) 473.
[60] A. Goriounova, P. Held, P. Becker, L. Bohatý, Acta Crystallogr. E 60 (2004) 1131.
[61] A. Goriounova, P. Held, P. Becker, L. Bohatý, Acta Crystallogr. E 59 (2003) i83.
[62] G. Canneri, Gazz. Chim. Ital. 56 (1926) 450.
[63] P. Becker, R. Fröhlich, Cryst. Res. Tech. 43 (2008) 1240.
[64] J. Weidelt, Z. Anorg. Allg. Chem. 374 (1970) 26.
[65] I.V. Tananaev, B.F. Dzhurinskii, R.F. Chistova, Inorg. Mater. 11 (1975) 69.
[66] I.V. Tananaev, B.F. Dzhurinskii, I.M. Belyakov, Izv. Akad. Nauk SSSR Neorgan. Materialy 2 (1966) 1791.
[67] I.V. Tananaev, B.F. Dzhurinskii, Izv. Akad. Nauk SSSR Neorgan. Materialy 11 (1975) 165.
[68] G.K. Abdullaev, Kh.S. Mamedov, G.G. Dzhafarov, Sov. Phys. Crystallogr. 20 (1975) 161.
[69] C. Sieke, T. Nikelski, Th. Schleid, Z. Anorg. Allg. Chem. 628 (2002) 819.
[70] H. Emme, C. Despotopoulou, H. Huppertz, Z. Anorg. Allg. Chem. 630 (2004) 1717.
[71] H. Emme, C. Despotopoulou, H. Huppertz, Z. Anorg. Allg. Chem. 630 (2004) 2450.
[72] G. Heymann, T. Soltner, H. Huppertz, Solid State Sci. 8 (2006) 821.
[73] A. Haberer, G. Heymann, H. Huppertz, Z. Naturforsch. 62b (2007) 759.
[74] V.I. Pakhomov, G.B. Sil'nitskaya, A.V. Medvedev, B.F. Dzhurinskii, Inorg. Mater. 8 (1972) 1107.
[75] H. Emme, G. Heymann, A. Haberer, H. Huppertz, Z. Naturforsch. 62b (2007) 765.
[76] T. Nikelski, Th. Schleid, Z. Anorg. Allg. Chem. 629 (2003) 1017.
[77] T. Nikelski, M.C. Schäfer, H. Huppertz, Th. Schleid, Z. Kristallogr. NCS 223 (2008) 177.
[78] H. Emme, T. Nikelski, Th. Schleid, R. Pöttgen, M.H. Möller, H. Huppertz, Z. Naturforsch. 59b (2004) 202.
[79] M. Leskelä, L. Niinistö, in: K.A. Gschneider Jr., L. Eyring (Eds.), Handbook on the Physics and Chemistry of Rare-Earth, vol. 9, Elsevier Science, Amsterdam, 1986, p. 203.
[80] K.I. Machida, G.Y. Adachi, H. Hata, J. Shiokawa, Bull. Chem. Soc. Jpn. 54 (1981) 1052.
[81] J.H. Lin, M.Z. Su, K. Wurst, E. Schweda, J. Solid State Chem. 126 (1996) 287.
[82] S. Noirault, S. Celerier, O. Joubert, M.T. Caldes, Y. Piffard, Inorg. Chem. 46 (2007) 9961.
[83] J.H. Lin, S. Zhou, L.Q. Yang, G.Q. Yao, M.Z. Su, L.P. You, J. Solid State Chem. 134 (1997) 158.

4.1 Rare-Earth Borates

[84] J.H. Lin, L.P. You, G.X. Lu, L.Q. Yang, M.Z. Su, J. Mater. Chem. 8 (1998) 1051.
[85] B.F. Dzhurinskii, A.B. Ilyukhin, Kristallografiya 47 (2002) 442.
[86] L. Li, P. Lu, Y. Wang, X. Jin, G. Li, Y. Wang, L. You, J. Lin, Chem. Mater. 14 (2002) 4963.
[87] L. Li, X. Jin, G. Li, Y. Wang, F. Liao, G. Yao, J. Lin, Chem. Mater. 15 (2003) 2253.
[88] T. Nikelski, M.C. Schäfer, Th. Schleid, Z. Anorg. Allg. Chem. 634 (2008) 49.
[89] H. Huppertz, B. von der Eltz, J. Am. Chem. Soc. 124 (2002) 9376.
[90] H. Huppertz, Z. Naturforsch. 58b (2003) 278.
[91] H. Emme, H. Huppertz, Z. Anorg. Allg. Chem. 628 (2002) 2165.
[92] H. Emme, H. Huppertz, Acta Crystallogr. C 61 (2005) i29.
[93] H. Emme, H. Huppertz, Chem. Eur. J. 9 (2003) 3623.
[94] H. Huppertz, S. Altmannshofer, G. Heymann, J. Solid State Chem. 170 (2003) 320.
[95] H. Emme, H. Huppertz, Acta Crystallogr. C 61 (2005) i23.
[96] H. Emme, M. Valldor, R. Pöttgen, H. Huppertz, Chem. Mater. 17 (2005) 2707.
[97] A. Haberer, G. Heymann, H. Huppertz, J. Solid State Chem. 180 (2007) 1595.
[98] A. Haberer, G. Heymann, H. Huppertz, Z. Anorg. Allg. Chem. 632 (2006) 2079.
[99] D. Walker, M.A. Carpenter, C.M. Hitch, Am. Mineral 75 (1990) 1020.
[100] D. Walker, Am. Mineral 76 (1991) 1092.
[101] M.H. Farok, G.A. Saunders, W.A. Lambsom, R. Krüger, H.B. Senin, S. Bartlett, S. Takel, Phys. Chem. Glasses 37 (1996) 125.
[102] K. Schmetzer, Naturwissenschaften 65 (1978) 592.
[103] STOE WinXpow, V1.2, STOE & CIE GMbH (2001) Darmstadt, (Germany).
[104] Z. Otwinowski, W. Minor, Meth. Enzymol. 276 (1997) 307.
[105] G.M. Sheldrick, Shelxs-97 and Shelxl-97 — Program Suite for the Solution and Refinement of Crystal Structures. University of Göttingen, Göttingen (Germany), 1997.
[106] G.M. Sheldrick, Acta Crystallogr. A 64 (2008) 112.
[107] P. Bonazzi, S. Menchetti, N. Jahrb. Mineral. Monatsh. (1989) 69.
[108] J.H. Fang, R.E. Newnham, Mineral Mag. 35 (1965) 196.
[109] L. Pauling, J. Am. Chem. Soc. 69 (1947) 542.
[110] I.D. Brown, D. Altermatt, Acta Crystallogr. B 41 (1985) 244.
[111] N.E. Brese, M. O'Keeffe, Acta Crystallogr. B 47 (1991) 192.
[112] A. Trzesowska, R. Kruszynski, T. Bartczak, J. Acta Crystallogr. B 60 (2004) 174.
[113] R. Hoppe, Z. Kristallogr. 150 (1979) 23.
[114] R. Hoppe, S. Voigt, H. Glaum, J. Kissel, H.P. Müller, K.J. Bernet, J. Less Common Met. 156 (1989) 105.
[115] S. Hering, H. Huppertz, Z. Naturforsch. 64b (2009) 1032.
[116] R. Hoppe, Angew. Chem. 78 (1966) 52; Angew. Chem. Int. Ed. 5 (1966) 96.
[117] R. Hoppe, Angew. Chem. 82 (1970) 7; Angew. Chem. Int. Ed. 9 (1970) 25.
[118] R. Hübenthal, Maple — Program for the Calculation of Maple Values, Vers. 4, University of Gießen, Gießen (Germany), 1993.
[119] A. Bartos, K.P. Lieb, M. Uhrmacher, D. Wiarda, Acta Crystallogr. B 49 (1993) 165.
[120] C.T. Prewitt, R.D. Shannon, Acta Crystallogr. B 24 (1968) 869.
[121] A. Neuhaus, Chimia 18 (1964) 93.
[122] F.M. Mirabella Jr (Ed.), Internal Reflection Spectroscopy, Theory and Applications, Marcel Dekker, Inc., 1992, p. 276.
[123] J.P. Laperches, P. Tarte, Spectrochim. Acta 22 (1966) 1201.
[124] R. Böhlhoff, H.U. Bambauer, W. Hoffmann, Z. Kristallogr. 133 (1971) 386.
[125] W.C. Steele, J.C. Decius, J. Chem. Phys. 25 (1956) 1184.
[126] H.R. Hoekstra, Science 164 (1964) 1163.

4.1 Rare-Earth Borates

4.1.3 λ-PrBO$_3$

The rare-earth orthoborates REBO$_3$ exhibit a complex polymorphism. According to the nomenclature of Meyer [150-152], the polymorphs were designated with Greek letters (Table 4.1-1). Many of the orthoborates listed above were only obtained as powder samples, and some structural refinements were reported from powder diffraction data. For the case of λ-PrBO$_3$, only cell parameters were given by Meyer [150]. The high-pressure / high-temperature technique now yielded green single crystals at 3 GPa and 800 °C (Figure 4.1-2). The pressure-induced crystallization is a known phenomenon, which is especially of interest for borates, which are glass formers in general. In the following paper, the first single-crystal data of λ-PrBO$_3$ is given as another step towards the thorough investigation of the rare-earth orthoborates.

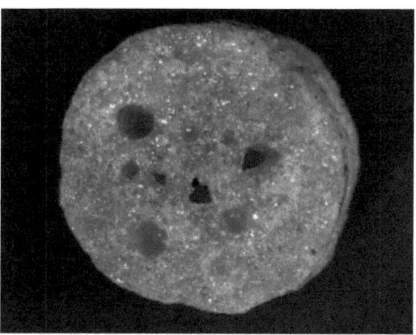

Figure 4.1-2
Crystalline sample of λ-PrBO$_3$

Synthesis and Crystal Structure of the Praseodymium Orthoborate λ-PrBO$_3$

Almut Haberer[a], Reinhard Kaindl[b], and Hubert Huppertz[a]

[a] Institut für Allgemeine, Anorganische und Theoretische Chemie, Leopold-Franzens-Universität Innsbruck, Innrain 52a, 6020 Innsbruck, Austria
[b] Institut für Mineralogie und Petrographie, Leopold-Franzens-Universität Innsbruck, Innrain 52, 6020 Innsbruck, Austria

Reprint requests to H. Huppertz. E-mail: Hubert.Huppertz@uibk.ac.at

Z. Naturforsch. 2010, 65b, 1206–1212; received May 18, 2010

The praseodymium orthoborate λ-PrBO$_3$ was synthesized from Pr$_6$O$_{11}$, B$_2$O$_3$, and PrF$_3$ under high-pressure / high-temperature conditions of 3 GPa and 800 °C in a Walker-type multianvil apparatus. The crystal structure was determined on the basis of single-crystal X-ray diffraction data, collected at room temperature. The title compound crystallizes in the orthorhombic aragonite-type structure, space group $Pnma$, with the lattice parameters a = 577.1(2), b = 506.7(2), c = 813.3(2) pm, and V = 0.2378(2) nm^3, with R_1 = 0.0400 and wR_2 = 0.0495 (all data). Within the trigonal-planar BO$_3$ groups, the average B–O distance is 137.2 pm. The praseodymium atoms are ninefold coordinated by oxygen atoms.

Key words: High Pressure, Crystal Structure, Multianvil, Orthoborate, Aragonite

Introduction

The rare earth orthoborates $REBO_3$ exhibit a complex polymorphism. Besides structural ambiguities, the inconsistent nomenclature of the different polymorphs has made this class of borates even less transparent. According to the nomenclature of Meyer [1–3], the first polymorphs were designated with Greek letters; we will stick to this representation in this article, if possible. Depending on the size of the rare earth cations, orthoborate modifications with CaCO$_3$-related structures are known. While the phases β-$REBO_3$ (RE = Sc, Er–Lu) [3–8] crystallize in the calcite-type structure, the borates λ-$REBO_3$ (RE = La–Eu) [1, 9–11] adopt the aragonite-type structure. Furthermore, there are a low-temperature modification π-$REBO_3$ (RE = Y, Nd–Lu) [7,9] and a high-temperature modification μ-$REBO_3$ (RE = Y, Sm–Gd, Dy–Lu) [12], which were suspected to have vaterite-related crystal structures [9]. In recent years, numerous studies concerning the structure and the space group of π- and μ-$REBO_3$ were performed. Several hexagonal or rhombohedral space groups were proposed [12–16], as well as a monoclinic unit cell with pseudohexagonal stacking of the borate building blocks [17, 18]. The discussion is still in progress; at least, a consensus has been reached on the constitution of these borates from B$_3$O$_9$ groups for π-$REBO_3$ (low-temperature phase), and from BO$_3$ groups for μ-$REBO_3$ (high-temperature form). Two other high-temperature phases termed H-$REBO_3$ (RE = La, Ce, Nd) [19, 20] with monoclinic crystal structures, and v-$REBO_3$ (RE = Ce–Nd, Sm–Dy) [1, 20–24] with triclinic crystal structures, are known. Additionally, high-pressure phases called χ-$REBO_3$ (RE = Dy–Er) [25, 26] were synthesized.

Many of the orthoborates listed above were only obtained as powder samples, and some structural refinements were reported from powder diffraction data. For the case of λ-PrBO$_3$, only cell parameters were given by Meyer [1]. Another paper refers to the IR properties of λ-PrBO$_3$, without giving any structural details [27]. In this article, we present the first crystal structure determination of λ-PrBO$_3$ from single crystals obtained by high-pressure / high-temperature synthesis.

Experimental Section

Synthesis

According to Meyer, the synthesis of λ-PrBO$_3$ was first carried out under ambient pressure conditions by sintering B$_2$O$_3$ and Pr$_6$O$_{11}$ at 1100 °C [1].

Single crystals of λ-PrBO$_3$ were now obtained under high-pressure / high-temperature conditions of 3 GPa

4.1 Rare-Earth Borates

and 800 °C. As the primary goal of the synthesis was a praseodymium fluoride borate, the starting reagents were Pr_6O_{11} (Strem Chemicals, 99.9 %), B_2O_3 (Strem Chemicals, 99.9+ %), and PrF_3 (Strem Chemicals, 99.9 %), which were ground together and filled into a boron nitride crucible (Henze BNP GmbH, HeBoSint® S100, Kempten, Germany) in the ratio $Pr_6O_{11} : B_2O_3 : PrF_3 = 11 : 6 : 27$. The boron nitride crucible was positioned inside the center of an 18/11 assembly, which was compressed by eight tungsten carbide cubes (TSM-10 Ceratizit, Reutte, Austria). A detailed description of the preparation of the assembly can be found in references [28–31]. The assembly was compressed to 3 GPa in 1.25 h, using a multianvil device, based on a Walker-type module and a 1000 t press (both devices from the company Voggenreiter, Mainleus, Germany). Having reached the reaction pressure, the sample was heated to 800 °C (cylindrical graphite furnace) in the following 10 min, kept there for 60 min, and cooled down to 600 °C within 20 min at constant pressure. After cooling down to r. t. by radiation heat loss, a decompression period of 3.75 h followed. A green crystalline sample could be separated from the surrounding boron nitride. Powder diffraction measurements showed that it consisted mainly of λ-$PrBO_3$, together with a praseodymium fluoride borate, which could not be fully characterized yet. Small green air- and water-resistant crystals of λ-$PrBO_3$ were mechanically separated from the product mixture.

IR spectroscopy

FTIR-ATR (*A*ttenuated *T*otal *R*eflection) spectra of the crystals were recorded with a Bruker Vertex 70 FT-IR spectrometer (resolution ~ 0.5 cm^{-1}), attached to a Hyperion 3000 microscope in a spectral range from 600–4000 cm^{-1}. A frustrum-shaped germanium ATR-crystal with a tip diameter of 100 μm was pressed onto the surface of the crystals with a power of 5 N, which let them crush into pieces of μm-size. 64 scans for sample and background were acquired. Besides spectra correction for atmospheric influences, an enhanced ATR-correction [32], using the OPUS 6.5 software, was performed. A mean refraction index of the sample of 1.6 was assumed for the ATR-correction. Background correction and peak fitting followed via polynomial and folded Gaussian-Lorentzian functions.

Raman spectroscopy

Confocal Raman spectra of single crystals were obtained with a Horiba Jobin Yvon LabRam-HR 800 Raman microspectrometer. The sample was excited by the 532 nm emission line of a 30 mW Nd-YAG-laser under an Olympus 100× objective (numerical aperture = 0.9). The size and power of the laser spot on the surface were approximately 1 μm and 5 mW. The scattered light was dispersed by a grating with 1800 lines mm^{-1} and collected by a 1024 × 256 open electrode CCD detector. The spectral resolution, determined by measuring the Rayleigh line, was about 1.4 cm^{-1}. The polynomial and convoluted Gauss-Lorentz functions were applied for background correction and band fitting. The wavenumber accuracy of about 0.5 cm^{-1} was achieved by adjusting the zero-order position of the grating and regularly checked by a neon spectral calibration lamp.

Crystal structure analysis

The powder diffraction pattern of the sample was collected with a Stoe Stadi P diffractometer, using monochromatized Mo$K_{\alpha 1}$ (λ = 70.93 pm) radiation. The main reflections of the powder pattern can be assigned to λ-$PrBO_3$ (Fig. 1), while additional reflections belong to a not yet fully characterized praseodymium fluoride borate. The diffraction pattern of λ-$PrBO_3$ was indexed with the program ITO [33] on the basis of an orthorhombic unit cell. The lattice parameters

Table 1. Crystal data and structure refinement for λ-$PrBO_3$.

Empirical formula	$PrBO_3$
Molar mass, g mol^{-1}	199.72
Crystal system	orthorhombic
Space group	$Pnma$ (no. 62)
Powder diffractometer	Stoe Stadi P
Radiation; λ, pm	Mo$K_{\alpha 1}$; 70.93
Powder diffraction data	
a, pm	576.3(2)
b, pm	506.3(1)
c, pm	812.7(2)
Volume, nm^3	0.2371(2)
Single-crystal diffractometer	Bruker AXS / Nonius Kappa CCD
Radiation; λ, pm	MoK_{α}; 71.073 (graphite monochromator)
Single-crystal data	
a, pm	577.1(2)
b, pm	506.7(2)
c, pm	813.3(2)
Volume, nm^3	0.2378(2)
Formula units per cell	$Z = 4$
Temperature, K	293(2)
Calculated density, g cm^{-3}	5.58
Crystal size, mm^3	0.03 × 0.03 × 0.025
Absorption coefficient, mm^{-1}	20.2
$F(000)$, e	352
θ range, deg	4.3–37.7
Range in hkl	±9, ±8, ±13
Total no. of reflections	3865
Independent reflections / R_{int}	687 / 0.0522
Reflections with $I \geq 2\sigma(I)$ / R_σ	579 / 0.0338
Data / ref. parameters	687 / 29
Absorption correction	Multi scan (SCALEPACK [35])
Final indices R_1 / wR_2 [$I \geq 2\sigma(I)$]	0.0283 / 0.0467
Indices R_1 / wR_2 (all data)	0.0400 / 0.0495
Goodness-of-fit on F^2	1.098
Larg. diff. peak / hole, e Å^{-3}	2.00 / −1.73

Table 2. Atomic coordinates and isotropic equivalent displacement parameters U_{eq} (Å^2) for λ-PrBO$_3$ (space group *Pnma*). U_{eq} is defined as one third of the trace of the orthogonalized U_{ij} tensor.

Atom	W. position	x	y	z	U_{eq}
Pr	4c	0.24225(4)	¼	0.58450(3)	0.00638(9)
B	4c	0.0841(9)	¼	0.2382(7)	0.0073(8)
O1	4c	0.0958(6)	¼	0.0696(4)	0.0107(7)
O2	8d	0.0864(5)	0.0165(5)	0.3231(3)	0.0099(5)

Table 3. Anisotropic displacement parameters U_{ij} (Å^2) for λ-PrBO$_3$ (space group *Pnma*).

Atom	U_{11}	U_{22}	U_{33}	U_{23}	U_{13}	U_{12}
Pr	0.0065(2)	0.0065(2)	0.0062(2)	0	0.00039(9)	0
B	0.007(2)	0.009(2)	0.006(2)	0	0.000(2)	0
O1	0.012(2)	0.013(2)	0.008(2)	0	−0.000(2)	0
O2	0.011(2)	0.008(2)	0.011(2)	0.0008(9)	−0.0018(9)	−0.0006(9)

Table 4. Interatomic distances (pm) and angles (deg), calculated with the single-crystal lattice parameters of λ-PrBO$_3$ with standard deviations in parentheses.

Pr–O1a	239.5(4)	B–O1	137.3(7)
Pr–O2a (2×)	244.7(3)	B–O2 (2×)	137.0(4)
Pr–O2b (2×)	256.3(3)		
Pr–O2c (2×)	259.4(3)	O1–B–O2 (2×)	120.2(2)
Pr–O1b (2×)	270.3(2)	O2–B–O2	119.5(4)

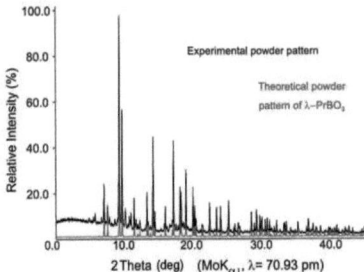

Fig. 1 (color online). Experimental powder pattern of the sample in comparison to the theoretical powder pattern of λ-PrBO$_3$ underneath, based on single-crystal diffraction data. Additional reflections of the sample are caused by a praseodymium fluoride borate not fully characterized yet.

$a = 576.3(2)$, $b = 506.3(1)$, and $c = 812.7(2)$ pm (Table 1) were gained from least-squares fits of the powder data. The correct indexing of the pattern was confirmed by intensity calculations [34], taking the atomic positions from the single-crystal structure refinements of λ-PrBO$_3$ (Table 2). The data are in good agreement with Meyer's parameters ($a = 576.9$, $b = 506.1$, and $c = 812.9$ pm) [1] and with the single-crystal data ($a = 577.1(2)$, $b = 506.7(2)$, and $c = 813.3(2)$ pm).

For the crystal structure analysis, small single crystals of λ-PrBO$_3$ were isolated by mechanical fragmentation and measured with an Enraf-Nonius Kappa CCD with graphite-monochromatized MoK$_\alpha$ (λ = 71.073 pm) radiation. Afterwards, a multi-scan absorption correction was applied to the data (SCALEPACK [35]). According to the systematic extinctions, the space groups $Pn2_1a$ (no. 33) and *Pnma* (no. 62) were derived. Structure solution and parameter refinement (full-matrix least-squares against F^2) were successfully performed in *Pnma*, using the SHELX-97 software suite [36, 37] with anisotropic displacement parameters for all atoms. All relevant details of the data collection and evaluation are listed in Table 1. The final difference Fourier synthesis did not reveal any significant residual peaks. Additionally, the positional parameters (Table 2), anisotropic displacement parameters (Table 3), interatomic distances, and angles (Table 4) are listed below.

Further details of the crystal structure investigation may be obtained from Fachinformationszentrum Karlsruhe, 76344 Eggenstein-Leopoldshafen, Germany (fax: +49-7247-808-666; e-mail: crysdata@fiz-karlsruhe.de, http://www.fiz-informationsdienste.de/en/DB/icsd/depot_anforderung.html) on quoting the deposition number CSD-421745.

Results and Discussion

λ-PrBO$_3$ crystallizes in an aragonite-type structure (space group *Pnma*), as predicted from powder diffraction data by Meyer [1]. The structure is composed of praseodymium atoms and isolated trigonal-planar BO$_3$ units (Fig. 2). Viewing along [100], the BO$_3$ groups are stacked in channels, surrounded by hexagonally arranged praseodymium cations in an antiprismatic arrangement, which is slightly off-centered, so that the boron atoms are incongruent (Fig. 3). The B–O bond lengths of the BO$_3$ groups of 137.0(4)

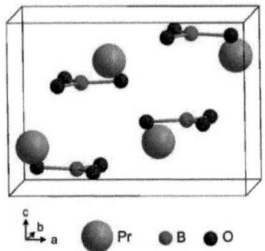

Fig. 2 (color online). Schematic drawing of the unit cell of λ-PrBO$_3$ with trigonal-planar BO$_3$ units.

4.1 Rare-Earth Borates

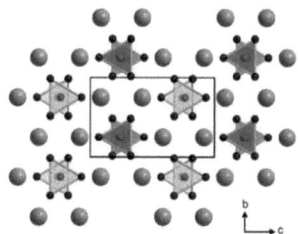

Fig. 3 (color online). View of the λ-PrBO$_3$ structure along [100]. The BO$_3$ groups are stacked in channels formed by hexagonally arranged praseodymium cations in a slightly off-centered antiprismatic arrangement.

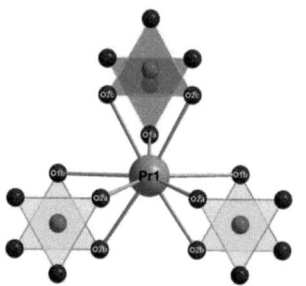

Fig. 4 (color online). Coordination sphere of the praseodymium cation in λ-PrBO$_3$. Six surrounding BO$_3$ groups donate either one or two oxygen atoms, leading to a ninefold coordination.

and 137.3(7) pm are typical for threefold coordinated boron and agree with the distances ascertained for λ-NdBO$_3$ (137.5 pm) [11]. The O–B–O angles are 119.5(4) and 120.2(2)°, as expected for a trigonal-planar geometry. Müller-Bunz et al. observed a 5.6 pm dislocation of the boron atom from the plane spanned by its oxygen ligands in λ-NdBO$_3$ [11]. For λ-PrBO$_3$, this value is reduced to 3.2 pm.

There is one crystallographically unique praseodymium atom in the structure. Six surrounding BO$_3$ groups donate either one or two oxygen atoms, leading to a ninefold coordination sphere of praseodymium (Fig. 4). The Pr–O distances are in the range between 239.5(4) and 270.3(2) pm, which is in agreement with the bond lengths of ninefold coordinated praseodymium in HP-PrO$_2$ (235.9(2) to 283.7(3) pm)

[38]. As expected from the smaller ionic radius of Nd^{3+} (130.3 pm) in comparison to Pr^{3+} (131.9 pm), the corresponding bond lengths in the isotypic phase λ-NdBO$_3$ are slightly shorter (238.6 to 269.3 pm) [11].

The calculation of the bond-valence sums for λ-PrBO$_3$ with the bond-length / bond-strength (BLBS) concept (VALIST: Bond Valence Calculation and Listing [39]) and the CHARDI concept (*cha*rge *di*stribution in solids according to Hoppe [40]) confirmed the formal ionic charges resulting from the single-crystal structure analysis (ΣV (BLBS): +3.02 (Pr), +3.00 (B), −1.93 (O1), −2.05 (O2) and ΣQ (CHARDI): +3.00 (Pr), +3.00 (B), −1.89 (O1), −2.05 (O2).

Furthermore, we calculated the MAPLE value (*Ma*delung *P*art of *L*attice *E*nergy according to Hoppe [41–43]) of λ-PrBO$_3$ in order to compare it with the sum of the MAPLE values for the high-pressure modification of Pr$_2$O$_3$ [44] and the ambient-pressure modification B$_2$O$_3$-I [45] [0.5 Pr$_2$O$_3$ (3457 kcal mol^{-1}) +0.5 B$_2$O$_3$-I (5237 kcal mol^{-1})]. The calculated value (4318 kcal mol^{-1}) for λ-PrBO$_3$ and the MAPLE value obtained from the sum of the binary oxides (4347 kcal mol^{-1}) tally well (deviation 0.7 %) and confirm the value of 4318 kcal mol^{-1} for the isotypic compound λ-NdBO$_3$ [11].

IR spectroscopy

Fig. 5 shows the FTIR-ATR spectrum of λ-PrBO$_3$ single crystals in the range between 600–1600 cm^{-1}. The IR absorption bands for aragonite-type orthoborates have been well discussed in the literature and can be divided into four different absorption areas: an in-

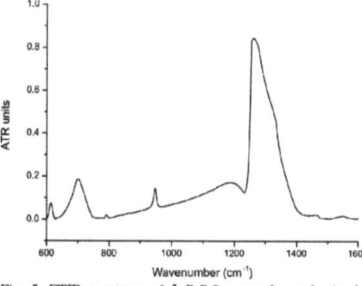

Fig. 5. FTIR spectrum of λ-PrBO$_3$ crystal powder in the range of 600–1600 cm^{-1}.

Table 5. Wavenumbers and possible assignment of ATR-FTIR absorption bands in the spectrum of λ-PrBO$_3$ single crystals (left) and of absorption bands reported from the bulk material [48].

Band	Single crystal IR Assignment	Bulk IR [48] Band
629	v_4	590, 610
700	v_2	713
790		788
1190	v_1	944
1260		1250, 1285
1463	v_3	

Table 6. Wavenumbers and possible assignment of Raman bands in the spectrum of λ-PrBO$_3$.

Band	Assignment	Band	Assignment
106		596	
113		607	pulse vibration
121		631	of BO$_3$
143		694	
180	Lattice, Pr–O,	950	symmetric stretching ?
193	bending/stretching	1240	
227	of BO$_3$	1242	asymmetric
252		1260	stretching of
330		1270	BO$_3$
373		1383	

plane bending (v_4) of the BO$_3$ groups results in bands between 600 and 670 cm^{-1}, out-of-plane bending (v_2) occurs near 740 cm^{-1}, symmetric stretching modes (v_1) between 940 and 1000 cm^{-1}, and asymmetric stretching vibrations (v_3) between 1300–1500 cm^{-1} [46, 47]. Bulk IR measurements of λ-PrBO$_3$ were previously performed by Laperches et al. [48]. Since their band assignment is not in agreement with the literature, Table 5 compares the positioning of the bands of bulk and single crystal λ-PrBO$_3$, but labels the absorption modes as shown above. There is good agreement between the absorptions of the FTIR-ATR spectrum of λ-PrBO$_3$ single crystals and the literature data. A comparable pattern of absorption bands is reported for aragonite (CaCO$_3$) [49].

Raman spectroscopy

The Raman spectrum of the λ-PrBO$_3$ single crystals shows several intense bands below 500 cm^{-1} (Fig. 6), which can be assigned to the *RE*–O bond bending and stretching vibrations, as well as to the lattice vibrations (Table 6). For aragonite (CaCO$_3$), also a large number of external Raman-active modes are reported below 500 cm^{-1} [49]. The bands observed between 600 and 700 cm^{-1} for λ-PrBO$_3$, correspond to the pulse vibration modes of CaCO$_3$ between 700 and 720 cm^{-1} (v_4 modes). The bands around 900 cm^{-1} in borates are usually assigned to the stretching modes of BO$_4$ tetrahedra, whereas BO$_3$ groups are expected > 1100 cm^{-1} [50–55]. Nevertheless, the most intense single band of λ-PrBO$_3$ occurs at 950 cm^{-1} and is also observed in the BO$_3$ group-containing phases $RE_5(BO_3)_2F_9$ (*RE* = Er, Yb) [56]. Aragonite-structured CaCO$_3$ shows it's most intense band at 1085 cm^{-1}, which is owed to the symmetric stretching vibration of the CO$_3$ group (v_1 mode) [49]. Above 1200 cm^{-1}, several broader bands for λ-PrBO$_3$ are observed and assigned to the asymmetric stretching vibrations (v_3) of the BO$_3$ groups [53]. In classical aragonite, these modes appear at higher wavenumbers.

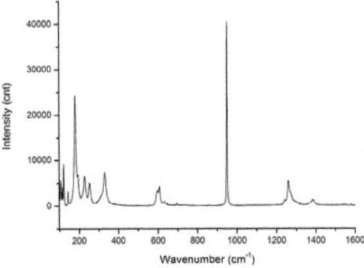

Fig. 6. Raman spectrum of a single crystal of λ-PrBO$_3$ in the range of 100–1600 cm^{-1}.

Conclusions

Under normal pressure conditions, glasses and powders are often the favored products of reactions in oxoborate chemistry. As demonstrated in this paper, high-pressure / high-temperature conditions can enforce the formation of a crystalline product, allowing the first single-crystal structure determination of λ-PrBO$_3$. Starting materials for the synthesis of λ-PrBO$_3$ were Pr$_6$O$_{11}$, B$_2$O$_3$, and PrF$_3$, resulting in a not yet fully characterized praseodymium fluoride borate as a side product. The formation of borates under the applied high-pressure/high-temperature conditions is often observed; they seem to be the most favorable reaction product even in the presence of rare earth fluorides. In these cases, the fluoride does either not participate in the reaction or humidity seems to lead to the

4.1 Rare-Earth Borates

formation of gaseous HF. An impact on the formation and stability range of the obtained borates has not been observed so far, the same reaction products occur without the fluoride.

Acknowledgements

We thank Dr. G. Heymann for collecting the single-crystal data. This work was financially supported by the Deutsche Forschungsgemeinschaft (HU966/2-3).

[1] H. J. Meyer, *Naturwissenschaften* **1969**, *56*, 458.
[2] H. J. Meyer, *Naturwissenschaften* **1972**, *59*, 215.
[3] H. J. Meyer, A. Skokan, *Naturwissenschaften* **1971**, *58*, 566.
[4] D. A. Keszler, H. Sun, *Acta Crystallogr.* **1988**, *C44*, 1505.
[5] H. Huppertz, *Z. Naturforsch.* **2001**, *56b*, 697.
[6] S. C. Abrahams, J. L. Bernstein, E. T. Keve, *J. Appl. Crystallogr.* **1971**, *4*, 284.
[7] S. Hosokawa, Y. Tanaka, S. Iwamoto, M. Inoue, *J. Mater. Sci.* **2008**, *43*, 2276.
[8] T. A. Bither, H. S. Young, *J. Solid State Chem.* **1973**, *6*, 502.
[9] E. M. Levin, R. S. Roth, J. B. Martin, *Am. Mineral.* **1961**, *46*, 1030.
[10] J. Weidelt, H. U. Bambauer, *Naturwissenschaften* **1968**, *55*, 342.
[11] H. Müller-Bunz, T. Nikelski, Th. Schleid, *Z. Naturforsch.* **2003**, *58b*, 375.
[12] R. E. Newnham, M. J. Redman, R. P. Santoro, *J. Am. Ceram. Soc.* **1963**, *46*, 253.
[13] W. F. Bradley, D. L. Graf, R. S. Roth, *Acta Crystallogr.* **1966**, *20*, 283.
[14] G. Chadeyron, M. El-Ghozzi, R. Mahiou, A. Arbus, J. C. Cousseins, *J. Solid State Chem.* **1997**, *128*, 261.
[15] M. Ren, J. H. Lin, Y. Dong, L. Q. Yang, M. Z. Su, L. P. You, *Chem. Mater.* **1999**, *11*, 1576.
[16] M. Th. Cohen-Adad, Ch. Kappenstein, O. Aloui-Lebbou, C. Goutaudier, G. Panczer, C. Dujardin, C. Pedrini, P. Florian, D. Massiot, F. Gerard, *J. Solid State Chem.* **2000**, *154*, 204.
[17] P. E. D. Morgan, P. J. Carroll, F. F. Lange, *Mater. Res. Bull.* **1977**, *12*, 251.
[18] J. H. Lin, S. Sheptyakov, Y. Wang, P. Allenspach, *Chem. Mater.* **2004**, *16*, 2418.
[19] R. Böhlhoff, H. U. Bambauer, W. Hoffmann, *Naturwissenschaften* **1970**, *57*, 129.
[20] R. Böhlhoff, H. U. Bambauer, W. Hoffman, *Z. Kristallogr.* **1971**, *133*, 386.
[21] K. K. Palkina, V. G. Kuznetsov, L. A. Butman, B. F. Dzhurinskii, *Acad. Sci. USSR* **1976**, *2*, 286.
[22] S. Noirault, O. Joubert, M. T. Caldes, Y. Piffard, *Acta Crystallogr.* **2006**, *E62*, i228.
[23] G. Corbel, M. Leblanc, E. Antic-Fidancev, M. Lemaître-Blaise, J. C. Krupa, *J. Alloys Compd.* **1999**, *287*, 71.

[24] H. Emme, H. Huppertz, *Acta Crystallogr.* **2004**, *C60*, i117.
[25] H. Huppertz, B. von der Eltz, R.-D. Hoffmann, H. Piotrowski, *J. Solid State Chem.* **2002**, *166*, 203.
[26] H. Huppertz, unpublished results.
[27] J. P. Laperches, P. Tarte, *Spectrochim. Acta* **1966**, *22*, 120.
[28] V. F. Ross, J. O. Edwards, *The Structural Chemistry of the Borates*, Wiley, New York, 1967.
[29] D. Walker, M. A. Carpenter, C. M. Hitch, *Am. Mineral.* **1990**, *75*, 1020.
[30] D. Walker, *Am. Mineral.* **1991**, *76*, 1092.
[31] H. Huppertz, *Z. Kristallogr.* **2004**, *219*, 330.
[32] F. M. Mirabella, Jr. (ed.), *Internal Reflection Spectroscopy, Theory and Applications*, Marcel Dekker, New York 1992, p. 276.
[33] J. W. Visser, *J. Appl. Crystallogr.* **1969**, *2*, 89.
[34] WINX^{POW} Software, Stoe & Cie GmbH, Darmstadt (Germany) 1998.
[35] Z. Otwinowski, W. Minor in *Methods in Enzymology*, Vol. 276, *Macromolecular Crystallography*, Part A (Eds.: C. W. Carter, Jr., R. M. Sweet), Academic Press, New York, 1997, pp. 307.
[36] G. M. Sheldrick, SHELXS-97 and SHELXL-97, Program Suite for the Solution and Refinement of Crystal Structures, University of Göttingen, Göttingen (Germany) 1997.
[37] G. M. Sheldrick, *Acta Crystallogr.* **1990**, *A46*, 467; *ibid.* **2008**, *A64*, 112.
[38] A. Haberer, H. Huppertz, *Z. Anorg. Allg. Chem.* **2010**, *636*, 363.
[39] A. S. Wills, VALIST (version 3.0.13), University College, London (U. K.), 1998 – 2008. Program available from: www.ccp14.ac.uk.
[40] R. Hoppe, S. Voigt, H. Glaum, J. Kissel, H. P. Müller, K. J. Bernet, *J. Less-Common Met.* **1989**, *156*, 105.
[41] R. Hoppe, *Angew. Chem.* **1966**, *78*, 52; *Angew. Chem., Int. Ed. Engl.* **1966**, *5*, 96.
[42] R. Hoppe, *Angew. Chem.* **1970**, *82*, 7; *Angew. Chem., Int. Ed. Engl.* **1970**, *9*, 25.
[43] R. Hübenthal, M. Serafin, R. Hoppe, MAPLE (version 4.0), Program for the Calculation of Distances, Angles, Effective Coordination Numbers, Coordination Spheres, and Lattice Energies, University of Gießen, Gießen (Germany) 1993.

[44] L. Eyring, N. C. Baenziger, *J. Appl. Phys.* **1962**, *33*, 428.
[45] G. E. Gurr, P. W. Montgomery, C. D. Knutson, B. T. Gorres, *Acta Crystallogr.* **1970**, *B26*, 906.
[46] C. E. Weir, R. A. Schroeder, *J. Research NBS-A. Phys. Chem. A* **1964**, *68*, 465.
[47] J. H. Denning, S. D. Ross, *Spectrochim. Acta* **1971**, *A28*, 1775.
[48] J. P. Laperches, P. Tarte, *Spectrochim. Acta* **1966**, *22*, 1201.
[49] R. Frech, E. C. Wang, J. B. Bates, *Spectrochim. Acta* **1980**, *A36*, 915.
[50] G. Corbel, R. Retoux, M. Leblanc, *J. Solid State Chem.* **1998**, *139*, 52.
[51] H. Huppertz, *J. Solid State Chem.* **2004**, *177*, 3700.
[52] G. Chadeyron, M. El-Ghozzi, R. Mahiou, A. Arbus, J. C. Cousseins, *J. Solid State Chem.* **1997**, *128*, 261.
[53] L. Jun, X. Shuping, G. Shiyang, *Spectrochim. Acta* **1995**, *A51*, 519.
[54] J. C. Zhang, Y. H. Wang, X. Guo, *J. Lumin.* **2007**, *122–123*, 980.
[55] G. Padmaja, P. Kistaiah, *J. Phys. Chem.* **2009**, *A113*, 2397.
[56] A. Haberer, R. Kaindl, J. Konzett, R. Glaum, H. Huppertz, *Z. Anorg. Allg. Chem.* **2010**, *636*, 1326.

4.1 Rare-Earth Borates

4.1.4 π-ErBO$_3$

While the crystal structures of most rare-earth orthoborate polymorphs REBO$_3$ seem to be fully solved, there are a low-temperature modification π-REBO$_3$ (RE = Y, Nd-Lu) [137-139, 146-148, 153] and a high-temperature modification μ-REBO$_3$ (RE = Y, Sm - Gd, Dy - Lu) [140, 142, 148, 153], which were originally suspected to have vaterite-related crystal structures. In recent years, numerous studies concerning the structure and the space group of π- and μ-REBO$_3$ were performed and only one of them was based on single-crystal data. Several hexagonal or rhombohedral space groups were proposed on the basis of powder diffraction data, as well as a monoclinic unit cell with pseudohexagonal stacking of the borate building blocks. The discussion is still in progress; at least, a consensus has been reached on the constitution of these borates from B$_3$O$_9$-groups for π-REBO$_3$ (low-temperature phase) and from BO$_3$-groups for μ-REBO$_3$ (high-temperature form).

The synthesis of π-ErBO$_3$, which was performed at 3 GPa and 800 °C (Figure 4.1-3) yielded pale pink crystals. Subsequently, a single-crystal structure determination was possible, shedding some light onto the real structural properties of π-REBO$_3$. The results are discussed in the following paper.

Figure 4.1-3
Crystalline sample of π-ErBO$_3$

Journal of Solid State Chemistry 184 (2011) 149–153

Contents lists available at ScienceDirect

Journal of Solid State Chemistry

journal homepage: www.elsevier.com/locate/jssc

The crystal structure of π-ErBO$_3$: New single-crystal data for an old problem

Almut Pitscheider [a], Reinhard Kaindl [b], Oliver Oeckler [c], Hubert Huppertz [a,*]

[a] *Institut für Allgemeine, Anorganische und Theoretische Chemie, Leopold-Franzens-Universität Innsbruck, Innrain 52a, A-6020 Innsbruck, Austria*
[b] *Institut für Mineralogie und Petrographie, Leopold-Franzens-Universität Innsbruck, Innrain 52, A-6020 Innsbruck, Austria*
[c] *Department Chemie, Ludwig-Maximilians-Universität München, Butenandtstrasse 5–13, D-81377 München, Germany*

ARTICLE INFO

Article history:
Received 24 September 2010
Received in revised form
8 November 2010
Accepted 9 November 2010
Available online 13 November 2010

Keywords:
Rare-earth
Orthoborate
High-pressure
Crystal structure

ABSTRACT

Single crystals of the orthoborate π-ErBO$_3$ were synthesized from Er$_2$O$_3$ and B$_2$O$_3$ under high-pressure/high-temperature conditions of 2 GPa and 800 °C in a Walker-type multianvil apparatus. The crystal structure was determined on the basis of single-crystal X-ray diffraction data, collected at room temperature. The title compound crystallizes in the monoclinic pseudowollastonite-type structure, space group $C2/c$, with the lattice parameters $a=1128.4(2)$ pm, $b=652.6(2)$ pm, $c=954.0(2)$ pm, and $\beta=112.8(1)°$ ($R_1=0.0124$ and $wR_2=0.0404$ for all data).

© 2010 Elsevier Inc. All rights reserved.

1. Introduction

The rare-earth orthoborates $REBO_3$ are of great scientific interest, due to their extraordinary optical properties like vacuum ultraviolet (VUV) transparency, outstanding optical damage threshold, and high luminescent efficiencies for Eu^{3+}-doped orthoborates. These compounds are thus attractive materials to be used in vacuum discharge lamps or screens and numerous studies have been performed on these borates in the past decades. The rare-earth orthoborates $REBO_3$ exhibit a complex polymorphism and most of them are designated with Greek letters according to the nomenclature of Meyer [1–3]. An overview of the known phases is given in Table 1.
While the crystal structures of most polymorphs have been determined unambiguously, there are a low-temperature modification and a high-temperature modification termed π-$REBO_3$ ($RE=$ Y, Nd, Sm–Lu) and μ-$REBO_3$ ($RE=$ Y, Sm–Gd, Dy–Lu), respectively, which are still the objects of lengthy discussions in the literature. In the following, a survey of the previous works is given:

(a) In 1961, Levin et al. [4] first described the orthoborates π-$REBO_3$ ($RE=$ Y, Nd, Sm–Lu) as pseudo-hexagonal vaterite-type borates with a possible boron coordination of more than three. A phase transition into μ-$REBO_3$ was observed, which was proposed to be nearly isostructural with vaterite, but also possibly pseudohexagonal.

(b) Newnham et al. [22] presented the first structure determination of π-YBO$_3$ powder in 1963 and proposed two possible hexagonal structure models with BO$_3$-groups. One was a disordered model with space group $P6_3/mmc$, the other one an ordered model with space group $P6_3/mcm$. Each rare-earth ion was said to be coordinated to eight oxygen atoms arranged in a distorted cube.

(c) In the following years, several IR, NMR, and Raman studies were performed on the π-orthoborates, all indicating a tetrahedral boron coordination in the form of B$_3$O$_9$-rings [29–33]. Bradley et al. [23] thus presented in 1966 the first structure determination from X-ray powder data, that accommodated these findings. For the low-temperature form π-$REBO_3$, the hexagonal space group $P\bar{6}c2$ was reported, based on Newnham's disordered model. Either a triangular boron coordination, analogous with vaterite, or a three-membered ring of boron tetrahedra were considered to be possible. For the high-temperature form μ-$REBO_3$, the hexagonal space group $P6_322$ with tilted triangular anions was derived.

(d) In 1977, Morgan et al. [24] picked up the thought of pseudohexagonality for the π-orthoborates, as indicated in the very beginning by Levin et al. Based on Bradley's model, they calculated a model in the monoclinic space group $C2/c$ with a pseudohexagonal stacking of the B$_3$O$_9$-rings. The compounds were assumed to crystallize in a pseudowollastonite-type structure rather than in a vaterite-type one.

(e) It was not until 1997, when the first single-crystal structure for π-YBO$_3$ was reported by Chadeyron et al. [25]. Here, the structure was solved in the hexagonal space group $P6_3/m$. The model implies the partial occupation of oxygen and boron

---

[*] Dedicated to Professor Peter Klüfers on the occasion of his 60th birthday.
[*] Corresponding author.
E-mail address: Hubert.Huppertz@uibk.ac.at (H. Huppertz).

0022-4596/$ - see front matter © 2010 Elsevier Inc. All rights reserved.
doi:10.1016/j.jssc.2010.11.018

4.1 Rare-Earth Borates

Table 1
Known polymorphs of $REBO_3$.

Polymorphs	RE	Comments
β-$REBO_3$	Sc, Er–Lu	Calcite structure [3–9]
λ-$REBO_3$	La–Nd, Sm, Eu	Aragonite structure [4,10–12]
χ-$REBO_3$	Dy–Er	Triclinic phases [13,14]
ν-$REBO_3$	Ce–Nd, Sm–Dy	Triclinic (H–$NdBO_3$) [2,15–18]
H-$REBO_3$	La, Ce, Nd	Monoclinic (H–$LaBO_3$) [19–21]
π-$REBO_3$	Y, Nd, Sm–Lu	Low-temperature phases [4,9,22–28]
μ-$REBO_3$	Y, Sm–Gd, Dy–Lu	High-temperature phases [3,23,26–28]

atoms and a possible rotational distortion of the B_3O_9-rings. BO_4-tetrahedra were identified at least as the main structural unit via IR and NMR investigations.

(f) Two years later, Ren et al. [26] brought two other space groups for π- and μ-$REBO_3$ into the discussion. On the basis of powder diffraction data, Chadeyron's hexagonal cell of π-YBO_3 was identified as a subcell of a rhombohedral structure. This way, a fully ordered structure in the rhombohedral space group $R32$ for π-$REBO_3$ was obtained, but unreasonable bond distances gave rise to doubts. The high-temperature polymorph μ-$REBO_3$ was said to crystallize in $P6_3/mmc$, the hexagonal space group previously identified for the low-temperature form by Newnham et al.

(g) Powder data of π-$GdBO_3$ from Cohen-Adad et al. [27] in 2000 were consistent with all proposed hexagonal space groups $P6_3/mmc$, $P\bar{6}c2$, $P6_3/mcm$, and $P6_3/m$, while the best agreement could be accounted in $P\bar{6}c2$. For μ-$GdBO_3$, the hexagonal space group $P6_322$ was assumed; both models are in agreement with the earlier findings of Bradley et al.

(h) A neutron diffraction study on $(Y_{0.92}Er_{0.08})BO_3$ powder samples was undertaken by Lin et al. [28] in 2004. The powder diffraction patterns of both the π- and the μ-orthoborate could only be indexed assuming a monoclinic cell in the space group $C2/c$, and the results thus support the earlier structure determination of Morgan et al. This monoclinic lattice was related to the rhombohedral lattice, observed by Ren et al. by removing the 3-fold axis.

(i) In 2008, Hosokawa et al. [9] presented another study on orthoborate powders synthesized by a glycothermal reaction. Here, the space group $P6_3/m$ was reported as previously determined from single-crystal data by Chadeyron et al.

It is clear from the literature that the crystal structure of the π- and the μ-orthoborate has to be clarified. The red $(Y,Gd)BO_3{:}Eu^{3+}$ phosphor is known to be one of the most familiar materials, being used for plasma panel displays, and its luminescent properties are often discussed on the basis of a hexagonal vaterite-type structure. Looking at the above-mentioned studies, this cannot be taken for granted. Because of its importance in application, it is highly desirable to refine this structure satisfactorily. Over the past 50 years, five different structural models were proposed, revised, doubted, and supported. Remarkably, only one single-crystal measurement on π-YBO_3 was reported, while all other models are based on powder measurements. Light pressure during solid state syntheses often yields coarse-crystalline samples of the products, due to pressure-induced crystallization [34]. The high-pressure/high-temperature synthesis now resulted in single crystals of π-$ErBO_3$, which made the first satisfying single-crystal structure determination possible.

2. Experimental section

2.1. Synthesis

A 1:1 mixture of Er_2O_3 (Strem Chemicals, 99.9%) and B_2O_3 (Strem Chemicals, 99.9+%) was ground and filled into a boron nitride crucible (Henze BNP GmbH, HeBoSint®S100, Kempten, Germany). This crucible was placed into the center of an 18/11-assembly, which was compressed by eight tungsten carbide cubes (TSM-10 Ceratizit, Reutte, Austria). The details of preparing the assembly can be found in Refs. [35–39]. Pressure was applied by a multianvil device, based on a Walker-type module and a 1000 ton press (both devices from the company Voggenreiter, Mainleus, Germany). The sample was compressed up to 2 GPa in 75 min, then heated to 800 °C in 10 min and kept there for 20 min. Afterwards, the sample was cooled down to 600 °C in 20 min and then to room temperature by switching off the heating. The decompression of the assembly required 3.75 h. The recovered MgO-octahedron (pressure transmitting medium, Ceramic Substrates & Components Ltd., Newport, Isle of Wight, UK) was broken apart revealing nearly phase pure π-$ErBO_3$, from which pink air- and water-resistant crystal platelets were obtained for the single-crystal structure determination.

2.2. Crystal structure analysis

The intensity data of a single-crystal of π-$ErBO_3$ were collected at room temperature, using a Nonius Kappa-CCD diffractometer with graded multilayer X-ray optics (Mo$K\alpha$ radiation, λ=71.073 pm). Afterwards, a multi-scan absorption correction was applied to the data [40]. All relevant details of the data collection and evaluation are listed in Table 2. According to the systematic extinctions, the monoclinic space groups Cc and $C2/c$ were derived. Structure solution and parameter refinement (full-matrix least-squares against F^2) were performed successfully in the space group $C2/c$, using the SHELX-97 software suite [41] with anisotropic atomic displacement parameters for all atoms. The positional parameters, interatomic distances, and interatomic angles are listed in Tables 3 and 4.

Further information of the crystal structure is available from the Fachinformationszentrum Karlsruhe (crysdata@fiz-karlsruhe.de), D-76344 Eggenstein-Leopoldshafen (Germany), quoting the Registry no. CSD-422094.

Table 2
Crystal data and structure refinement for π-$ErBO_3$.

Empirical formula	$ErBO_3$
Molar mass, g mol^{-1}	226.07
Crystal system	Monoclinic
Space group	$C2/c$ (No. 15)
Single-crystal diffractometer	Bruker AXS/Nonius Kappa CCD
Radiation	Mo$K\alpha$ (λ=71.073 pm)
	(graphite monochromator)
a, pm	1128.4(2)
b, pm	652.6(2)
c, pm	954.0(2)
β, deg.	112.8(1)
Volume, nm^3	0.6476(2)
Formula units per cell	Z=12
Temperature, K	293(2)
Calculated density, g cm^{-3}	6.96
Crystal size, mm^3	0.04 × 0.04 × 0.02
Absorption coefficient, mm^{-1}	38.5
F(0 0 0), e	1164
θ range, deg.	3.7–30.0
Range in hkl	$\pm$15, $\pm$9, $\pm$13
Total number of reflections	3552
Independent reflections/R_{int}	946/0.0202
Reflections with $I \geq 2\sigma(I)/R_\sigma$	914/0.0170
Data/ref. parameters	946/72
Absorption correction	semiempirical [40]
Goodness-of-fit on F^2	1.016
Final indices R_1/wR_2 [$I \geq 2\sigma(I)$]	0.0120/0.0401
Indices R_1/wR_2 (all data)	0.0124/0.0404
Larg. diff. peak/hole, e Å^{-3}	1.06/−0.77

2.3. IR spectroscopy

FTIR-attenuated total reflection (ATR) spectra of the crystals were recorded with a Bruker Vertex 70 FT-IR spectrometer (resolution ~0.5 cm^{-1}), attached to a Hyperion 3000 microscope in a spectral range from 600–4000 cm^{-1}. A frustrum shaped germanium ATR-crystal with a tip diameter of 100 μm was pressed on the surface of the crystals with a power of 5 N, which let them crush into pieces of μm-size. Sixty-four scans for sample and background were acquired. Beside spectra correction for atmospheric influences, an enhanced ATR-correction, [42] using the OPUS 6.5 software, was performed. A mean refraction index of the sample of 1.6 was assumed for the ATR-correction.

2.4. Raman spectroscopy

Confocal Raman spectra of single crystals were gained with a HORIBA JOBIN YVON LabRam-HR 800 Raman micro-spectrometer. The sample was excited, using the 633 nm emission line of a 17 mW He–Ne-laser and an OLYMPUS 100× objective (numerical aperture=0.9). The size and power of the laser spot on the surface were approximately 1 μm and 5 mW, respectively. The scattered light was dispersed by a grating with 1800 lines/mm and collected by a 1024 × 256 open electrode CCD detector. The spectral resolution, determined by measuring the Rayleigh line, was about 1.4 cm^{-1}. The wavenumber accuracy of about 0.5 cm^{-1} was achieved by adjusting the zero-order position of the grating and regularly checked by a Neon spectral calibration lamp.

3. Results and discussion

The single-crystal structure of π-ErBO$_3$ could be solved and refined in the space group $C2/c$ and is thus isotypic to the pseudowollastonite-type CaSiO$_3$ [43] and not to the vaterite-type CaCO$_3$ [44]. The results agree with the findings of Morgan et al. [24] and the neutron diffraction study of Lin et al. [28]. From the single-crystal data, we unambiguously derived a monoclinic unit cell with the parameters a=1128.4(2) pm, b=652.6(2) pm, c=954.0(2) pm,

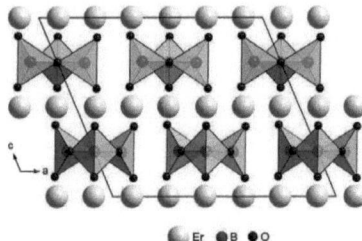

Fig. 1. Crystal structure of π-ErBO$_3$ along [0 1 0], displaying isolated B$_3$O$_9$-rings of BO$_4$-tetrahedra.

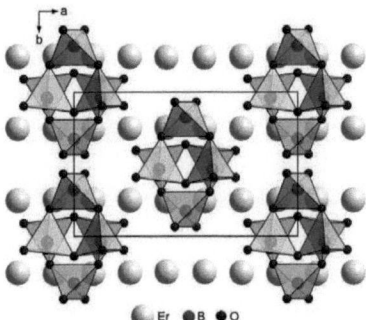

Fig. 2. Crystal structure of π-ErBO$_3$ along [0 0 1], displaying the staggered stacking of the B$_3$O$_9$-rings.

Table 3
Atomic coordinates and isotropic equivalent displacement parameters U_{eq} (Å^2) for π-ErBO$_3$ (space group $C2/c$).

Atom	Wyckoff position	x	y	z	U_{eq}
Er1	4e	1/4	1/4	0	0.00399(7)
Er2	8f	0.085415(7)	0.25566(2)	0.499356(7)	0.00376(7)
B1	8f	0.1201(2)	0.0379(3)	0.2469(2)	0.0058(3)
B2	4e	0	0.6752(4)	1/4	0.0057(4)
O1	8f	0.1255(2)	0.0920(2)	0.1020(2)	0.0062(2)
O2	8f	0.2229(2)	0.0932(2)	0.3887(2)	0.0061(2)
O3	8f	0.0484(2)	0.5664(2)	0.3923(2)	0.0061(2)
O4	8f	0.3914(2)	0.3082(2)	0.2517(2)	0.0056(2)
O5	4e	0	0.1350(2)	1/4	0.0052(3)

U_{eq} is defined as one third of the trace of the orthogonalized U_{ij} tensor.

Table 4
Interatomic distances (pm) and angles (deg.), calculated with the single-crystal lattice parameters of π-ErBO$_3$ with standard deviations in parentheses.

Er1–O1	224.6(2) 2×	Er2–O3	223.6(2)	B1–O2	144.5(2)	B2–O3	143.9(2) 2×
		Er2–O2	223.6(2)	B1–O1	145.1(2)		
Er1–O4	234.3(2) 2×	Er2–O5	232.4(2)	B1–O4	150.5(2)	B2–O4	150.7(2) 2×
		Er2–O3	233.0(2)	B1–O5	150.7(2)		
Er1–O3	242.0(2) 2×	Er2–O3	243.0(2)	∅=147.7		∅=147.3	
		Er2–O2	243.3(2)				
Er1–O2	244.6(2) 2×	Er2–O1	244.1(2)	O2–B1–O1	121.1(2)	O3–B2–O3	120.9(2)
		Er2–O1	244.3(2)	O2–B1–O4	106.3(2)	O3–B2–O4	106.4(2) 2×
∅=236.4		∅=235.9		O1–B1–O4	106.6(2)		
				O2–B1–O5	106.2(2)	O3–B2–O4	106.6(2) 2×
				O1–B1–O5	106.7(2)		
				O4–B1–O5	109.7(2)	O4–B2–O4	109.6(2)

4.1 Rare-Earth Borates

and $\beta=112.8(1)°$ (Table 2). The parameters of the neutron powder diffraction experiments are similar with $a=1131.38(3)$ pm, $b=654.03(2)$ pm, $c=954.99(2)$ pm, and $\beta=112.902(1)°$ [28], while no cell parameters were given by Morgan et al.

The crystal structure of π-ErBO$_3$ has got two crystallographically independent boron cations (Table 3), which are fourfold coordinated by oxygen anions. Connecting via common corners, three BO$_4$-tetrahedra (two around B1 and one around B2) form isolated B$_3$O$_9$-rings (Fig. 1). The tetrahedra are regular with a displacement of the central boron atoms towards the non-bridging oxygen atoms, resulting in two shorter and two larger B–O bonds for each boron atom (Table 4). A similar distortion for BO$_4$-tetrahedra with two bridging and two non-bridging ligands was observed previously, e.g. in the fluoride borate La$_4$B$_4$O$_{11}$F$_2$ [45]. With an average B–O bond lengths of 147.7 and 147.3 pm, they go together with the known average value of 147.6 pm for BO$_4$-tetrahedra in borates [46,47]. The bond lengths are comparable, but more consistent than those obtained by Lin et al. via powder neutron diffraction [28]. For one of the hexagonal structure models, Chadeyron et al. report B–O bond lengths between 137 and 192 pm [25], which are rather doubtful for BO$_4$-tetrahedra. The distortions of the BO$_4$-tetrahedra in π-ErBO$_3$ also cause enlarged O–B–O angles of $\sim$120° between the non-bridging oxygen atoms.

Viewing along [0 0 1], the B$_3$O$_9$-rings are stacked in a shifted arrangement (Fig. 2). Earlier structure determinations assumed the presence of rare-earth cations inside of the B$_3$O$_9$-rings [23], which can be now excluded from the single-crystal structure determination. Due to the staggered arrangement and the distortion of the BO$_4$-tetrahedra, three- or sixfold rotational axes do not exist in the structure of π-ErBO$_3$. Therefore, all hexagonal and rhombohedral models have to take disorder or partially occupied atomic positions into account, which is not the case for the monoclinic model, stated earlier by Lin et al. [28].

The two erbium cations in the structure are both eightfold coordinated (Fig. 3). The Er–O distances are in a narrow range between 223.6(2) and 244.6(2) pm. They are thus much more uniformly distributed than the bond lengths reported for π-YBO$_3$ by Lin et al., which range between 225.7(8)–295.0(8) pm [28].

The calculations of the bond-valence sums for π-ErBO$_3$ with the bond-length/bond-strength (BLBS) concept [48] and the charge distribution concept (CHARDI) in solids according to Hoppe [49–51] were performed and confirmed the formal ionic charges of the single-crystal structure analysis (Table 5).

Furthermore, we calculated the Madelung Part of Lattice Energy value (MAPLE) according to Hoppe [52–54] of π-ErBO$_3$ in order to compare it with the sum of the MAPLE values for the high-pressure modification of Er$_2$O$_3$ [55] and that of B$_2$O$_3$-II [56] [0.5 Er$_2$O$_3$ (15,283 kJ mol^{-1})+0.5 B$_2$O$_3$-II (21,938 kJ mol^{-1})]. The calculated value (18,739 kJ mol^{-1}) for π-ErBO$_3$ and the MAPLE value of the sum of the binary oxides (18,611 kJ mol^{-1}) tally well (deviation 0.7%).

Fig. 4 shows the FTIR-ATR measurement of the sample between 600 and 2000 cm^{-1}. The main absorption bands between 750 and 1250 cm^{-1} are those typical for the tetrahedral borate group [BO$_4$]$^{5-}$ as in π-GdBO$_3$, π-YBO$_3$, or TaBO$_4$ [26,31,57]. Fig. 5 shows the Raman spectrum of π-ErBO$_3$ in the range 100–1500 cm^{-1}. In previous Raman measurements of borates, the area below 650 cm^{-1} could be assigned to lattice and RE–O modes. Therefore, the area between 700 and 1100 cm^{-1} shows probably the bands evoked by the BO$_4$-groups. The positions of these Raman bands (Fig. 5) are compared to those reported for π-GdBO$_3$ in the literature (Table 6) [27].

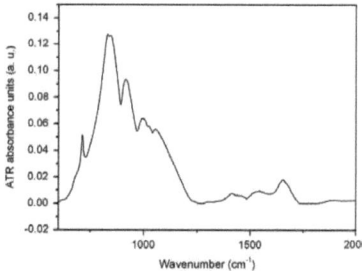

Fig. 4. FTIR-ATR spectrum of π-ErBO$_3$ in the range 600–2000 cm^{-1}.

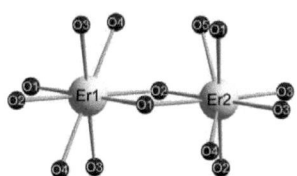

Fig. 3. Coordination spheres of the erbium cations in the crystal structure of π-ErBO$_3$.

Table 5
Charge distribution in π-ErBO$_3$, calculated with VaList (ΣV) and the CHARDI concept (ΣQ).

	Er1	Er2	B1	B2	O1	O2	O3	O4	O5
ΣV	+2.97	+3.01	+3.01	+3.05	−1.89	−1.92	−1.96	−2.18	−2.18
ΣQ	+3.03	+3.02	+2.99	+2.95	−1.91	−1.94	−1.98	−2.11	−2.12

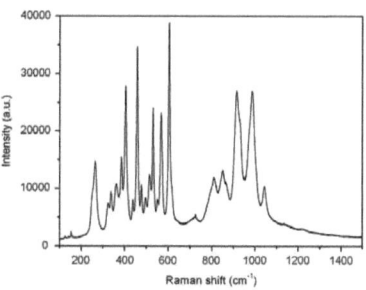

Fig. 5. Raman spectrum of π-ErBO$_3$ in the range 100–1500 cm^{-1}.

Table 6
Wavenumbers of Raman bands in the spectrum of π-ErBO$_3$ single crystals (left) and of bands of π-GdBO$_3$, reported for the bulk material (right) [27].

Single-crystal Raman bands (π-ErBO$_3$)	Bulk Raman bands (π-GdBO$_3$)
724	
810	825
849	843
865	881
913	918
986	996
1043	1016

4. Conclusions

In this paper, we presented the first satisfying single-crystal structure determination of π-ErBO$_3$, which sheds additional light on the extensively discussed structure of π-orthoborates. The application of light pressure (2 GPa) during the solid state synthesis yielded a coarse-crystalline sample of this ambient-pressure phase, due to the pressure-induced crystallization. Based on the single-crystal data, we can support the monoclinic structural model, introduced by Morgan et al. [24] and Lin et al. [28]. Due to the staggered arrangement and the distortion of the BO$_4$-tetrahedra, no three- or sixfold rotational axes can be applied to the structure of π-ErBO$_3$. All hexagonal and rhombohedral models have to take disorder or partially occupied atomic positions into account, thus representing subcells and/or incorrectly averaged structural models. The neglecting of several very weak reflections and the splitting of appearing single reflections in the powder pattern led to numerous hexagonal cells. The new single-crystal data should end the discussion about the correct indexing of the powder patterns. The luminescence properties for the red (Y,Gd)BO$_3$:Eu^{3+} phosphor should be discussed on the basis of a monoclinic rather than a hexagonal structure.

For the future, single crystals of the high-temperature polymorph μ-REBO$_3$ will be of great interest.

Appendix A. Supplementary materials

Supplementary data associated with this article can be found in the online version at doi:10.1016/j.jssc.2010.11.018.

References

[1] H.J. Meyer, Naturwissenschaften 56 (1969) 458.
[2] H.J. Meyer, Naturwissenschaften 59 (1972) 215.
[3] H.J. Meyer, A. Skokan, Naturwissenschaften 58 (1971) 566.
[4] E.M. Levin, R.S. Roth, J.B. Martin, Am. Mineral. 46 (1961) 1030.
[5] S.C. Abrahams, J.L. Bernstein, E.T. Keve, J. Appl. Crystallogr. 4 (1971) 284.
[6] T.A. Bither, H.S. Young, J. Solid State Chem. 6 (1973) 502.
[7] D.A. Keszler, H. Sun, Acta Crystallogr. C 44 (1988) 1505.
[8] H. Huppertz, Z. Naturforsch. 56b (2001) 697.
[9] S. Hosokawa, Y. Tanaka, S. Iwamoto, M. Inoue, J. Mater. Sci. 43 (2008) 2276.
[10] J. Weidelt, H.U. Bambauer, Naturwissenschaften 55 (1968) 342.
[11] G.K. Abdullaev, Kh.S. Mamedov, G.G. Dzhafarov, Azerb. Khim. Zh. (1976) 117.
[12] H. Müller-Bunz, T. Nikelski, Th. Schleid, Z. Naturforsch. 58b (2003) 375.
[13] H. Huppertz, B. von der Eltz, R-D. Hoffmann, H. Piotrowski, J. Solid State Chem. 166 (2002) 203.
[14] H. Huppertz, Unpublished results.
[15] K.K. Palkina, V.G. Kuznetsov, L.A. Butman, B.F. Dzhurinskii, Acad. Sci. USSR 2 (1976) 286.
[16] G. Corbel, M. Leblanc, E. Antic-Fidancev, M. Lemaître-Blaise, J.C. Krupa, J. Alloys Compd. 287 (1999) 71.
[17] H. Emme, H. Huppertz, Acta Crystallogr. C 60 (2004) i117.
[18] S. Noirault, O. Joubert, M.T. Caldes, Y. Piffard, Acta Crystallogr. E 62 (2006) i228.
[19] R. Böhlhoff, H.U. Bambauer, W. Hoffmann, Naturwissenschaften 57 (1970) 129.
[20] R. Böhlhoff, H.U. Bambauer, W. Hoffmann, Z. Kristallogr. 133 (1971) 386.
[21] S. Lemanceau, G. Bertrand-Chadeyron, R. Mahiou, M. El-Ghozzi, J.C. Cousseins, P. Conflant, R.N. Vannier, J. Solid State Chem. 148 (1999) 229.
[22] R.E. Newnham, M.J. Redman, R.P. Santoro, J. Am. Ceram. Soc. 46 (1963) 253.
[23] W.F. Bradley, D.L. Graf, R.S. Roth, Acta Crystallogr. 20 (1966) 283.
[24] P.E.D. Morgan, P.J. Carroll, F.F. Lange, Mater. Res. Bull. 12 (1977) 251.
[25] G. Chadeyron, M. El-Ghozzi, R. Mahiou, A. Arbus, J.C. Cousseins, J. Solid State Chem. 128 (1997) 261.
[26] M. Ren, J.H. Lin, Y. Dong, L.Q. Yang, M.Z. Su, L.P. You, Chem. Mater. 11 (1999) 1576.
[27] M.Th. Cohen-Adad, Ch. Kappenstein, O. Aloui-Lebbou, C Goutaudier, G. Panczer, C. Dujardin, C. Pedrini, P. Florian, D. Massiot, F. Gerard, J. Solid State Chem. 154 (2000) 204.
[28] J.H. Lin, S. Sheptyakov, Y. Wang, P. Allenspach, Chem. Mater. 16 (2004) 2418.
[29] C.E. Weir, E.R. Lippincott, J. Res. Natl. Bur. Stand. A 65 (1961) 173.
[30] C.E. Weir, R.A. Schroeder, J. Res. Natl. Bur. Stand. A 68 (1964) 465.
[31] J.P. Laperches, P. Tarte, Spectrochim. Acta 22 (1966) 1201.
[32] H.M. Kriz, P.J. Bray, J. Chem. Phys. 51 (1969) 3642.
[33] J.H. Denning, S.D. Ross, Spectrochim. Acta A 28 (1972) 1775.
[34] H. Huppertz, Chem. Commun. (2011) 10.1039/C0CC02715D.
[35] D. Walker, M.A. Carpenter, C.M. Hitch, Am. Mineral. 75 (1990) 1020.
[36] D. Walker, Am. Mineral. 76 (1991) 1092.
[37] H. Huppertz, Z. Kristallogr. 219 (2004) 330.
[38] D.C. Rubie, Phase Transitions 68 (1999) 431.
[39] N. Kawai, S. Endo, Rev. Sci. Instrum. 8 (1970) 1178.
[40] Z. Otwinowski, W. Minor, Methods Enzymol. 276 (1997) 307.
[41] G.M. Sheldrick, Acta Crystallogr. A 64 (2008) 112.
[42] F.M. Mirabella Jr. (Ed.), Principles, Theory, and Practice of Internal Reflection Spectroscopy in Internal Reflection Spectroscopy, Theory and Applications, Marcel Dekker Inc., New York, 1993, pp. 17–53.
[43] H. Yang, C.T. Prewitt, Am. Mineral. 84 (1999) 929.
[44] H.J. Meyer, Z. Kristallogr. 128 (1969) 183.
[45] A. Haberer, R. Kaindl, O. Oeckler, H. Huppertz, J. Solid State Chem. 183 (2010) 1970.
[46] F.C. Hawthorne, P.C. Burns, J.D. Grice, Reviews in Mineralogy (Chapter 2), in: Boron: Mineralogy, Petrology, and Geochemistry, vol. 33, Mineralogical Society of America, Washington, DC (1996).
[47] E. Zobetz, Z. Kristallogr. 191 (1990) 45.
[48] A.S. Wills, VaList Version 3.0.13, University College London, UK, 1998–2008 Program available from <www.ccp14.ac.uk>.
[49] R. Hoppe, S. Voigt, H. Glaum, J. Kissel, H.P. Müller, K.J. Bernet, J. Less-Common Met. 156 (1989) 105.
[50] I.D. Brown, D. Altermatt, Acta Crystallogr. B 41 (1985) 244.
[51] N.E. Brese, M. O'Keeffe, Acta Crystallogr. B 47 (1991) 192.
[52] R. Hoppe, Angew. Chem. 78 (1966) 52;
R. Hoppe, Angew. Chem. Int. Ed. 5 (1966) 96.
[53] R. Hoppe, Angew. Chem. 82 (1970) 7;
R. Hoppe, Angew. Chem. Int. Ed. 9 (1970) 25.
[54] R. Hübenthal, M. Serafin, R. Hoppe, Maple (version 4.0), Program for the Calculation of Distances, Angles, Effective Coordination Numbers, Coordination Spheres, and Lattice Energies, University of Gießen, Gießen (Germany), (1993).
[55] H.R. Hoekstra, Inorg. Chem. 5 (1966) 754.
[56] C.T. Prewitt, R.D. Shannon, Acta Crystallogr. B 24 (1968) 869.
[57] G. Blasse, G.P.M. van den Heuvel, Phys. Status Solidi 19 (1973) 111.

4.2 Rare-Earth Fluoride and Fluorido Borates

The main goal of this thesis was the establishment of the high-pressure / high-temperature technique on the field of rare-earth fluoride and fluorido borates. The difference between fluorido borates and fluoride borates lies in the coordination sphere of boron. While borates, that contain only isolated fluoride anions, are generally termed "fluoride borates", "fluorido borates" contain fluorine atoms covalently bound to the boron atoms. On the one hand, fluoroborate glasses have been object of extensive research [203-205]. The doping with rare-earth element cations leads to interesting luminescence, fluorescence, and even dielectric properties of the glasses [206-209]. It is known that fluoroborate glasses, as well as borate glasses, are UV-transmitting; the addition of fluorine enlarges the optical gap [210, 211]. On the other hand, crystalline fluorido and fluoride borates are not that well studied, excluding natural minerals with more than two cations. Table 4.2-1 displays all known quaternary and quinary crystalline fluorido and fluoride borates, synthesized under ambient pressure. Among them, one finds the nonlinear-optical material $KBe_2BO_3F_2$ (KBBF), showing second harmonic generator properties in the vacuum ultraviolet region [212]. In $M_3B_7O_{13}F$ (M = Mg, Cr, Mn, Fe, Zn), BO_4-tetrahedra are present in the boracite structure-type arrangement [213], whereas in $Ba_5(B_2O_5)_2F_2$, a connection of two BO_3-groups *via* common corners is found, leading to a typical pyroborate unit [214]. Furthermore, there probably exist three fluorido borates with mixed B–O/F coordination, namely MB_6O_9F [215], $BaBOF_3$ [216], and BiB_2O_4F [217] (highlighted with grey in Table 4.2-1). The presence of BO_3F-tetrahedra was confirmed for the former compounds and is discussed for the latter compounds. All other compounds in Table 4.2-1 exhibit isolated trigonal-planar BO_3-groups, which seem to represent the main structural motif of the fluoride borates. In the field of crystalline rare-earth fluorido and fluoride borates, only the compounds $RE_3(BO_3)_2F_3$ (RE = Sm, Eu, Gd) [218] and $Gd_2(BO_3)F_3$ [219] were known at the beginning of our investigations. They were synthesized by heating a stoichiometric mixture of RE_2O_3, B_2O_3, and REF_3 under ambient pressure conditions. From this enumeration, it is obvious that in comparison to the extensive synthetic and structural chemistry of oxoborates, next to nothing was known about fluorido and fluoride borates.

Table 4.2-1
Overview of existing quaternary and quinary fluorido and fluoride borates

Compound	Metal M	Comment	Authors
Quaternary			
$M_2(BO_3)F$	Be	$C2/c$	Baidina et al. [220]
$\alpha\text{-}M_2(BO_3)F$	Mg	$Pna2_1$	Brovkin et al. [221]
$\beta\text{-}M_2(BO_3)F$	Mg	$Pnam$	Nikishova et al. [222]
$M_2(BO_3)F_3$	Gd		Müller-Bunz et al. [219]
$M_3(BO_3)F_3$	Mg		dal Negro et al. [223]
$M_3(BO_3)_2F_3$	Sm, Eu, Gd		Corbel et al. [218]
$M_3B_7O_{13}F$	Mg, Cr, Mn, Fe, Zn	Fluoroboracites	Bither et al. [213]
$M_5(BO_3)_3F$	Ca, Mg, Sr, Eu	Monoclinic (Cm) ? Orthorhombic ($Pnma$)	Fletcher et al. [224], Brovkin et al. [225], Alekel et al. [226], Kazmierczak et al. [227]
$M_5(B_2O_5)_2F_2$	Ba	Pyroborate fluoride	Alekel et al. [214]
$M_6(BO_3)_5F_3$	Al	Jeremejevite	Rodellas et al. [228]
MB_6O_9F	Li		Cakmak et al. [215]
$MBOF_3$	Ba		Chackraburtty [216]
MB_2O_4F	Bi		Li et al. [217]
$B(OMF_5)_3$	Te	Boron tris(penta fluoro oxotellurate)	Sawyer et al. [229]
Quinary			
$MBa(BO_3)F_2$	Al, Ga		Park et al. [230]
$MBe_2(BO_3)F_2$	Na, K, Rb, Cs	NLO materials	Mei et al. [231,232], Baidina et al. [233]

4.2 Rare-Earth Fluoride and Fluorido Borates

Considering the extensive research on the syntheses of oxoborates under high-pressure/high-temperature conditions in our group in the last decade, it was tempting to extend our investigations into the field of fluorido and fluoride borates. Under extreme synthetic conditions, the number and variety of oxoborates could be greatly extended, revealing fascinating structures and attracting material properties (Sections 1.3 and 4.1). Due to the structural similarities to fluorido and fluoride borates, it was more than likely that numerous new compounds and crystal structures of this field could be obtained under pressure, as well.

Among the existing fluorido and fluoride borates, there are mainly trigonal BO_3-groups, mostly connected *via* common corners. Consequently, the field of fluorido and fluoride borates has an enormous potential for high-pressure investigations. The connection of existing building blocks or the increase of coordination numbers, as discussed in Section 1.3 for oxoborates, implies a tremendous number of possible, higher condensed structures. Furthermore, the tendency of glass-formation in the class of fluorido and fluoride borates can presumably be suppressed *via* pressure induced crystallization.

During this thesis, the successful syntheses and characterizations of the compounds $RE_5(BO_3)_2F_9$ (RE = Er - Yb), $Pr_4B_3O_{10}F$, and the non-isotypic fluoride borates $RE_4B_4O_{11}F_2$ (RE = Eu, Gd, Dy) and $La_4B_4O_{11}F_2$ were performed. In the Sections 4.2.1 - 4.2.7, the papers concerning these compounds are presented. Furthermore, it was possible to derive the crystal structures of the compounds $RE_3(BO_3)_2F_3$ (RE = Gd, Dy), $RE_{12}B_{11}O_{31}F_7$ (RE = La, Pr, Nd, Sm), and of the first apatite-structured fluorine-containing borates "$RE_5(BO_{3.66}F_{0.34})_3F$" ($RE$ = Gd, Yb). At the printing of this work, some investigations concerning these phases were still in progress, so that in the Sections 4.2.8 - 4.2.10 intermediate results are presented.

4.2 Rare-Earth Fluoride and Fluorido Borates

4.2.1 $Yb_5(BO_3)_2F_9$

The first high-pressure / high-temperature syntheses were based on previous fluoride borate syntheses under ambient pressure. Here, rare-earth oxides, the corresponding rare-earth fluorides and boron oxide were mixed in simple ratios. Syntheses of mere rare-earth fluoride with boron oxide led to no reaction, even though all elements were present in the reaction mixture. This leads to the assumption that a reaction of the rare-earth oxides with the boron oxide might be a starting step for the following formation of fluorine-containing borates. However, only syntheses during which all three compound classes were present, led to a reaction.

The first rare-earth fluoride borate obtained under high-pressure / high-temperature conditions was $Yb_5(BO_3)_2F_9$ (Figure 4.2-1). Its crystal structure is closely related to those of the previously known ambient-pressure compounds $RE_3(BO_3)_2F_3$ (RE = Sm, Eu, Gd) [218] and $Gd_2(BO_3)F_3$ [219]. All the details concerning the synthesis, crystal structure, and structural relationship can be found in the following publication.

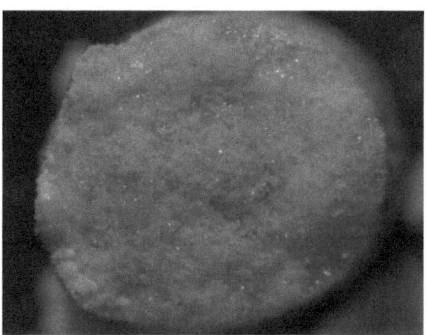

Figure 4.2-1
Crystalline sample of $Yb_5(BO_3)_2F_9$

Journal of Solid State Chemistry 182 (2009) 888–895

Contents lists available at ScienceDirect

Journal of Solid State Chemistry

journal homepage: www.elsevier.com/locate/jssc

High-pressure synthesis, crystal structure, and structural relationship of the first ytterbium fluoride borate $Yb_5(BO_3)_2F_9$

Almut Haberer, Hubert Huppertz *

Institut für Allgemeine, Anorganische und Theoretische Chemie, Leopold-Franzens-Universität Innsbruck, Innrain 52a, A-6020 Innsbruck, Austria

ARTICLE INFO

Article history:
Received 20 December 2008
Received in revised form
20 January 2009
Accepted 22 January 2009
Available online 30 January 2009

Keywords:
High pressure
Multianvil
Crystal structure
Fluoride borate

ABSTRACT

$Yb_5(BO_3)_2F_9$ was synthesized under high-pressure/high-temperature conditions in a Walker-type multianvil apparatus at 7.5 GPa and 1100 °C, representing the first known ytterbium fluoride borate. The compound exhibits isolated BO_3-groups next to ytterbium cations and fluoride anions, showing a structure closely related to the other known rare-earth fluoride borates $RE_3(BO_3)_2F_3$ (RE = Sm, Eu, Gd) and $Gd_2(BO_3)F_3$. Monoclinic $Yb_5(BO_3)_2F_9$ crystallizes in space group $C2/c$ with the lattice parameters $a = 2028.2(4)$ pm, $b = 602.5(2)$ pm, $c = 820.4(2)$ pm, and $\beta = 100.63(3)°$ ($Z = 4$). Three different ytterbium cations can be identified in the crystal structure, each coordinated by nine fluoride and oxygen anions. None of the five crystallographically independent fluoride ions is coordinated by boron atoms, solely by trigonally-planar arranged ytterbium cations. In close proximity to the above mentioned compounds $RE_3(BO_3)_2F_3$ (RE = Sm, Eu, Gd) and $Gd_2(BO_3)F_3$, $Yb_5(BO_3)_2F_9$ can be described via alternating layers with the formal compositions "$YbBO_3$" and "YbF_3" in the bc-plane.

© 2009 Elsevier Inc. All rights reserved.

1. Introduction

In the past years, the field of rare-earth borates could be extended by the application of high-pressure/high-temperature techniques, leading to a large variety of new compounds, for example the rare-earth borates $RE_4B_6O_{15}$ (RE = Dy, Ho) [1–3], α-$RE_2B_4O_9$ (RE = Sm–Ho) [4–6], β-$RE_2B_4O_9$ (RE = Gd [7], Dy [8]), $Pr_4B_{10}O_{21}$ [9], and the meta-borates δ-$RE(BO_2)_3$ (RE = La, Ce) [10,11].

Borates, being glass formers in general, show an increasing willingness to crystallize under pressure. This kind of pressure-induced crystallization can be observed in β-SnB_4O_7 [12], β-ZrB_2O_5 [13], or β-HfB_2O_5 [14]. Also fluoroborates tend to form glasses with interesting optical properties. While borate glasses can be used for vacuum ultraviolet (VUV) optics, the addition of fluorine enlarges the optical gap [15]. We start now to investigate fluoro- and fluoride borates under high-pressure/high-temperature conditions, because they are likely yielding crystalline fluoroborate phases with interesting optical properties.

Until now, only the rare-earth fluoride borates $RE_3(BO_3)_2F_3$ (RE = Sm, Eu, Gd) [16] and $Gd_2(BO_3)F_3$ [17] are known, which were synthesized by heating a stoichiometric mixture of RE_2O_3, B_2O_3, and REF_3 under ambient pressure conditions. The structure of $Gd_2(BO_3)F_3$ was solved from X-ray powder diffraction data, as suitable single crystals could not be found in the sample. In 2000, luminescence studies on $Eu_3(BO_3)_2F_3$ led to a disordered model of the structure [18]. In $Eu_3(BO_3)_2F_3$, three fluoride anions in the structure were supposed to be partially replaced by oxoborate anions BO_3^{3-}, resulting in the formula $Eu_3(BO_3)_{2+x}F_{3-3x}$. This affects each of the three crystallographically different fluorides in the structure. In 2002, Müller-Bunz et al. [17] in vain tried to reproduce the isotypic compound $Gd_3(BO_3)_2F_3$. Instead of $Gd_3(BO_3)_2F_3$, the group yielded crystals of $Gd_2(BO_3)F_3$. But they found a close relationship between the two structures, both of which can be described via alternating layers of the formal compositions "$REBO_3$" and "REF_3" in the bc-plane. It should be emphasized that the crystal structures of the actual compounds $REBO_3$ and REF_3 cannot be compared with the structures of the layers, so that "$REBO_3$" and "REF_3" only stand for the formal compositions of the layers. The detection of disorder in crystals of $Gd_2(BO_3)F_3$ prompted to a new model of disorder, also applicable to $Eu_3(BO_3)_2F_3$.

In this article, we present the first ytterbium fluoride borate $Yb_5(BO_3)_2F_9$, obtained by high-pressure/high-temperature synthesis. The crystal structure of $Yb_5(BO_3)_2F_9$ can be developed from the structures mentioned above, showing alternating layers of the formal compositions "$YbBO_3$" and "YbF_3". We found disorder in parts of the structure, too, which might be explained along the model of disorder, proposed by Müller-Bunz et al. In the following, synthesis, structural details, and structural relationships of the new compound $Yb_5(BO_3)_2F_9$ are reported.

2. Experimental section

According to Eq. (1), the synthesis of $Yb_5(BO_3)_2F_9$ happened under high-pressure/high-temperature conditions, starting from

---

* Corresponding author. Fax: +435125072934.
E-mail address: hubert.huppertz@uibk.ac.at (H. Huppertz).

0022-4596/$ - see front matter © 2009 Elsevier Inc. All rights reserved.
doi:10.1016/j.jssc.2009.01.023

4.2 Rare-Earth Fluoride and Fluorido Borates

the binary oxides Yb_2O_3 and B_2O_3, as well as YbF_3:

$$Yb_2O_3 + B_2O_3 + 3\,YbF_3 \xrightarrow{7.5\,GPa,1100\,°C} Yb_5(BO_3)_2F_9 \qquad (1)$$

A mixture of Yb_2O_3 (Smart Elements, 99.99%), B_2O_3 (Strem Chemicals, 99.9+%), and YbF_3 (Strem Chemicals, 99.9%) at a molar ratio of 1:1:3 (Eq. (1)) was ground up and filled into a boron nitride crucible (Henze BNP GmbH, HeBoSint® S10, Kempten, Germany). This crucible was placed into the center of an 18/11-assembly, which was compressed by eight tungsten carbide cubes (TSM-10 Ceratizit, Reutte, Austria). The details of preparing the assembly can be found in Refs. [19–23]. Pressure was applied by a multianvil device, based on a Walker-type module, and a 1000 ton press (both devices from the company Voggenreiter, Mainleus, Germany). The sample was compressed up to 7.5 GPa for 3 h, then heated to 1100 °C in 15 minutes and kept there for 20 minutes. Afterwards, the sample was cooled down to 850 °C in 20 minutes, followed by natural cooling down to room temperature after switching off heating. The decompression required 9 h. The recovered experimental MgO-octahedron (pressure transmitting medium, Ceramic Substrates & Components Ltd., Newport, Isle of Wight, UK) was broken apart and the sample carefully separated from the surrounding boron nitride crucible, obtaining colorless, air- and water-resistant, irregularly shaped crystals of $Yb_5(BO_3)_2F_9$.

3. Crystal structure analysis

The sample was characterized by powder X-ray diffraction, which was performed in transmission geometry on a flat sample of the reaction product, using a STOE STADI P powder diffractometer with MoKα_1 radiation (Ge monochromator, $\lambda = 71.073$ pm). Fig. 1 shows a powder pattern of the sample (top), exhibiting $Yb_5(BO_3)_2F_9$, as well as reflections of another still unknown side product, marked with lines. The experimental powder pattern tallies well with the theoretical pattern (bottom), simulated from single-crystal data. Indexing the reflections of the ytterbium fluoride borate, we got the parameters $a = 2028.0(2)$ pm, $b = 602.5(3)$ pm, and $c = 821.5(5)$ pm, with $\beta = 100.62(5)°$ and a volume of $986.6(7)\,Å^3$. This confirms the lattice parameters, obtained from single-crystal X-ray diffraction (Table 1).

The intensity data of a single crystal of $Yb_5(BO_3)_2F_9$ were collected at room temperature by use of a Kappa CCD diffractometer (Bruker AXS/Nonius, Karlsruhe), equipped with a Miracol Fiber Optics Collimator and a Nonius FR590 generator (graphite-monochromatized MoKα_1 radiation, $\lambda = 71.073$ pm). Additionally, the data set was subjected to a numerical absorption correction (HABITUS [24]). All relevant details of the data collection and evaluation are listed in Table 1.

Structure solution and parameter refinement (full-matrix least-squares against F^2) were successfully performed, using the SHELX-97 software suite [25,26] with anisotropic atomic displacement parameters for all atoms. According to the systematic extinctions, the monoclinic space groups $C2/c$ and Cc were derived. The structure solution in $C2/c$ (no. 15) succeeded. The final difference Fourier syntheses did not reveal any significant residual peaks in all refinements. The positional parameters of the refinements, anisotropic displacement parameters, interatomic distances, and interatomic angles are listed in the Tables 2–5. Further information of the crystal structure is available from the Fachinformationszentrum Karlsruhe (crysdata@fiz-karlsruhe.de), D-76344 Eggenstein-Leopoldshafen (Germany), quoting the Registry no. CSD-420182.

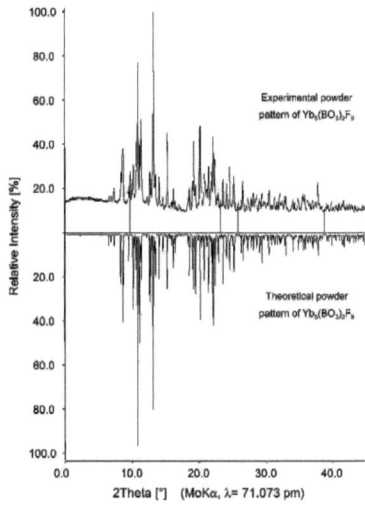

Fig. 1. Top: Experimental powder pattern of $Yb_5(BO_3)_2F_9$; reflections of an unknown phase are indicated with lines. Bottom: theoretical powder pattern of $Yb_5(BO_3)_2F_9$, based on single-crystal diffraction data.

4. Results and discussion

4.1. Crystal structure of $Yb_5(BO_3)_2F_9$

The structure of $Yb_5(BO_3)_2F_9$ consists of isolated BO_3-groups, ytterbium cations, and fluoride anions (Fig. 2). As shown in Fig. 3, the structure can be described via alternating layers of the formal compositions "$YbBO_3$" and "YbF_3", spreading into the bc-plane. The layers are labeled in a way, that reveals their relationship to other rare-earth fluoride borates, as explained below.

There are three crystallographically independent Yb^{3+} ions in the structure, which are nine-fold coordinated by oxygen and fluorine (Fig. 4). Yb1 and Yb3 have got four oxide and five fluoride ions in the coordination sphere, whereas Yb2 is surrounded by three oxide and six fluoride ions. The average interatomic distance Yb–O with 234.2 pm is in the same range, but shorter than the average Gd–O distance of ninefold coordinated Gd^{3+} in $Gd_2[BO_3]F_3$ (242.2 pm [17]), as we would expect from the smaller ionic radius of Yb^{3+}.

Each one of the five fluoride ions in $Yb_5(BO_3)_2F_9$ is coordinated by three ytterbium ions (Fig. 5). The bond lengths in Table 4 range between 218.6(4) and 294.2(5) pm, being in the same region as the bond lengths of threefold-coordinated fluoride ions in YbF_3 (203.9–284.7 pm) [27]. The Yb–F angles sum up to ~120° (Table 5), as expected from trigonal-planar geometry.

Regarding the BO_3-group in the structure, the average B–O distance is 142.7 pm (Table 4). This is fairly large for interatomic distances in BO_3-groups, which are usually in a range

Table 1
Crystal data and structure refinement of $Yb_5(BO_3)_2F_9$.

Empirical formula	$Yb_5(BO_3)_2F_9$
Molar mass (g mol^{-1})	1153.82
Crystal system	Monoclinic
Space group	$C2/c$ (No. 15)
Lattice parameters from powder data	
Powder diffractometer	Stoe Stadi P
Radiation	MoKα_1 ($\lambda = 71.073$ pm)
a (pm)	2028.0(2)
b (pm)	602.5(3)
c (pm)	821.5(5)
β (deg.)	100.62(5)
Volume (Å^3)	986.6(7)
Single-crystal data	
Single-crystal diffractometer	Bruker AXS/Nonius Kappa CCD
Radiation	MoKα_1 ($\lambda = 71.073$ pm)
a (pm)	2028.2(4)
b (pm)	602.5(2)
c (pm)	820.4(2)
β (deg.)	100.63(3)
Volume (Å^3)	985.3(3)
Formula units per cell	$Z = 4$
Temperature (K)	293(2)
Calculated density (g cm^{-3})	7.778
Crystal size (mm^3)	$0.04 \times 0.03 \times 0.02$
Absorption coefficient (mm^{-1})	47.163
F (000)	1956
θ range (deg.)	$2.04 < \theta < 32.49$
Range in $h\,k\,l$	$-28/30, \pm 9, \pm 12$
Total no. of reflections	13200
Independent reflections	1784 ($R_{int} = 0.1157$)
Reflections with $I > 2\sigma(I)$	1604 ($R_\sigma = 0.0494$)
Data/parameters	1784/102
Absorption correction	numerical (HABITUS [24])
Transm. ratio (min/max)	0.1987/0.3881
Goodness-of-fit (F^2)	1.065
Final R indices ($I > 2\sigma(I)$)	$R1 = 0.0294$
	$wR2 = 0.0638$
R indices (all data)	$R1 = 0.0352$
	$wR2 = 0.0658$
Largest differ. peak, deepest hole (e/Å^3)	$2.07/-2.58$

Table 2
Atomic coordinates and isotropic equivalent displacement parameters (U_{eq}/Å^2) for $Yb_5(BO_3)_2F_9$ (space group: $C2/c$).

Atom	Wyckoff site	x	y	z	U_{eq}
Yb1	8f	0.30685(2)	0.12006(2)	0.18004(3)	0.00705(9)
Yb2	8f	0.39063(2)	0.38939(4)	0.59241(3)	0.00896(9)
Yb3	4e	1/2	0.11101(7)	1/4	0.0088(2)
B1	8f	0.3880(5)	0.905(2)	0.437(2)	0.020(2)
O1	8f	0.4094(2)	0.7587(8)	0.5663(5)	0.0071(8)
O2	8f	0.3375(3)	0.072(2)	0.4630(6)	0.014(2)
O3	8f	0.4077(3)	0.1053(8)	0.7801(6)	0.0106(9)
F1	8f	0.2897(2)	0.4241(7)	0.0198(5)	0.0109(7)
F2	8f	0.3689(2)	0.4250(7)	0.3138(5)	0.0125(8)
F3	4e	1/2	0.489(2)	1/4	0.013(2)
F4	8f	0.2735(2)	0.7819(8)	0.2172(5)	0.0145(8)
F5	8f	0.4690(2)	0.1870(9)	0.5119(6)	0.021(2)

U_{eq} is defined as one-third of the trace of the orthogonalized U_{ij} tensor.

Table 3
Anisotropic displacement parameters (U_{ij}/Å^2) for $Yb_5(BO_3)_2F_9$ (space group: $C2/c$).

Atom	U_{11}	U_{22}	U_{33}	U_{12}	U_{13}	U_{23}
Yb1	0.0088(2)	0.0047(2)	0.0075(2)	0.00016(9)	0.00110(9)	$-0.00031(8)$
Yb2	0.0160(2)	0.0048(2)	0.0058(2)	0.00200(9)	0.00142(9)	$-0.00021(8)$
Yb3	0.0061(2)	0.0095(2)	0.0106(2)	0	0.0010(2)	0
B1	0.020(4)	0.023(4)	0.017(4)	$-0.002(3)$	0.001(3)	0.001(3)
O1	0.010(2)	0.006(2)	0.005(2)	$-0.000(2)$	0.000(2)	$-0.001(2)$
O2	0.011(2)	0.022(3)	0.008(2)	$-0.011(2)$	$-0.005(2)$	0.003(2)
O3	0.015(2)	0.009(2)	0.006(2)	0.004(2)	$-0.001(2)$	$-0.001(2)$
F1	0.012(2)	0.010(2)	0.010(2)	$-0.002(2)$	$-0.001(2)$	$-0.001(2)$
F2	0.018(2)	0.012(2)	0.008(2)	$-0.004(2)$	0.003(2)	$-0.005(2)$
F3	0.013(3)	0.012(3)	0.014(3)	0	$-0.001(2)$	0
F4	0.013(2)	0.013(2)	0.017(2)	$-0.004(2)$	0.001(2)	0.005(2)
F5	0.018(2)	0.019(2)	0.026(2)	$-0.002(2)$	0.006(2)	$-0.008(2)$

Table 4
Interatomic distances (pm) in $Yb_5(BO_3)_2F_9$, calculated with the single-crystal lattice parameters.

Yb1–O2a	230.2(5)	Yb2–O1	227.4(5)	Yb3–O1	228.9(4) (2×)
Yb1–O2b	230.8(5)	Yb2–O3	228.6(5)	Yb3–O3	233.1(5) (2×)
Yb1–O3	246.6(5)	Yb2–O2	235.2(5)	Yb3–F3	227.8(6)
Yb1–O1	253.9(5)	Yb2–F5	219.9(5)	Yb3–F5a	239.2(5) (2×)
Yb1–F4a	218.6(4)	Yb2–F2a	224.4(4)	Yb3–F5b	264.2(5) (2×)
Yb1–F4b	219.7(4)	Yb2–F2b	225.7(4)		
Yb1–F1a	224.4(4)	Yb2–F1	231.5(4)		
Yb1–F1b	232.6(4)	Yb2–F3	246.4(2)		
Yb1–F2	237.8(4)	Yb2–F4	294.2(5)		
B1–O1	138.7(10)				
B1–O3	141.5(11)				
B1–O2	148.0(12)	$\varnothing = 142.7$			
F1–Yb1a	224.4(5)	F2–Yb2a	224.4(5)	F3–Yb3	227.8(7)
F1–Yb2	231.5(5)	F2–Yb2b	225.7(4)	F3–Yb2a	246.4(2) (2×)
F1–Yb1b	232.6(5)	F2–Yb1	237.8(4)		
F4–Yb1a	218.6(5)	F5–Yb2	219.9(5)		
F4–Yb2	219.7(5)	F5–Yb2b	239.2(5)		
F4–Yb1b	294.2(5)	F5–Yb1	264.2(6)		

Table 5
Interatomic angles (deg.) in $Yb_5(BO_3)_2F_9$, calculated with the single-crystal lattice parameters.

O1–B1–O3	124.6(8)	Yb1a–F1–Yb2	102.2(2)	Yb2a–F2–Yb2b	146.7(2)
O1–B1–O2	116.5(7)	Yb1a–F1–Yb1b	109.8(2)	Yb2a–F2–Yb1	100.2(2)
O3–B1–O2	118.7(7)	Yb2–F1–Yb1b	145.4(3)	Yb2b–F2–Yb1	112.4(2)
	$\varnothing = 119.9$		$\varnothing = 119.1$		$\varnothing = 119.8$
Yb3–F3–Yb2a	107.3(2)	Yb1a–F4–Yb1b	137.6(2)	Yb2–F5–Yb3a	134.5(2)
Yb3–F3–Yb2b	107.3(2)	Yb1a–F4–Yb2	89.3(2)	Yb2–F5–Yb3b	104.0(2)
Yb2a–F3–Yb2b	145.4(3)	Yb1b–F4–Yb2	133.0(2)	Yb3a–F5–Yb3b	117.9(2)
	$\varnothing = 120.0$		$\varnothing = 120.0$		$\varnothing = 118.8$

around 137 pm, e.g. in borates with calcite structure (AlBO$_3$ (137.96(4) pm) [28], β-YbBO$_3$ (137.8(4) pm) [29], and FeBO$_3$ (137.9(2) pm) [30]). The bond lengths B–O1 (138.7(10) pm) and B–O3 (141.5(11) pm) show high standard deviations, but normal values. So the large average B–O distance is mainly due to the large distance B–O2 with 148.0(12) pm, exhibiting a high standard deviation, too. This holds also true for the O–B–O angles, where the large angle O1–B1–O3 (124.6(8)°) underlines the displacement of the boron atom from the center of the BO$_3$ triangle. High standard deviations can be observed, too, for the isotropic and anisotropic displacement parameters of B1 and O2. All these findings might be evoked by a possible disorder in the BO$_3$-group (Table 3). This was also observed in Gd$_3$(BO$_3$)$_2$F$_3$ [16] and Gd$_2$(BO$_3$)F$_3$ [17] and will be discussed in the following section.

4.2 Rare-Earth Fluoride and Fluorido Borates

The formal layers in the bc-plane of $Yb_5(BO_3)_2F_9$ (Fig. 3) show the formal compositions "YbBO$_3$" and "YbF$_3$". In detail, the layer B in Fig. 3 formally comprises Yb3 in strings of YbFF$_{4/2}$, running along [001], as shown in Fig. 6. Layer C has the formal composition (Yb2)BO$_3$, depicted in Fig. 7 (left). Layer E (Fig. 7, right) has got the same composition, but a shifted arrangement by $\frac{1}{2}$ along the b-axis compared to layer C. A corrugated sheet with Yb1 in the formal composition YbF$_{8/2}$ builds up layer D (Fig. 8). The arrangement of the Yb cations in the layers explains the similar coordination spheres of Yb1 and Yb3 (4 O, 5 F), and the differing coordination of Yb2 (3 O, 6 F). Simply adding the formal constitutions of the layers, illustrated in Fig. 3, (B+B' = 2 YbF$_3$, C+C'+E+E' = 4 YbBO$_3$, D+D' = 4 YbF$_3$), we come to the formula 4 YbBO$_3$ · 6 YbF$_3$ = 2 Yb$_5$(BO$_3$)$_2$F$_9$.

The calculations of the charge distribution of the atoms in Yb$_5$(BO$_3$)$_2$F$_9$ via bond valence sums (ΣV) with ValList (bond valence calculation and listing) [31] and with the CHARDI concept (ΣQ) [32–34] confirm the formal valence states in the fluoride borate (Table 6). The low value of 2.59 for B1 in the ValList calculation is caused by the extraordinary long B–O2 bond, which is not adequately considered in the bond-length/bond-strength calculations. CHARDI calculations provide the expected value of 2.99 for B1.

Furthermore, we calculated the Madelung Part of Lattice Energy (MAPLE) values [35–37] for Yb$_5$(BO$_3$)$_2$F$_9$ in order to compare them with the MAPLE values of the high-pressure modification of Yb$_2$O$_3$ [38], of the ambient-pressure modification of B$_2$O$_3$ (B$_2$O$_3$–I) [39], and of YbF$_3$ [27]. The additive potential of the MAPLE values allows the calculation of hypothetical values for Yb$_5$(BO$_3$)$_2$F$_9$, starting from binary oxides and fluorides. As a result, we obtained a value of 53 819 kJ/mol in comparison to 53 661 kJ/mol (deviation: 0.3%), starting from the binary components [1 × Yb$_2$O$_3$ (15 590 kJ/mol) + 1 × B$_2$O$_3$–I (20 626 kJ/mol) + 3 × YbF$_3$ (17 445 kJ/mol)].

4.2. Comparison of $Yb_5(BO_3)_2F_9$ to other rare-earth fluoride borates

Besides Yb$_5$(BO$_3$)$_2$F$_9$, the only other known rare-earth fluoride borates are RE_3(BO$_3$)$_2$F$_3$ (RE = Sm, Eu, Gd) [16] and Gd$_2$(BO$_3$)F$_3$ [17]. Both structures show a quite similar constitution of formal rare-earth borate and fluoride layers (Fig. 9 left and middle). Müller-Bunz et al. realized that the duplicating of the formula Gd$_2$(BO$_3$)F$_3$ leads to Gd$_4$(BO$_3$)$_2$F$_6$, which can be written as Gd$_3$(BO$_3$)$_2$F$_3$·GdF$_3$ [17]. This illustrates the relationship of the two structures: by inserting the GdF$_3$ layer D in layer A' of Gd$_3$(BO$_3$)$_2$F$_3$, there emerge the layers C and C'.

Looking at Yb$_5$(BO$_3$)$_2$F$_9$, we found that it can be written as Yb$_3$(BO$_3$)$_2$F$_3$ · 2 YbF$_3$. In analogy to the insertion step mentioned above, an additional YbF$_3$ layer D can be found in the structure, splitting the remaining layer A of Gd$_2$(BO$_3$)$_2$F$_3$ (Fig. 9 middle) into the layers E and E' in Yb$_5$(BO$_3$)$_2$F$_9$ (Fig. 9 right).

The insertion of REF_3 layers into the structure of Gd$_3$(BO$_3$)$_2$F$_3$ stretches the a-axis, ranging from 1253.4(1) pm in Gd$_3$(BO$_3$)$_2$F$_3$ via

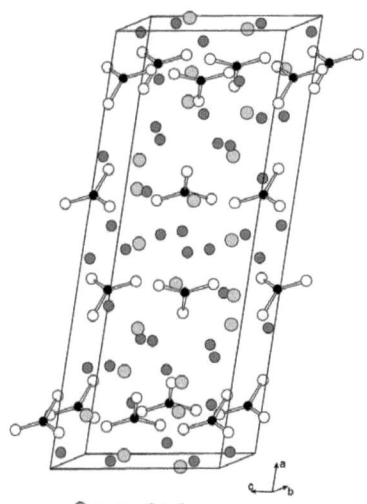

Fig. 2. Crystal structure of Yb$_5$(BO$_3$)$_2$F$_9$, showing isolated BO$_3$-groups.

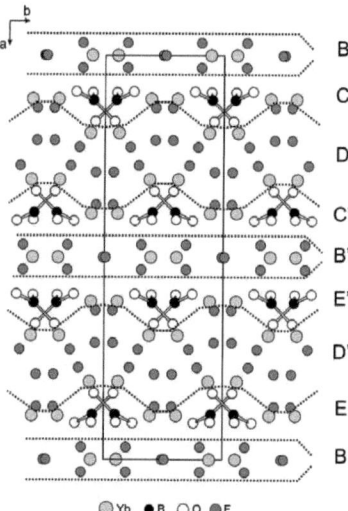

Fig. 3. Structure of Yb$_5$(BO$_3$)$_2$F$_9$, depicting alternating layers in the bc-plane with the formal compositions "YbBO$_3$" and "YbF$_3$".

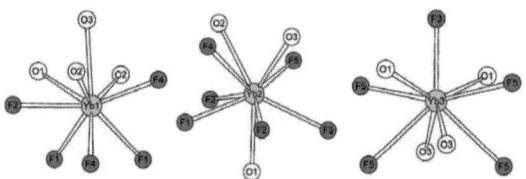

Fig. 4. Coordination spheres of the three Yb^{3+} ions in Yb$_5$(BO$_3$)$_2$F$_9$.

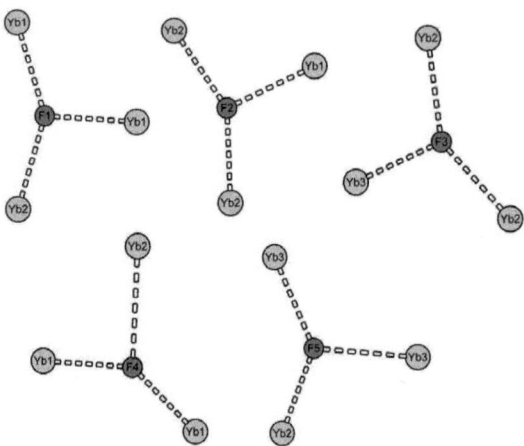

Fig. 5. Trigonal coordination of fluoride ions in Yb$_5$(BO$_3$)$_2$F$_9$.

1637.2(1) pm in Gd$_2$(BO$_3$)F$_3$ to 2028.2(4) pm in Yb$_5$(BO$_3$)$_2$F$_9$. Compared to the a-axis, the other cell edges differ only marginally, as depicted in Fig. 10. Due to the smaller ionic radius of Yb^{3+}, they shrink slightly in Yb$_5$(BO$_3$)$_2$F$_9$.

The boron position in Gd$_2$(BO$_3$)F$_3$ could not be refined by Corbel et al. [16]. Studying the fluorescence properties of Eu$_3$(BO$_3$)$_2$F$_3$, Antic-Fidancev et al. [18] found some evidence for a disorder in the BO$_3$-group of the structure. When measuring crystals of Gd$_2$(BO$_3$)F$_3$, Müller-Bunz et al. discovered that there is more than one possible position for the boron atoms in the structure: if you switch the positions of an oxygen ion of the BO$_3$-group and an adjacent fluoride ion, the location of the BO$_3$-group changes. A more detailed description of the model of disorder can be found in Ref. [17]. We simulated this for Yb$_5$(BO$_3$)$_2$F$_9$, splitted the boron position manually and resulted in acceptable B2–O distances, even though we did not find evidence for a second boron position in our crystals. So, probably, the same kind of disorder, found in Gd$_2$(BO$_3$)F$_3$ and postulated for Gd$_3$(BO$_3$)$_2$F$_3$, exists (at least to a small amount) in Yb$_5$(BO$_3$)$_2$F$_9$ as well. This would explain the large standard deviations of the structural parameters in the BO$_3$-group.

4.3. IR spectroscopy

The infrared spectrum of Yb$_5$(BO$_3$)$_2$F$_9$ was recorded on a Nicolet 5700 FT-IR spectrometer, scanning a range from 400 to 4000 cm^{-1}. Before measuring, the sample was thoroughly dried under high vacuum for several days. Fig. 10 shows the complete spectral region between 400 and 4000 cm^{-1}. The absorptions in

4.2 Rare-Earth Fluoride and Fluorido Borates

the area of 2000 cm^{-1} are evoked by the diamond window of the spectrometer and thus part of the background of the measurement. The absorptions between 1200 and 1400 cm^{-1}, between 600 and 800 cm^{-1} and below 500 cm^{-1} characterize triangular BO$_3$-groups as in λ-LaBO$_3$ [40], H-LaBO$_3$ [41], or EuB$_2$O$_4$ [42]. In the region of 3000–3500 cm^{-1}, absorption peaks could not be detected. Peaks at those wavelengths can be assigned to OH-groups and typically reveal water-containing borates. On the basis of this IR measurement, we can exclude OH-groups in Yb$_5$(BO$_3$)$_2$F$_9$ (Fig. 11).

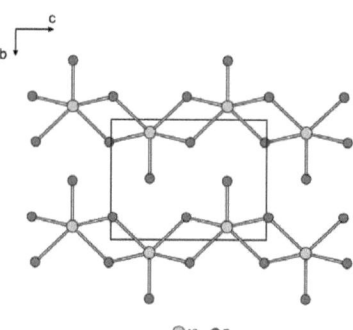

Fig. 6. View of layer B in the bc-plane of Yb$_5$(BO$_3$)$_2$F$_9$, comprising Yb3 in Yb–F-strings with the formal composition YbF$_{6/2}$.

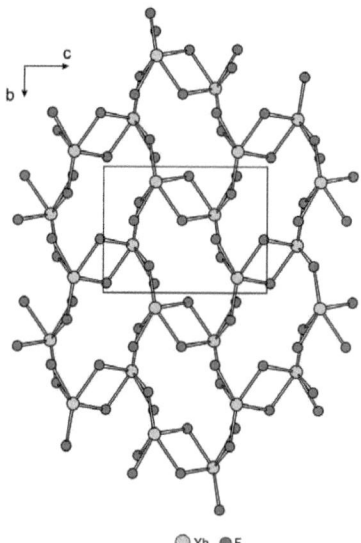

Fig. 8. Corrugated sheet with Yb1 in the formal composition YbF$_{8/2}$ in the bc-plane, building up layer D in Yb$_5$(BO$_3$)$_2$F$_9$.

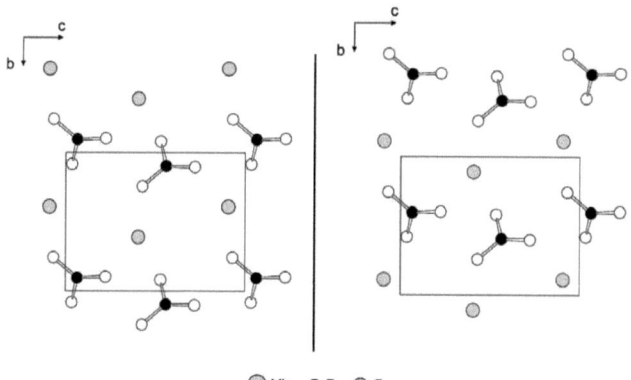

Fig. 7. Layers C (left) and E (right) in the bc-plane of Yb$_5$(BO$_3$)$_2$F$_9$, exhibiting the formal composition (Yb2)BO$_3$.

Table 6
Charge distribution in $Yb_5(BO_3)_2F_9$, calculated with ValList (ΣV) [31] and the CHARDI concept (ΣQ) [32–34].

	Yb1	Yb2	Yb3	B1				
ΣV	3.04	2.91	2.66	2.59				
ΣQ	3.05	2.94	3.05	2.99				
	O1	O2	O3	F1	F2	F3	F4	F5
ΣV	−2.02	−1.89	−1.94	−0.97	−0.98	−0.74	−0.91	−0.79
ΣQ	−2.15	−1.79	−2.03	−1.10	−1.13	−0.87	−0.92	−0.94

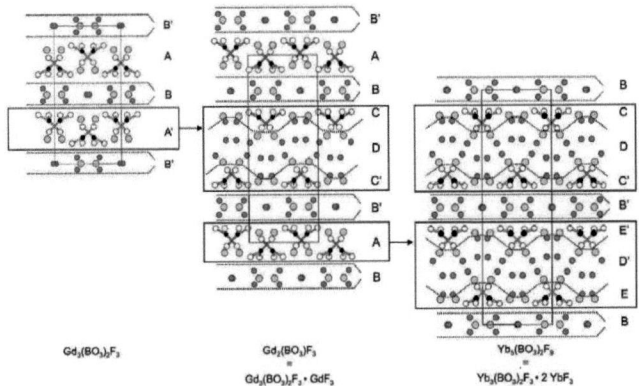

Fig. 9. Comparison of the structures of $Gd_3(BO_3)_2F_3$ (left), $Gd_2(BO_3)F_3$ (middle) and $Yb_5(BO_3)_2F_9$ (right). The splitting of layer A' into layers C, D, and C' is depicted in the left part; splitting of layer A into layers E, D', and E' is shown on the right (RE = light gray spheres, B = black spheres, O = hollow spheres, F = dark gray spheres).

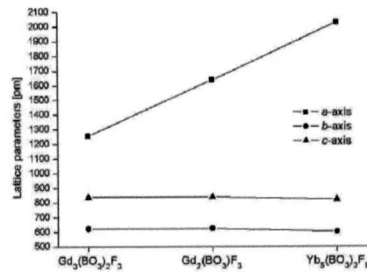

Fig. 10. Lattice parameters of $Gd_3(BO_3)_2F_3$ (left), $Gd_2(BO_3)F_3$ (middle), and $Yb_5(BO_3)_2F_9$ (right).

5. Conclusions

In this article, we described the high-pressure synthesis and crystal structure of the first ytterbium fluoride borate $Yb_5(BO_3)_2F_9$. It shows a structure closely related to the known gadolinium fluoride borates $Gd_3(BO_3)_2F_3$ and $Gd_2(BO_3)F_3$. We found some evidence of disorder in the BO_3-group of $Yb_5(BO_3)_2F_9$; nevertheless, we can neither confirm nor refute the disorder model existing for the gadolinium fluoride borates. The synthesis of similar structures, starting from oxides and fluorides of different rare-earth cations, can—as we hope—make understand the actual formation of the structure and will be the object of our forthcoming studies. Investigations concerning optical properties of our fluoride borate, especially a possible enlargement of the optical gap, will be performed in the future.

Another point, stimulating our interest, are higher pressures on $Yb_5(BO_3)_2F_9$ to transform the BO_3-groups of the structure into BO_4-tetrahedra. This already succeeded and led to a new structure exhibiting very interesting properties [43].

4.2 Rare-Earth Fluoride and Fluorido Borates

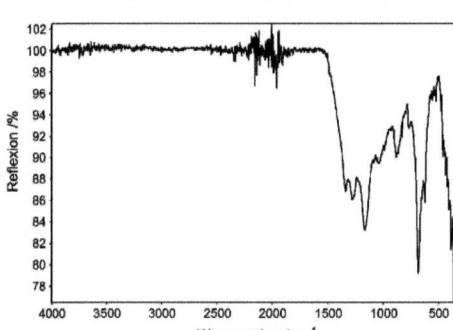

Fig. 11. IR spectrum of $Yb_5(BO_3)_2F_9$.

Acknowledgments

We thank Dr. Klaus Wurst for collecting the single-crystal data. Special thanks go to Prof. Dr. W. Schnick (LMU München) for his continuous support of these investigations. This work was financially supported by the Fonds der Chemischen Industrie.

Appendix A. Supplementary materials

Supplementary data associated with this article can be found in the online version at: doi:10.1016/j.jssc.2009.01.023.

References

[1] H. Huppertz, B. von der Eltz, J. Am. Chem. Soc. 124 (2002) 9376.
[2] H. Huppertz, Z. Naturforsch. B 58 (2003) 278.
[3] H. Huppertz, H. Emme, J. Phys. Condens. Matter 16 (2004) 1283.
[4] H. Emme, H. Huppertz, Z. Anorg. Allg. Chem. 628 (2002) 2165.
[5] H. Emme, H. Huppertz, Chem. Eur. J. 9 (2003) 3623.
[6] H. Emme, H. Huppertz, Acta Crystallogr. C 61 (2005) i29.
[7] H. Emme, H. Huppertz, Acta Crystallogr. C 61 (2005) i23.
[8] H. Huppertz, S. Altmannshofer, G. Heymann, J. Solid State Chem. 170 (2003) 320.
[9] A. Haberer, G. Heymann, H. Huppertz, J. Solid State Chem. 180 (2007) 1595.
[10] G. Heymann, T. Soltner, H. Huppertz, Solid State Sci. 8 (2006) 821.
[11] A. Haberer, G. Heymann, H. Huppertz, Z. Naturforsch. B 62 (2007) 759.
[12] J.S. Knyrim, F.M. Schappacher, R. Pöttgen, J. Schmedt auf der Günne, D. Johrendt, H. Huppertz, Chem. Mater. 19 (2007) 254.
[13] J.S. Knyrim, H. Huppertz, Z. Naturforsch. 63b (2008) 707.
[14] J.S. Knyrim, H. Huppertz, J. Solid State Chem. 180 (2007) 742.
[15] T. Suzuki, M. Hirano, H. Hosono, J. Appl. Phys. 91 (2002) 4149.
[16] G. Corbel, R. Retoux, M. Leblanc, J. Solid State Chem. 139 (1998) 52.
[17] H. Müller-Bunz, Th. Schleid, Z. Anorg. Allg. Chem. 628 (2002) 2750.
[18] E. Antic-Fidancev, G. Corbel, N. Mercier, M. Leblanc, J. Solid State Chem. 153 (2000) 270.
[19] D. Walker, M.A. Carpenter, C.M. Hitch, Am. Mineral 75 (1990) 1020.
[20] D. Walker, Am. Mineral 76 (1991) 1092.
[21] H. Huppertz, Z. Kristallogr. 219 (2004) 330.
[22] D.C. Rubie, Phase Transitions 68 (1999) 431.
[23] N. Kawai, S. Endo, Rev. Sci. Instrum. 8 (1970) 1178.
[24] W. Herrendorf, H. Bärnighausen, HABITUS, Universities of Karlsruhe and Giessen, Germany, 1993/1997.
[25] G.M. Sheldrick, SHELXS-97 and SHELXL-97, Program suite for the solution and refinement of crystal structures, University of Göttingen, Göttingen, Germany, 1997.
[26] G.M. Sheldrick, Acta Crystallogr. A 64 (2008) 112.
[27] A. Zalkin, D.H. Templeton, J. Am. Chem. Soc. 75 (1953) 2453.
[28] A. Vegas, Acta Crystallogr. B 33 (1977) 3607.
[29] H. Huppertz, Z. Naturforsch. B 56 (2001) 697.
[30] R. Diehl, Solid State Commun. 17 (1975) 743.
[31] A.S. Wills, ValList Version 3.0.13, University College London, UK, 1998–2008. Program available from ⟨www.ccp14.ac.uk⟩.
[32] I.D. Brown, D. Altermatt, Acta Crystallogr. B 41 (1985) 244.
[33] N.E. Brese, M. O'Keeffe, Acta Crystallogr. B 47 (1991) 192.
[34] R. Hoppe, S. Voigt, H. Glaum, J. Kissel, H.P. Müller, K.J. Bernet, Less-Common Met. 156 (1989) 105.
[35] R. Hoppe, Angew. Chem. 78 (1966) 52;
R. Hoppe, Angew. Chem. Int. Ed. 5 (1966) 96.
[36] R. Hoppe, Angew. Chem. 82 (1970) 7;
R. Hoppe, Angew. Chem. Int. Ed. 9 (1970) 25.
[37] R. Hübenthal, MAPLE Program for the Calculation of MAPLE Values, Version 4, University of Gießen, Gießen, Germany, 1993.
[38] H.R. Hoekstra, Inorg. Chem. 5 (1966) 754.
[39] S.V. Berger, Acta Crystallogr. 5 (1952) 389.
[40] J.P. Laperches, P. Tarte, Spectrochim. Acta 22 (1966) 1201.
[41] H. Böhlhoff, U. Bambauer, W. Hoffmann, Z. Kristallogr. 133 (1971) 386.
[42] K. Machida, H. Hata, K. Okuna, G. Adachi, J. Shiokawa, J. Inorg. Nucl. Chem. 41 (1979) 1425.
[43] A. Haberer, H. Huppertz, unpublished results.

4.2 Rare-Earth Fluoride and Fluorido Borates

4.2.2 Er$_5$(BO$_3$)$_2$F$_9$

The successful synthesis of Yb$_5$(BO$_3$)$_2$F$_9$ was the starting point for the research on isotypic phases. Due to the comparable ionic radius of the cations, neighboring elements in the periodic system are likely to crystallize in isotypic structures, while larger size differences often result in different crystal structures. An attempt with a mixture of erbium oxide, erbium fluoride and boron oxide was made and pink crystals of Er$_5$(BO$_3$)$_2$F$_9$ could be obtained (Figure 4.2-2). We performed single-crystal IR and Raman measurements on both isotypic compounds in order to compare their spectroscopic properties. The results and the structural data for Er$_5$(BO$_3$)$_2$F$_9$ are presented in the following publication.

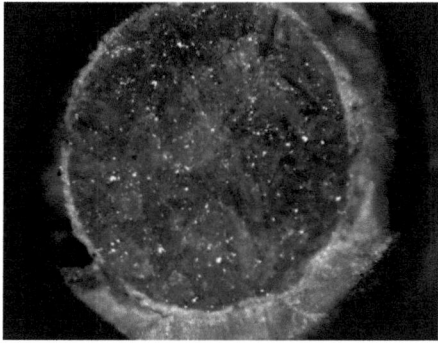

Figure 4.2-2
Crystalline sample of Er$_5$(BO$_3$)$_2$F$_9$

ARTICLE

DOI: 10.1002/zaac.201000085

Synthesis of Er$_5$(BO$_3$)$_2$F$_9$ and Properties of RE_5(BO$_3$)$_2$F$_9$ (RE = Er, Yb)

Almut Haberer,[a] Reinhard Kaindl,[b] Jürgen Konzett,[b] Robert Glaum,[c] and Hubert Huppertz*[a]

Dedicated to Professor Rüdiger Kniep on the Occasion of His 65th Birthday

Keywords: Rare earths; Fluorine; Oxides; Boron; High-pressure; X-ray diffraction

Abstract. Er$_5$(BO$_3$)$_2$F$_9$ was synthesised under conditions of 3 GPa and 800 °C in a Walker-type multianvil apparatus. The crystal structure was determined on the basis of single-crystal X-ray diffraction data, collected at room temperature. Er$_5$(BO$_3$)$_2$F$_9$ is isotypic to the recently synthesised Yb$_5$(BO$_3$)$_2$F$_9$ and crystallises in $C2/c$ with the lattice parameters a = 2031.2(4) pm, b = 609.5(2) pm, c = 824.6(2) pm, and β = 100.29(3)°. The physical properties of RE_5(BO$_3$)$_2$F$_9$ (RE = Er, Yb) including high temperature behaviour and single crystal IR- / Raman spectroscopy were investigated.

Introduction

In the past years, the field of rare-earth fluorido- and fluoride borates could be greatly extended by the application of high-pressure / high-temperature techniques, leading to the new compounds Yb$_5$(BO$_3$)$_2$F$_9$ [1], Pr$_4$B$_3$O$_{10}$F [2], and Gd$_4$B$_4$O$_{11}$F$_2$ [3].

Before we started our high-pressure / high-temperature research, rare-earth fluorido- and fluoride borates were only represented by the compounds RE_3(BO$_3$)$_2$F$_3$ (RE = Sm, Eu, Gd) [4] and Gd$_2$(BO$_3$)F$_3$ [5], synthesised by heating a stoichiometric mixture of RE_2O$_3$, B$_2$O$_3$, and REF$_3$ under ambient pressure conditions. There is a close relationship between these two structure types and the recently discovered high-pressure ytterbium fluoride borate Yb$_5$(BO$_3$)$_2$F$_9$ [1]. All three compounds can be described by alternating layers of the formal compositions "REBO$_3$" and "REF$_3$" in the bc-plane. It should be emphasised that the crystal structures of the actual compounds REBO$_3$ and REF$_3$ can not be compared with the structures of the layers, so that "REBO$_3$" and "REF$_3$" only stand for the formal composition of the layers. We are now trying to substitute the Yb^{3+} ions of Yb$_5$(BO$_3$)$_2$F$_9$ with other rare-earth cations to yield isotypic RE_5(BO$_3$)$_2$F$_9$ phases, different structures, or new compositions dependent on the size of the rare-earth ion. E.g., during our studies of high-pressure phases of the rare-earth borates, the compounds RE_4B$_6$O$_{15}$ (RE = Dy, Ho) [6–8] were intensively investigated. Nevertheless, it was not possible to obtain isotypical compounds with RE = Er, Tm, Yb, or Lu. For these rare-earth elements, high-pressure / high-temperature experiments always yielded the compounds RE_3B$_5$O$_{12}$ (RE = Er–Lu) [9].

In analogy to these investigations of the rare-earth borates, we are trying to define the formation region for the composition RE_5(BO$_3$)$_2$F$_9$. Whereas Yb$_5$(BO$_3$)$_2$F$_9$ was synthesised at 7.5 GPa [1], an isotypic erbium fluoride borate Er$_5$(BO$_3$)$_2$F$_9$ could be obtained now at a pressure of 3 GPa. This indicates that there is a large pressure region for the formation of these high-pressure phases and the application of similar synthetic conditions to neighbouring rare-earth elements could thus lead to additional isotypic compounds. Since it was possible to obtain Er$_5$(BO$_3$)$_2$F$_9$, the existence of an isotypic thulium compound is more than likely. Future studies will show, if the synthesis with larger rare-earth cations is possible as well and at which point we observe the formation of another structure type as in the borate case. However, experiments on Yb$_5$(BO$_3$)$_2$F$_9$ at higher pressures (10 GPa) resulted already in the formation of an apatite-structured compound, wherein all boron atoms showed a fourfold coordination [10].

In this paper, we present the synthesis and structural details of Er$_5$(BO$_3$)$_2$F$_9$ in comparison to the isotypic phase Yb$_5$(BO$_3$)$_2$F$_9$. Furthermore, several physical properties of the compounds RE_5(BO$_3$)$_2$F$_9$ (RE = Er, Yb) are reported.

Experimental Section

To synthesise the compound Er$_5$(BO$_3$)$_2$F$_9$, high-pressure / high-temperature conditions of 3 GPa and 800 °C were applied, starting from Er$_2$O$_3$ (Strem Chemicals, 99.9 %), B$_2$O$_3$ (Strem Chemicals, 99.9+%),

---

* Prof. Dr. H. Huppertz
  E-Mail: Hubert.Huppertz@uibk.ac.at
[a] Institut für Allgemeine, Anorganische und Theoretische Chemie
    Leopold-Franzens-Universität Innsbruck
    Innrain 52a
    6020 Innsbruck, Austria
[b] Institut für Mineralogie und Petrographie
    Leopold-Franzens-Universität Innsbruck
    Innrain 52
    6020 Innsbruck, Austria
[c] Institut für Anorganische Chemie
    Universität Bonn
    Gerhard-Domagk-Str. 1
    53121 Bonn, Germany

4.2 Rare-Earth Fluoride and Fluorido Borates

Synthesis of Er$_5$(BO$_3$)$_2$F$_9$ and Properties of RE_5(BO$_3$)$_2$F$_9$ (RE = Er, Yb)

and ErF$_3$ (Strem Chemicals, 99.+%) at equal molar ratios. The starting materials were ground up and filled into a boron nitride crucible (Henze BNP GmbH, HeBoSint® S100, Kempten, Germany). The compounds were compressed and heated by a multianvil assembly, based on a Walker-type module and a 1000 ton press (both devices from the company Voggenreiter, Mainleus, Germany). Eight tungsten carbide cubes (TSM-20, Ceratizit, Reutte, Austria) compressed the boron nitride crucible, which was positioned inside the centre of an 18/11 assembly. A detailed description of the assembly can be found in reference [11].

The 18/11 assembly was compressed up to 3 GPa in 1.25 h and heated to 800 °C (cylindrical graphite furnace) in the following 10 min, kept there for 20 min, and cooled down to 600 °C in 20 min at constant pressure. After natural cooling down to room temperature by switching off the heating, a decompression period of 3.75 h was required. Then the recovered octahedral pressure medium (MgO, Ceramic Substrates & Components Ltd., Newport, Isle of Wight, UK) was broken apart and the sample carefully separated from the surrounding graphite and boron nitride. Pink air- and water-resistant crystals of Er$_5$(BO$_3$)$_2$F$_9$ were obtained.

Crystal Structure Analysis

The sample was characterised by powder X-ray diffraction, which was performed in transmission geometry on a flat sample of the reaction product, using a STOE STADI P powder diffractometer with Mo-$K_{\alpha1}$ radiation (Ge monochromator, λ = 71.073 pm). The powder pattern showed reflections of Er$_5$(BO$_3$)$_2$F$_9$ and of μ-ErBO$_3$ [12] as a by-product of the synthesis (Figure 1). The experimental powder pattern tallied well with the theoretical pattern of Er$_5$(BO$_3$)$_2$F$_9$, simulated from single-crystal data. Indexing the reflections of the erbium fluoride borate, we got the parameters a = 2033.0(3) pm, b = 610.6(3) pm, and c = 825.5(4) pm, with β = 100.29(6)° and a volume of 1008.2(9) Å^3. This confirmed the lattice parameters, obtained from single-crystal X-ray diffraction (Table 1).

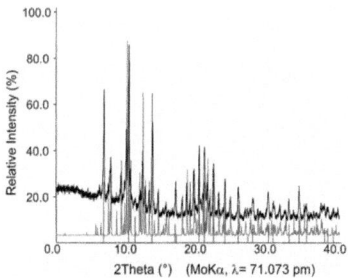

Figure 1. Experimental powder pattern of Er$_5$(BO$_3$)$_2$F$_9$ in comparison to the theoretical powder pattern beneath, based on single-crystal X-ray diffraction data. Reflections of μ-ErBO$_3$ are marked with lines.

Intensity data of a single-crystal of Er$_5$(BO$_3$)$_2$F$_9$ were collected at room temperature by use of a Kappa CCD diffractometer (Bruker AXS / Nonius, Karlsruhe), equipped with a Miracol Fiber Optics Collimator and a Nonius FR590 generator (graphite-monochromatised Mo-$K_{\alpha1}$ ra-

Table 1. Crystal data and structure refinement of Er$_5$(BO$_3$)$_2$F$_9$

Empirical Formula	Er$_5$(BO$_3$)$_2$F$_9$
Molar mass /g·mol^{-1}	1124.92
Crystal system	monoclinic
Space group	C2/c (No. 15)
Lattice parameters from powder data	
Powder diffractometer	STOE STADI P
Radiation	Mo-$K_{\alpha1}$ (λ = 71.073 pm)
a /pm	2033.0(3)
b /pm	610.6(3)
c /pm	825.5(4)
β /deg	100.29(6)
Volume /Å^3	1008.2(9)
Single-crystal data	
Single-crystal diffractometer	Bruker AXS / Nonius Kappa CCD
Radiation	Mo-$K_{\alpha1}$ (λ = 71.073 pm)
a /pm	2031.2(4)
b /pm	609.5(2)
c /pm	824.6(2)
β /deg	100.29(3)
Volume /Å^3	1004.4(3)
Formula units per cell	4
Temperature /K	293(2)
Calculated density /g·cm^{-3}	7.44
Crystal size /mm	0.03 × 0.02 × 0.02
Absorption coefficient /mm^{-1}	41.49
F(000)	1916
θ range /deg	2.0 ≤ θ ≤ 37.9
Range in $h\ k\ l$	−33/34, ±10, −14/12
Total no. reflections	7955
Independent reflections	2700 (R_{int} = 0.0382)
Reflections with $I > 2\sigma(I)$	2315 (R_σ = 0.0389)
Data / parameters	2700 / 102
Absorption correction	Multi-scan [13]
Goodness-of-fit (F^2)	1.062
Final R indices [$I > 2\sigma(I)$]	$R1$ = 0.0289
	$wR2$ = 0.0530
R indices (all data)	$R1$ = 0.0384
	$wR2$ = 0.0558
Largest diff. peak / deepest hole / e·Å^{-3}	2.71 / −2.55

diation, λ = 71.073 pm). An absorption correction, based on multiscans [13], was applied to the data set. All relevant details of the data collection and evaluation are listed in Table 1.

The structure solution and the parameter refinement (full-matrix least-squares against F^2) were successfully performed, using the SHELX-97 software tool [14, 15] with anisotropic atomic displacement parameters for all atoms. According to the systematic extinctions, the orthorhombic space groups C2/c and Cc were derived. The structure solution in C2/c (no. 15) succeeded. The final difference Fourier syntheses did not reveal any significant residual peaks in all refinements. The positional parameters of the refinements, anisotropic displacement parameters, interatomic distances, and interatomic angles are listed in Table 2, Table 3, Table 4, and Table 5. Further information of the crystal structure is available from the Fachinformationszentrum Karlsruhe (crysdata@fiz-karlsruhe.de), 76344 Eggenstein-Leopoldshafen, Germany, by quoting the Registry No. CSD-421469.

ARTICLE

Table 2. Atomic coordinates and isotropic equivalent displacement parameters (U_{eq} /Å^2) for Er$_5$(BO$_3$)$_2$F$_9$ (space group: C2/c). Wyckoff sites are 4e for Er3 and F3 and 8f for all other atoms. U_{eq} is defined as one-third of the trace of the orthogonalised U_{ij} tensor.

Atom	x	y	z	U_{eq}
Er1	0.30661(2)	0.11935(3)	0.18057(2)	0.00619(5)
Er2	0.39031(2)	0.38886(3)	0.59380(2)	0.00770(5)
Er3	½	0.11138(5)	¼	0.00783(6)
B1	0.3884(3)	0.902(2)	0.4368(8)	0.014(2)
O1	0.4087(2)	0.7605(6)	0.5638(4)	0.0080(6)
O2	0.3392(2)	0.0678(7)	0.4620(4)	0.0123(7)
O3	0.4082(2)	0.1054(6)	0.7829(4)	0.0081(6)
F1	0.2890(2)	0.4245(5)	0.0195(4)	0.0089(5)
F2	0.3683(2)	0.5761(5)	0.8143(4)	0.0107(5)
F3	½	0.4892(7)	¼	0.0134(8)
F4	0.2752(2)	0.7781(5)	0.2150(4)	0.0128(6)
F5	0.4693(2)	0.1849(6)	0.5136(5)	0.0192(7)

Table 3. Interatomic distances /pm in Er$_5$(BO$_3$)$_2$F$_9$, calculated with the single-crystal lattice parameters.

Er1–O2a	231.8(4)	Er2–O3	231.2(3)	Er3–O1	232.2(4) (2×)
Er1–O2b	232.7(4)	Er2–O1	231.6(4)		
Er1–O3	249.6(4)	Er2–O2	238.4(4)	Er3–O3	234.0(3) (2×)
Er1–O1	254.5(4)	Er2–F5	222.2(4)		
Er1–F4a	220.9(3)	Er2–F2a	225.8(3)	Er3–F3	230.3(5)
Er1–F4b	222.4(3)	Er2–F2b	227.8(3)	Er3–F5a	240.8(4)
Er1–F1a	227.6(3)	Er2–F1	233.6(3)		(2×)
Er1–F1b	232.8(3)	Er2–F3	248.0(2)	Er3–F5b	264.9(4)
Er1–F2	239.6(3)	Er2–F4	288.8(4)		(2×)
B1–O1	136.4(7)				
B1–O3	139.8(7)				
B1–O2	146.1(8)	Ø ≈ 140.8			
F1–Er1a	227.6(3)	F2–Er2a	225.8(3)	F3–Er3	230.3(5)
F1–Er2	233.6(3)	F2–Er2b	227.8(3)	F3–Er2	248.0(2)
F1–Er1b	232.8(3)	F2–Er1	239.6(3)		(2×)
F4–Er1a	220.9(3)	F5–Er2	222.2(4)		
F4–Er1b	222.4(3)	F5–Er3a	239.2(5)		
F4–Er2	288.8(4)	F5–Er3b	264.9(4)		

Table 4. Interatomic angles /deg in Er$_5$(BO$_3$)$_2$F$_9$, calculated with the single-crystal lattice parameters.

O1–B1–O3	125.4(5)	Er1a–F1–Er2	101.9(2)
O1–B1–O2	116.7(5)	Er1a–F1–Er1b	110.1(2)
O3–B1–O2	117.9(5)	Er2–F1–Er1b	145.7(2)
	Ø = 120.0		Ø = 119.2
Er2a–F2–Er2b	146.2(2)	Er3–F3–Er2a	107.4(2)
Er2a–F2–Er1	100.6(2)	Er3–F3–Er2b	107.4(2)
Er2b–F2–Er1	112.6(2)	Er2a–F3–Er2b	145.1(2)
	Ø = 119.8		Ø = 120.0
Er1a–F4–Er1b	135.4(2)	Er2–F5–Er3a	133.6(2)
Er1a–F4–Er2	91.0(2)	Er2–F5–Er3b	104.5(2)
Er1b–F4–Er2	133.4(2)	Er3a–F5–Er3b	118.2(2)
	Ø = 199.9		Ø = 118.8

Results and Discussion

Crystal Structure of Er$_5$(BO$_3$)$_2$F$_9$

The structure of Er$_5$(BO$_3$)$_2$F$_9$ consists of isolated BO$_3$-groups, erbium cations, and fluoride anions (Figure 2). The arrangement can be described by alternating layers of the formal compositions "ErBO$_3$" and "ErF$_3$", spreading into the bc-plane. For a more detailed description of the structure, including a comparison to the related structures of RE_3(BO$_3$)$_2$F$_3$ (RE = Sm, Eu, Gd) [5] and Gd$_2$(BO$_3$)F$_3$ [6], the reader is referred to the description of the isotypic compound Yb$_5$(BO$_3$)$_2$F$_9$ (ref. [1]). In this paper, we briefly compare the main features of the compounds.

Table 5. Charge distribution in Er$_5$(BO$_3$)$_2$F$_9$, calculated with VaList (ΣV) [20] and the CHARDI concept (ΣQ) [21–23].

	Er1	Er2	Er3	B1
ΣV	3.09	2.94	2.70	2.73
ΣQ	3.06	2.92	3.05	3.00

	O1	O2	O3	F1
ΣV	–2.06	–1.94	–1.99	–0.96
ΣQ	–2.20	–1.80	–1.95	–1.10

	F2	F3	F4	F5
ΣV	–1.01	–0.76	–0.93	–0.81
ΣQ	–1.14	–0.88	–0.91	–0.95

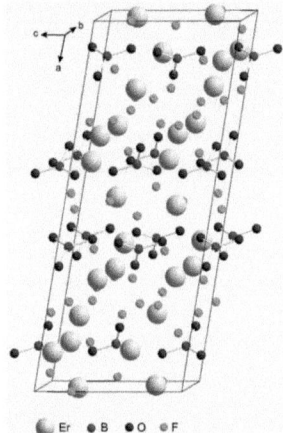

Figure 2. Crystal structure of Er$_5$(BO$_3$)$_2$F$_9$, showing isolated BO$_3$-groups.

With $a = 2031.4(2)$ pm, $b = 609.5(2)$ pm, and $c = 824.6(2)$ pm, the lattice parameters of Er$_5$(BO$_3$)$_2$F$_9$ are slightly larger than those of Yb$_5$(BO$_3$)$_2$F$_9$ ($a = 2028.0(2)$, $b = 602.5(3)$, and $c = 821.5(5)$ pm). This can be ascribed to the slightly larger size of Er^{3+} (ionic radius = 120 pm for a ninefold coordination [16]), compared to Yb^{3+} (ionic radius = 118 pm for coordina-

Synthesis of Er$_5$(BO$_3$)$_2$F$_9$ and Properties of RE_5(BO$_3$)$_2$F$_9$ (RE = Er, Yb)

tion number nine [16]). Since the size difference is marginally, no greater deviances of the bond lengths and angles are observed. In both compounds, each of the three crystallographically independent rare-earth cations exhibits a coordination number of nine. In Er$_5$(BO$_3$)$_2$F$_9$, we find Er–O/F distances in the range of 231.2(3)–288.8(4) pm, which tally well with the values obtained for the Yb–O/F distances in Yb$_5$(BO$_3$)$_2$F$_9$ (227.4(5) to 294.2(5) pm). Focussing on the coordination spheres of the fluoride ions, we find a coordination number of three for all fluoride ions in the two isotypic compounds. In Yb$_5$(BO$_3$)$_2$F$_9$, the F–Yb distances are slightly shorter than the F–Er bond lengths in Er$_5$(BO$_3$)$_2$F$_9$, which agrees with the slight size difference of the cations.

Inside the BO$_3$-groups, the B–O distances of Er$_5$(BO$_3$)$_2$F$_9$ vary between 136.4(7) and 146.1(8) pm (Table 3). The B–O bond lengths tally with those in Yb$_5$(BO$_3$)$_2$F$_9$ (between 138.7(10) and 148.0(12) pm). The unusually large B–O2 bond of 148 pm, observed in Yb$_5$(BO$_3$)$_2$F$_9$, is slightly shorter in the erbium compound (146.1(8) pm), but still exceeds typical bond lengths in BO$_3$-groups, which are usually in a range around 137 pm, e.g. in borates with a calcite structure (AlBO$_3$ (137.96(4) pm) [17], β-YbBO$_3$ (137.8(4) pm) [18], and FeBO$_3$ (137.9(2) pm) [19]).

In analogy to Yb$_5$(BO$_3$)$_2$F$_9$, we have calculated the charge distribution of the atoms in Er$_5$(BO$_3$)$_2$F$_9$ via bond valence sums (ΣV) with VaList (Bond Valence Calculation and Listing) [20] and with the CHARDI concept (ΣQ) [21–23], confirming the formal valence states in the fluoride borate. Table 5 displays the results of the formal ionic charges of the atoms, which are in agreement within the limits of both concepts. In Yb$_5$(BO$_3$)$_2$F$_9$, the bond valence sums for the boron atom were extraordinarily low (ΣV = 2.59), due to the large B–O2 bond present in the structure. Since the boron bond lengths are shorter in Er$_5$(BO$_3$)$_2$F$_9$, we now obtain a better value of ΣV = 2.73 for B1. Furthermore, we have calculated the MAPLE value (*M*adelung *P*art of *L*attice *E*nergy according to Hoppe [24–26]) of Er$_5$(BO$_3$)$_2$F$_9$ to compare it with the sum of the MAPLE values for the underlying binary components, namely the high-pressure modification of Er$_2$O$_3$ [27], the ambient-pressure modification of B$_2$O$_3$ (B$_2$O$_3$–I) [28], and ErF$_3$ [29]. We obtained a value of 53879 kJ·mol^{-1} in comparison to 53240 kJ·mol^{-1} (deviation: 1.2 %), starting from the binary components [1 × Er$_2$O$_3$ (15283 kJ·mol^{-1}) + 1 × B$_2$O$_3$–I (20626 kJ·mol^{-1}) + 3 × ErF$_3$ (17331 kJ·mol^{-1})].

Elemental Analyses

Since it is nearly impossible to distinguish between fluoride ions and hydroxyl groups by means of electron density or bond lengths, we performed elemental analyses on our samples of RE_5(BO$_3$)$_2$F$_9$ (RE = Er, Yb) to assure the atom assignments in both structures. EDX measurements were performed on a JEOL JSM-6500F scanning electron microscope, equipped with a field emission gun at an acceleration voltage of 16 kV. Samples were prepared by placing single crystals of both phases on adhesive conductive pads and subsequently coating them with a thin conductive carbon film. Each EDX spectrum (Oxford Instruments) was recorded with the analysed area limited on one single crystal to avoid the influence of possibly contaminating phases. The crystals of Er$_5$(BO$_3$)$_2$F$_9$ showed average atomic compositions (theoretical values in brackets) of Er 63.1 (74.3) wt-%; B 8.9 (1.9) wt-%; O 9.3 (8.5) wt-%; F 18.7 (15.2) wt-%. For Yb$_5$(BO$_3$)$_2$F$_9$, average compositions (wt-%) of Yb 70.7 (75.0) wt-%; B 4.6 (1.9) wt-%; O 8.4 (8.3) wt-%; F 16.3 (14.8) wt-% were obtained. Due to the light weight of boron, measurements have to be taken with caution, but still, these results confirm the presence of all elements and the composition, obtained from the single crystal structure determination.

Furthermore, the composition of Yb$_5$(BO$_3$)$_2$F$_9$ was confirmed by chemical analysis with a JEOL JXA-8100 electron microprobe. The sample was prepared by embedding bulk material in epoxy resin and polishing with diamond polishing paste. Analytical conditions were a 15 kV acceleration voltage and a 20 nA beam current with measurement times of 40 sec on the peaks and 20 sec on the backgrounds of the O-K_α, F-K_α, and Yb-L_α X-ray lines. SiO$_2$, Al$_2$SiO$_4$F$_2$, and Yb$_5$P$_5$O$_{14}$ were used as standards for O, F, and Yb, respectively, and the raw counts were corrected with the ZAF correction procedure. Boron could not be determined quantitatively with the electron microprobe due to its low concentration. An average of 3 analyses on three different crystals yielded (theoretical values in brackets): Yb 74.4 ± 0.2 (75.0) wt-%; O 8.6 ± 0.4 (8.3) wt-%; F 14.5 ± 0.6 (14.8) wt-%.

Physical Properties of RE_5(BO$_3$)$_2$F$_9$ (RE = Er, Yb)

UV/Vis Spectroscopy of Yb$_5$(BO$_3$)$_2$F$_9$

The electronic absorption spectrum of Yb$_5$(BO$_3$)$_2$F$_9$ was measured at 293 K in the NIR/Vis/UV (5800–28000 cm^{-1}, step-width $\Delta\lambda$ (UV/Vis) = 1 nm, $\Delta\lambda$ (NIR) = 2 nm), using a

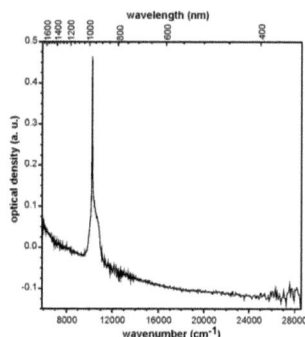

Figure 3. Electronic absorption spectrum of a polycrystalline agglomerate of Yb$_5$(BO$_3$)$_2$F$_9$.

ARTICLE

A. Haberer, R. Kaindl, J. Konzett, R. Glaum, H. Huppertz

strongly modified CARY 17 microcrystal spectrophotometer (Spectra Services, ANU Canberra, Australia). Details of the spectrometer have already been described in literature [30, 31]. A polycrystalline agglomerate with dimensions 0.2 × 0.2 × 0.05 mm was selected for the investigation. The noisy spectrum (Figure 3) results from the small size and rather modest optical quality of the agglomerate. The reference intensity was measured with a pinhole instead of the crystal mounted on an aperture. The electronic absorption spectrum of a polycrystalline agglomerate of $Yb_5(BO_3)_2F_9$ shows the transition $^2F_{7/2} \rightarrow {}^2F_{5/2}$, expected for Yb^{3+} at $\tilde{v} \approx 10400$ cm^{-1} [32]. Due to weak ligand-field interaction, the signal is split into a sharp peak at $\tilde{v} = 10236$ cm^{-1} and a broader shoulder at $\tilde{v} = 10660$ cm^{-1}.

IR Spectroscopy

FTIR absorbance spectra of single crystals of $RE_5(BO_3)_2F_9$ (RE = Er, Yb) were measured in transmitted, polarised light on a BaF$_2$ plate using a BRUKER VERTEX 70 spectrometer, attached to a HYPERION 3000 microscope, a MIR light source, and a LN-MCT detector. Spectral resolution was ~4 cm^{-1}. Minimum-maximum normalisation was done by the OPUS 6.5 software (BRUKER). In Figure 4, the spectra of both compounds are displayed. The absorptions of $Yb_5(BO_3)_2F_9$ (Figure 4, bottom) correspond to the absorptions measured on the bulk material (Ref. [1]). In the IR spectrum of the isotypic erbium compound (Figure 4, top), the absorption bands are slightly shifted to higher wavelengths. The absorption patterns of the IR spectra are typical for borates exhibiting triangular BO$_3$-groups [1, 33, 34]. Below 600 cm^{-1} for $Yb_5(BO_3)_2F_9$ and below 650 cm^{-1} for $Er_5(BO_3)_2F_9$, bands are evoked by the in-plane bending (v_4) of the BO$_3$-groups. The strong and usually sharp absorptions derived from the out-of-plane bending (v_2) of the trigonal ion occur in the range of 600–780 cm^{-1} for $Yb_5(BO_3)_2F_9$ and 650–850 cm^{-1} for $Er_5(BO_3)_2F_9$. According to the crystalline environment, the absorption bands between 800–950 cm^{-1} for $Yb_5(BO_3)_2F_9$ and 900–1000 cm^{-1} for $Er_5(BO_3)_2F_9$ can be classified as symmetric stretching vibrations (v_1). Asymmetric stretching vibrations (v_3) of the BO$_3$-groups are assigned in the areas of 1150–1400 cm^{-1} for $Yb_5(BO_3)_2F_9$ and of 1200–1500 cm^{-1} for $Er_5(BO_3)_2F_9$. Interestingly, two bands between 3400 and 3600 cm^{-1} were detected in the single crystal spectra of $RE_5(BO_3)_2F_9$ (RE = Er, Yb) (Figure 5), which were not present in the corresponding spectrum of dried bulk $Yb_5(BO_3)_2F_9$ [1]. Bands in this range usually indicate OH-groups in water-containing borates. The substitution of fluoride with hydroxyl groups is a problem commonly known from fluoride borates [3, 4]. The single crystals, which were examined by IR spectroscopy, were exposed to air and humidity and were not dried under high vacuum like the bulk material of $Yb_5(BO_3)_2F_9$. Thus, the exchange of fluoride with OH-groups occurs due to hygroscopic alteration. In apatites, the exchange of hydroxyl and fluoride ions was previously observed and studied [35, 36].

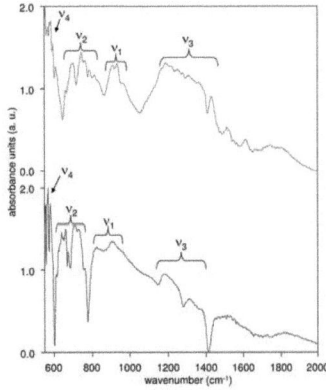

Figure 4. FTIR absorbance spectra of single crystals of $Er_5(BO_3)_2F_9$ (top) and $Yb_5(BO_3)_2F_9$ (bottom) in the range 550–2000 cm^{-1}.

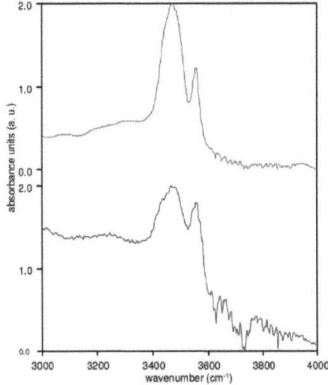

Figure 5. FTIR absorbance spectra on single crystals of $Er_5(BO_3)_2F_9$ (top) and $Yb_5(BO_3)_2F_9$ (bottom) in the range 3000–4000 cm^{-1}.

Raman Spectroscopy

Raman spectra of single crystals of $RE_5(BO_3)_2F_9$ (RE = Er, Yb) were measured without polarisers in the range 100–1600 cm^{-1}, using a HORIBA LABRAM HR-800 confocal micro-spectrometer and a 785 nm diode laser. Laser focus and power on the sample surface was ~1 μm and ~10 mW, respec-

Synthesis of Er$_5$(BO$_3$)$_2$F$_9$ and Properties of RE_5(BO$_3$)$_2$F$_9$ (RE = Er, Yb)

tively. Depth resolution was ~4 μm, spectral resolution ~6 cm^{-1}. Wavenumber accuracy of ~3 cm^{-1} was achieved by regularly adjusting the zero-order position of the grating and checked by a Neon calibration lamp. Third order polynomial background subtraction, normalisation, and band fitting by Gauss-Lorentz functions were done by the LABSPEC 5 software (HORIBA). Some glue contamination cannot be excluded.

The Raman spectra of the isotypic compounds RE_5(BO$_3$)$_2$F$_9$ (RE = Er, Yb) are quite similar with the most intense bands < 500 cm^{-1} and the narrow, "isolated" band at ~940 cm^{-1} (Figure 6). Above 1000 cm^{-1}, one broad band for Yb$_5$(BO$_3$)$_2$F$_9$ and several broader bands for Er$_5$(BO$_3$)$_2$F$_9$ are observed. Bands around 900 cm^{-1} in borates are usually assigned to stretching modes of BO$_4$-tetrahedra whereas BO$_3$-groups are expected > 1100 cm^{-1} [4, 37–42]. However, in hydrated monoborates very intense bands at 882 and 501 cm^{-1} were observed and assigned to symmetric stretching vibrations of the isolated BO$_3$-groups [39]. The unusually large variation of B–O distances inside the BO$_3$-groups (see above) might account for the comparatively low wavenumber and the large wavenumber range of these bands. Bands < 500 cm^{-1} can be assigned to bending and stretching vibrations of the isolated BO$_3$-groups, RE–O, RE–F, and F–O bonds, as well as lattice vibrations.

room temperature to 500 °C in 100 °C steps, and from 500 °C to 1100 °C in 50 °C steps. The heating rate was set to 40 °C·min^{-1}. Afterwards, the sample was cooled down to 500 °C in 50 °C steps, and from 500 °C to room temperature in 100 °C steps (cooling rate: 50 °C·min^{-1}). After each heating step, a diffraction pattern was recorded over the angular range $3° \leq 2\theta \leq 40°$. The temperature-dependent X-ray powder diffraction experiment on Yb$_5$(BO$_3$)$_2$F$_9$ indicated a decomposition and subsequent phase changes of the decomposition products (Figure 7). Successive heating of Yb$_5$(BO$_3$)$_2$F$_9$ in the range of 650–850 °C resulted in a decomposition into a mixture of π-YbBO$_3$ [43] and another still unidentified phase. Further heating led to a transformation of π-YbBO$_3$ into the high-temperature polymorph μ-YbBO$_3$ [43] in the range of 900–1100 °C. Additionally, reflections of YbF$_3$ and a formation of the silicate Yb$_2$Si$_2$O$_7$ [44] were observed. The latter was formed by a reaction with the silica capillary, which is known to happen occasionally. Subsequent cooling showed a complete retransformation into the low-temperature polymorph π-YbBO$_3$ below 500 °C. The formation and transformation of π-YbBO$_3$ into the high-temperature polymorph μ-YbBO$_3$ and back were often observed in temperature-dependent X-ray powder diffraction experiments of ytterbium borates, e.g. with Yb$_3$B$_5$O$_{12}$ [9].

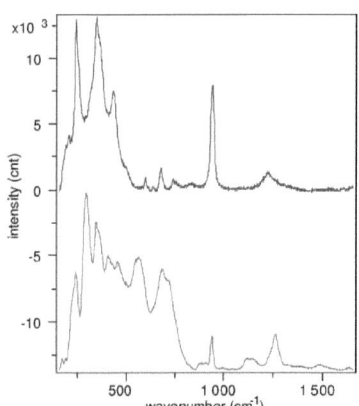

Figure 6. Raman-spectra on single crystals of Yb$_5$(BO$_3$)$_2$F$_9$ (top) and Er$_5$(BO$_3$)$_2$F$_9$ (bottom) in the range 100–1600 cm^{-1}.

Thermal Behaviour of Yb$_5$(BO$_3$)$_2$F$_9$

In situ X-ray powder diffraction experiments were done with a STOE STADI P powder diffractometer [Mo-$K_{\alpha 1}$ radiation (λ = 71.073 pm)] with a computer controlled STOE furnace: The sample was enclosed in a silica capillary and heated from

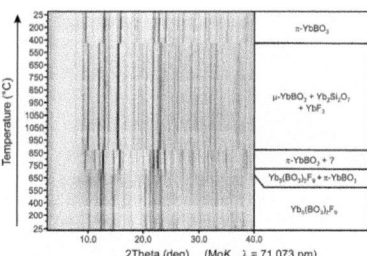

Figure 7. Temperature-dependent X-ray powder patterns, following the decomposition reaction of Yb$_5$(BO$_3$)$_2$F$_9$.

Conclusions

The successful syntheses of Yb$_5$(BO$_3$)$_2$F$_9$ (7.5 GPa) and the isotypic erbium fluoride borate Er$_5$(BO$_3$)$_2$F$_9$ at a pressure of 3 GPa suggest a large formation region for these high pressure phases. The application of similar synthetic conditions to neighbouring rare-earth elements could lead to additional isotypic compounds and will be studied in the future. However, in the case of RE^{3+} = Y^{3+}, experiments at higher pressures (10 GPa) resulted in the formation of an apatite-structured compound, wherein all boron atoms showed a fourfold coordination [10].

Acknowledgement

We thank *Dr. Gunter Heymann* for collecting the single-crystal data. Special thanks go to *Christian Minke* (LMU Munich) for the EDX analyses.

References

[1] A. Haberer, H. Huppertz, *J. Solid State Chem.* **2009**, *182*, 888.
[2] A. Haberer, H. Huppertz, *Solid State Sci.* **2010**, *12*, 515.
[3] A. Haberer, R. Kaindl, H. Huppertz, *J. Solid State Chem.* **2010**, *183*, 471.
[4] G. Corbel, R. Retoux, M. Leblanc, *J. Solid State Chem.* **1998**, *139*, 52.
[5] H. Müller-Bunz, Th. Schleid, *Z. Anorg. Allg. Chem.* **2002**, *628*, 2750.
[6] H. Huppertz, B. von der Eltz, *J. Am. Chem. Soc.* **2002**, *124*, 9376.
[7] H. Huppertz, *Z. Naturforsch.* **2003**, *58b*, 278.
[8] H. Huppertz, H. Emme, *J. Phys.: Condens. Matter* **2004**, *16*, 1283.
[9] H. Emme, M. Valldor, R. Pöttgen, H. Huppertz, *Chem. Mater.* **2005**, *17*, 2707.
[10] A. Haberer, H. Huppertz, unpublished results.
[11] H. Huppertz, *Z. Kristallogr.* **2004**, *219*, 330.
[12] R. E. Newnham, M. J. Redman, R. P. Santoro, *J. Am. Ceram. Soc.* **1963**, *46*, 253.
[13] Z. Otwinowski, W. Minor, *Methods Enzymol.* **1997**, *276*, 307.
[14] G. M. Sheldrick, *SHELXS-97* and *SHELXL-97*, Program Suite for the Solution and Refinement of Crystal Structures, University of Göttingen, Göttingen, Germany, **1997**.
[15] G. M. Sheldrick, *Acta Crystallogr., Sect. A* **2008**, *64*, 112.
[16] R. D. Shannon, *Acta Crystallogr., Sect. A* **1976**, *32*, 751.
[17] A. Vegas, *Acta Crystallogr., Sect. B* **1977**, *33*, 3607.
[18] H. Huppertz, *Z. Naturforsch.* **2001**, *56b*, 697.
[19] R. Diehl, *Solid State Commun.* **1975**, *17*, 743.
[20] A. S. Wills, *VaList Version 3.0.13*, University College London, UK, **1998–2008**; Program available from www.ccp14.ac.uk.
[21] I. D. Brown, D. Altermatt, *Acta Crystallogr., Sect. B* **1985**, *41*, 244.
[22] N. E. Brese, M. O'Keeffe, *Acta Crystallogr., Sect. B* **1991**, *47*, 192.
[23] R. Hoppe, S. Voigt, H. Glaum, J. Kissel, H. P. Müller, K. J. Bernet, *J. Less-Common Met.* **1989**, *156*, 105.
[24] R. Hoppe, *Angew. Chem.* **1966**, *78*, 52; *Angew. Chem. Int. Ed. Engl.* **1966**, *5*, 95.
[25] R. Hoppe, *Angew. Chem.* **1970**, *82*, 7; *Angew. Chem. Int. Ed. Engl.* **1970**, *9*, 25.
[26] R. Hübenthal, *MAPLE (version 4)*, Program for The Calculation of MAPLE Values, University of Gießen, Gießen, Germany, **1993**.
[27] H. R. Hoekstra, *Inorg. Chem.* **1966**, *5*, 754.
[28] S. V. Berger, *Acta Crystallogr.* **1952**, *5*, 389.
[29] A. Zalkin, D. H. Templeton, *J. Am. Chem. Soc.* **1953**, *75*, 2453.
[30] E. Krausz, *Aust. J. Chem.* **1993**, *46*, 1041.
[31] E. Krausz, *AOS News* **1998**, *12*, 21.
[32] B. Henderson, G. F. Imbusch, *Optical Spectroscopy of Solids*, Oxford University Press, **1989**.
[33] J. P. Laperches, P. Tarte, *Spectrochim. Acta* **1966**, *22*, 1201.
[34] G. Heymann, K. Beyer, H. Huppertz, *Z. Naturforsch.* **2004**, *59b*, 1200.
[35] A. Knappwost, *Naturwissenschaften* **1959**, *46*, 555.
[36] V. P. Orlovskii, S. P. Ionov, T. V. Belyaevskaya, S. M. Barinov, *Inorg. Mater.* **2002**, *38*, 182.
[37] H. Emme, H. Huppertz, *Chem. Eur. J.* **2003**, *9*, 3623.
[38] H. Huppertz, *J. Solid State Chem.* **2004**, *177*, 3700.
[39] L. Jun, X. Shuping, G. Shiyang, *Spectrochim. Acta A* **1995**, *51*, 519.
[40] G. Chadeyron, M. El-Ghozzi, R. Mahiou, A. Arbus, J. C. Cousseins, *J. Solid State Chem.* **1997**, *128*, 261.
[41] J. C. Zhang, Y. H. Wang, X. Guo, *J. Lumin.* **2007**, *122–123*, 980.
[42] G. Padmaja, P. Kistaiah, *J. Phys. Chem. A* **2009**, *113*, 2397.
[43] W. F. Bradley, D. L. Graf, R. S. Roth, *Acta Crystallogr.* **1966**, *20*, 284.
[44] Y. I. Smolin, Y. F. Shepelev, *Acta Crystallogr., Sect. B* **1970**, *26*, 484.

Received: February 10, 2010
Published Online: May 3, 2010

4.2 Rare-Earth Fluoride and Fluorido Borates

4.2.3 Tm$_5$(BO$_3$)$_2$F$_9$

Since the compounds Er$_5$(BO$_3$)$_2$F$_9$ and Yb$_5$(BO$_3$)$_2$F$_9$ were synthesized at 3 GPa and 7.5 GPa, respectively, it was most likely to obtain the intermediate phase Tm$_5$(BO$_3$)$_2$F$_9$ at intermediate pressures. Indeed, the synthesis at 5 GPa was succesful and colorless crystals of Tm$_5$(BO$_3$)$_2$F$_9$ were obtained (Figure 4.2-3). This was achieved in cooperation with M. Enders during his Master's thesis in our working group. All relevant details on this phase can be found in the following publication.

Figure 4.2-3
Crystalline sample of Tm$_5$(BO$_3$)$_2$F$_9$

High-pressure Synthesis and Characterization of the Fluoride Borate $Tm_5(BO_3)_2F_9$

Almut Haberer[a], Michael Enders[a], Reinhard Kaindl[b], and Hubert Huppertz[a]

[a] Institut für Allgemeine, Anorganische und Theoretische Chemie, Leopold-Franzens-Universität Innsbruck, Innrain 52a, 6020 Innsbruck, Austria

[b] Institut für Mineralogie und Petrographie, Leopold-Franzens-Universität Innsbruck, Innrain 52, 6020 Innsbruck, Austria

Reprint requests to H. Huppertz. E-mail: Hubert.Huppertz@uibk.ac.at

Z. Naturforsch. 2010, 65b, 1213–1218; received June 17, 2010

The rare earth fluoride borate $Tm_5(BO_3)_2F_9$ was synthesized from Tm_2O_3, B_2O_3, and TmF_3 under high-pressure/high-temperature conditions of 5 GPa and 900 °C in a Walker-type multianvil apparatus. The single-crystal structure determination revealed that $Tm_5(BO_3)_2F_9$ is isotypic to the compounds $RE_5(BO_3)_2F_9$ (RE = Er, Yb). $Tm_5(BO_3)_2F_9$ crystallizes in the space group $C2/c$ (Z = 4) with the parameters a = 2030.9(4), b = 606.2(2), c = 822.6(2) pm, β = 100.5(1)°, V = 995.7(3) Å^3, R_1 = 0.0341, and wR_2 = 0.0724 (all data). The structure is composed of isolated BO_3 groups, ninefold coordinated thulium cations, and fluoride anions. Infrared and Raman spectroscopic data of $Tm_5(BO_3)_2F_9$ are compared to the data of $RE_5(BO_3)_2F_9$ (RE = Er, Yb).

Key words: Rare Earth, Fluoride, Borate, High Pressure, Crystal Structure

Introduction

In the chemistry of rare-earth fluoride borates, the first known compounds $RE_3(BO_3)_2F_3$ (RE = Sm, Eu, Gd) [1, 2] and $Gd_2(BO_3)_2F_3$ [3] were obtained through the heating of stoichiometric mixtures of RE_2O_3, B_2O_3, and REF_3 under ambient-pressure conditions. In the last two years, the implementation of high-pressure/high-temperature techniques have given access to four new structure types with the compositions $RE_5(BO_3)_2F_9$ (RE = Er, Yb) [4, 5], $La_4B_4O_{11}F_2$ [6], $Gd_4B_4O_{11}F_2$ [7], and $Pr_4B_3O_{10}F$ [8]. Recently, the first divalent rare-earth fluoride borate $Eu_5(BO_3)_3F$ [9] was synthesized by Kazmierczak *et al.* under ambient-pressure conditions. The successful syntheses of the phases $RE_5(BO_3)_2F_9$ (RE = Er, Yb) [4, 5] indicated that most likely this structure type may also exist with the rare-earth cation Tm^{3+}. While $Er_5(BO_3)_2F_9$ and $Yb_5(BO_3)_2F_9$ were synthesized at 3 GPa and 7.5 GPa, respectively, now, the isotypic thulium fluoride borate $Tm_5(BO_3)_2F_9$ could be synthesized successfully at a pressure of 5 GPa. In this paper, we report the high-pressure synthesis of $Tm_5(BO_3)_2F_9$, including a comparison with the isotypic compounds $RE_5(BO_3)_2F_9$ (RE = Er, Yb) with respect to structural and spectroscopic data.

Experimental Section

Synthesis

The synthesis of $Tm_5(BO_3)_2F_9$ was achieved under high-pressure/high-temperature conditions of 5 GPa and 900 °C. A stoichiometric mixture of B_2O_3 (Strem Chemicals, 99.9+ %), Tm_2O_3 (Strem Chemicals, 99.9+ %), and TmF_3 (Alfa Aesar GmbH & Co KG, 99+ %) was pestled in a glove box and filled into a boron nitride crucible (Henze BNP GmbH, HeBoSint® S100, Kempten, Germany). The latter was positioned into an 18/11 assembly, which was compressed by eight tungsten carbide cubes (TSM-20, Ceratizit, Reutte, Austria) in a Walker-type multianvil device and a 1000 ton press (both devices from the company Voggenreiter, Mainleus, Germany). A detailed description of the assembly can be found in the references [10–14].

The 18/11 assembly was compressed to 5 GPa in 135 min, heated to 900 °C (cylindrical graphite furnace) within 15 min, kept at this temperature for 20 min, and cooled down to 700 °C during another 20 min. After cooling to r. t. by radiation heat loss, the decompression of the assembly lasted 7 h. The recovered octahedral pressure medium (MgO, Ceramic Substrates & Components Ltd., Newport, Isle of Wight, UK) was broken apart and the surrounding boron nitride and graphite were separated from the sample. This procedure yielded colorless air- and water-resistant crystals of the compound $Tm_5(BO_3)_2F_9$.

4.2 Rare-Earth Fluoride and Fluorido Borates

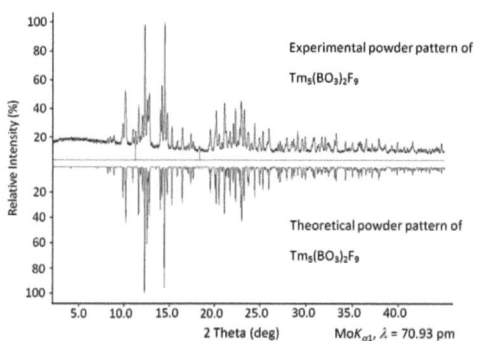

Fig. 1 (color online). Top: experimental powder pattern of $Tm_5(BO_3)_2F_9$; reflections of an unknown phase are indicated with lines. Bottom: theoretical powder pattern of $Tm_5(BO_3)_2F_9$ based on single-crystal diffraction data.

Table 1. Crystal data and structure refinement of $Tm_5(BO_3)_2F_9$ (standard deviations in parentheses).

Empirical formula	$Tm_5(BO_3)_2F_9$
Molar mass, g mol^{-1}	1133.27
Crystal system	monoclinic
Space group	$C2/c$
Powder diffractometer	Stoe STADI P
Radiation	$MoK_{\alpha1}$ (λ = 70.93 pm)
Powder data	
a, pm	2030.1(8)
b, pm	605.8(4)
c, pm	822.7(5)
β, deg	100.6(1)
V, Å^3	994.7(7)
Single-crystal diffractometer	Nonius Kappa CCD
Radiation	MoK_α (λ = 71.073 pm)
Crystal size, mm^3	0.04 × 0.04 × 0.06
Single-crystal data	
a, pm	2030.9(4)
b, pm	606.2(2)
c, pm	822.6(2)
β, deg	100.5(1)
V, Å^3	995.7(3)
Formula units per cell Z	4
Calculated density, g cm^{-3}	7.56
Temperature, K	293(2)
Absorption coefficient, mm^{-1}	44.3
$F(000)$, e	1936
θ range, deg	2.0 – 32.5
Range in hkl	±30, ±9, ±12
Total no. of reflections	6212
Independent reflections / R_{int} / R_σ	1809 / 0.0583 / 0.0422
Reflections with $I \geq 2\sigma(I)$	1643
Data / ref. parameters	1809 / 102
Absorption correction	multi-scan
Final $R1/wR2$ [$I \geq 2\sigma(I)$]	0.0296 / 0.0702
$R1/wR2$ (all data)	0.0341 / 0.0724
Goodness-of-fit on F^2	1.058
Largest diff. peak and hole, e Å^{-3}	3.7 / −3.2

Table 2. Atomic coordinates and isotropic equivalent displacement parameters U_{eq} (Å^2) for $Tm_5(BO_3)_2F_9$ (space group: $C2/c$) (standard deviations in parentheses). U_{eq} is defined as one third of the trace of the orthogonalized U_{ij} tensor.

Atom	W.-position	x	y	z	U_{eq}
Tm1	8f	0.30677(2)	0.11941(4)	0.18042(3)	0.00654(9)
Tm2	8f	0.39044(2)	0.38928(4)	0.59287(3)	0.0083(2)
Tm3	4e	1/2	0.11053(6)	1/4	0.0082(2)
B1	8f	0.3885(5)	0.904(2)	0.437(2)	0.018(3)
O1	8f	0.4094(2)	0.7583(7)	0.5648(5)	0.0077(8)
O2	8f	0.3385(2)	0.0698(9)	0.4625(6)	0.0123(9)
O3	8f	0.4075(2)	0.1050(7)	0.7813(6)	0.0094(9)
F1	8f	0.2891(3)	0.4227(7)	0.0198(5)	0.0105(7)
F2	8f	0.3683(2)	0.5758(7)	0.8134(5)	0.0119(7)
F3	4e	1/2	0.4879(9)	1/4	0.014(2)
F4	8f	0.2740(2)	0.7807(7)	0.2167(5)	0.0135(8)
F5	8f	0.4688(2)	0.1851(6)	0.5136(6)	0.0209(9)

Crystal structure analysis

$Tm_5(BO_3)_2F_9$ was identified by X-ray powder diffraction on a flat sample of the reaction product, using a Stoe Stadi P powder diffractometer with $MoK_{\alpha1}$ radiation (transmission geometry, Ge monochromator, λ = 70.93 pm). Fig. 1 shows the powder pattern displaying reflections of $Tm_5(BO_3)_2F_9$ as well as a still unknown side product (marked with lines in Fig. 1). The experimental powder pattern (top) is in good agreement with the theoretical pattern (bottom), simulated from the single-crystal data. Indexing the reflections of the thulium fluoride borate, we derived the parameters a = 2030.1(8), b = 605.8(4), c = 822.7(5) pm, β = 100.6(1)° and a volume of 994.7(7) Å^3. This validated the lattice parameters received from the single-crystal X-ray diffraction data (Table 1). Intensity data of a single crystal of $Tm_5(BO_3)_2F_9$ were gathered at r.t. by use of a Kappa CCD diffractometer (Bruker AXS/Nonius, Karlsruhe), equipped with a Miracol fiber optics collimator and a Nonius FR590

A. Haberer et al. · Fluoride Borate Tm$_5$(BO$_3$)$_2$F$_9$

Table 3. Anisotropic displacement parameters U_{ij} (Å^2) for Tm$_5$(BO$_3$)$_2$F$_9$ (space group $C2/c$).

Atom	U_{11}	U_{22}	U_{33}	U_{12}	U_{13}	U_{23}
Tm1	0.0080(2)	0.0047(2)	0.0069(2)	−0.00038(8)	0.0013(2)	0.00022(8)
Tm2	0.0148(2)	0.0045(2)	0.0055(2)	−0.00022(8)	0.0015(2)	0.00286(8)
Tm3	0.0057(2)	0.0091(2)	0.0097(2)	0	0.0012(2)	0
B1	0.023(4)	0.012(3)	0.019(4)	−0.001(3)	0.006(3)	−0.003(3)
O1	0.011(2)	0.005(2)	0.008(2)	0.000(2)	0.002(2)	−0.001(2)
O2	0.009(2)	0.020(2)	0.007(2)	0.005(2)	−0.002(2)	−0.007(3)
O3	0.011(2)	0.010(2)	0.007(2)	0.000(2)	0.001(2)	0.000(2)
F1	0.006(2)	0.013(2)	0.012(3)	0.001(2)	0.002(2)	−0.001(2)
F2	0.021(2)	0.007(2)	0.008(2)	0.001(2)	0.003(2)	0.007(3)
F3	0.014(3)	0.010(3)	0.017(3)	0	0.000(2)	0
F4	0.015(2)	0.012(3)	0.013(2)	0.001(2)	0.002(2)	−0.006(2)
F5	0.016(2)	0.022(2)	0.026(2)	−0.009(3)	0.007(2)	−0.003(2)

Table 4. Interatomic distances (pm) in Tm$_5$(BO$_3$)$_2$F$_9$ (space group: $C2/c$), calculated with the single-crystal lattice parameters (standard deviations in parentheses).

Tm1–O3	247.0(5)	Tm2–O3	230.2(5)	Tm3–O3	234.1(5)
Tm1–O1	255.4(5)	Tm2–O1	228.9(5)		2 ×
Tm1–O2a	231.2(5)	Tm2–O2	236.7(6)	Tm3–O1	230.6(5)
Tm1–O2b	231.6(5)	Tm2–F1	233.4(4)		2 ×
Tm1–F1a	225.5(4)	Tm2–F3	247.6(3)	Tm3–F3	228.7(6)
Tm1–F1b	232.3(4)	Tm2–F5	220.7(5)	Tm3–F5a	241.0(5)
Tm1–F4a	219.5(4)	Tm2–F4	292.7(4)		2 ×
Tm1–F4b	220.7(4)	Tm2–F2a	225.1(4)	Tm3–F5b	263.5(5)
Tm1–F2	238.2(4)	Tm2–F2b	227.1(4)		2 ×
⌀ = 233.5		⌀ = 238.0		⌀ = 240.8	
B1–O3	140.6(11)				
B1–O1	137.9(10)				
B1–O2	146.9(10)				
⌀ = 141.8					
F1–Tm1a	225.5(4)	F2–Tm1	238.2(4)	F3–Tm2	247.5(3)
F1–Tm1b	232.3(4)	F2–Tm2a	225.1(4)		2 ×
F1–Tm2	233.4(4)	F2–Tm2b	227.1(4)	F3–Tm3	228.7(6)
⌀ = 230.4		⌀ = 230.1		⌀ = 241.2	
F4–Tm1a	219.5(4)	F5–Tm2	220.7(5)		
F4–Tm1b	220.7(4)	F5–Tm3a	241.0(5)		
F4–Tm2	292.7(4)	F5–Tm3b	263.5(5)		
⌀ = 244.3		⌀ = 241.7			

Table 5. Interatomic angles (deg) in Tm$_5$(BO$_3$)$_2$F$_9$ (space group: $C2/c$), calculated with the single-crystal lattice parameters (standard deviations in parentheses).

O1–B1–O3	124.8(7)	Tm1a–F1–Tm1b	110.4(2)
O1–B1–O2	117.1(7)	Tm1a–F1–Tm2	102.0(2)
O3–B1–O2	117.9(7)	Tm1b–F1–Tm2	145.2(3)
⌀ = 119.9		⌀ = 119.2	
Tm3–F3–Tm2a	107.5(2)	Tm2–F5–Tm3a	133.9(2)
Tm3–F3–Tm2b	107.5(2)	Tm2–F5–Tm3b	104.6(3)
Tm2a–F3–Tm2b	145.0(3)	Tm3a–F5–Tm3b	117.9(3)
⌀ = 120.0		⌀ = 118.8	
Tm1a–F4–Tm1b	137.0(2)	Tm2a–F2–Tm2b	146.1(3)
Tm1a–F4–Tm2	89.8(3)	Tm2a–F2–Tm1	100.6(2)
Tm1b–F4–Tm2	132.9(3)	Tm2–F2–Tm1	112.5(2)
⌀ = 119.9		⌀ = 119.7	

generator (graphite-monochromatized MoK_α radiation, λ = 71.073 pm). An absorption correction, based on multi-scans,

Table 6. Comparison of the lattice parameters (pm, deg) and cell volumes V (Å^3) of RE_5(BO$_3$)$_2$F$_9$ (RE = Er, Tm, Yb) (standard deviations in parentheses).

Compound	a	b	c	β	V
Er$_5$(BO$_3$)$_2$F$_9$	2031.2(4)	609.5(2)	824.6(2)	100.3(1)	1004.4(3)
Tm$_5$(BO$_3$)$_2$F$_9$	2030.9(4)	606.2(2)	822.6(2)	100.5(1)	995.7(3)
Yb$_5$(BO$_3$)$_2$F$_9$	2028.2(4)	602.5(2)	820.4(2)	100.6(1)	985.3(3)

was performed with SCALEPACK [15]. All significant details of the data collection and analyses are listed in Table 1. For the structure refinement, the positional parameters of the isotypic compound Er$_5$(BO$_3$)$_2$F$_9$ were used as starting values [4]. The parameter refinement (full-matrix least-squares against F^2) was achieved by using SHELXL-97 [16, 17]. All atoms were refined with anisotropic atomic displacement parameters. The final difference Fourier syntheses did not reveal any significant residual peaks in all refinements. The positional parameters of the atom refinements, anisotropic displacement parameters, interatomic distances, and interatomic angles are listed in the Tables 2 – 5.

Further details of the crystal structure investigation may be obtained from the Fachinformationszentrum Karlsruhe, 76344 Eggenstein-Leopoldshafen, Germany (fax: +49-7247-808-666; e-mail: crysdata@fiz-karlsruhe.de, http://www.fiz-informationsdienste.de/en/DB/icsd/depot_anforderung.html) on quoting the deposition number CSD-421889.

Results and Discussion

Crystal structure of Tm$_5$(BO$_3$)$_2$F$_9$

The structure of Tm$_5$(BO$_3$)$_2$F$_9$ consists of isolated BO$_3$ groups, ninefold coordinated thulium cations, and fluoride anions (Fig. 2). For a detailed depiction of the structure, the reader is referred to the description of the isotypic compound Yb$_5$(BO$_3$)$_2$F$_9$ [5]. In this paper, a comparison of the three isotypic compounds RE_5(BO$_3$)$_2$F$_9$ (RE = Er, Tm, Yb) is given.

Table 6 shows the values of the lattice parameters of Er$_5$(BO$_3$)$_2$F$_9$ [4], Tm$_5$(BO$_3$)$_2$F$_9$, and Yb$_5$(BO$_3$)$_2$F$_9$ [5]. The differences correspond to the decreasing

4.2 Rare-Earth Fluoride and Fluorido Borates

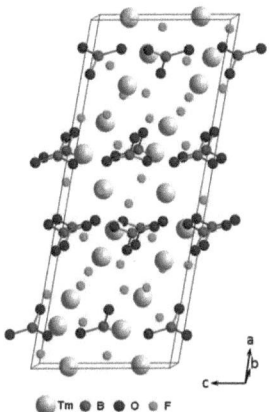

Fig. 2 (color online). Crystal structure of $Tm_5(BO_3)_2F_9$, showing isolated BO_3 groups.
(legend: Tm ● B ● O ● F)

Table 7. Charge distribution in $Tm_5(BO_3)_2F_9$, calculated with VALIST (ΣV) [19] and the CHARDI concept (ΣQ) [22].

	Tm1	Tm2	Tm3	B1
ΣV	2.95	2.79	2.63	2.66
ΣQ	3.04	2.94	3.05	3.00
	O1	O2	O3	F1
ΣV	−2.01	−1.99	−2.03	−0.87
ΣQ	−2.14	−1.81	−2.01	−1.09
	F2	F3	F4	F5
ΣV	−0.88	−0.66	−0.81	−0.71
ΣQ	−1.13	−0.88	−0.93	−0.95

Additionally, we calculated the MAPLE value (*M*adelung *P*art of *L*attice *E*nergy according to Hoppe [23–25]) of $Tm_5(BO_3)_2F_9$, and checked it against the sum of the MAPLE values received from the binary compounds Tm_2O_3 [26], B_2O_3-I [27], and TmF_3 [28]. We obtained a value of 53842 kJ mol^{-1} for $Tm_5(BO_3)_2F_9$, compared to 53544 kJ mol^{-1} (deviation: 0.6 %), starting from the binary components [1 × Tm_2O_3 (15484 kJ mol^{-1}) + 1 × B_2O_3-I (20669 kJ mol^{-1}) + 3 × TmF_3 (5797 kJ mol^{-1})].

ionic radii (lanthanoide contraction) of the ninefold coordinated rare-earth cations Er^{3+} (ionic radius = 120 pm [18]), compared to Tm^{3+} (119 pm [18]), and Yb^{3+} (118 pm [18]). Because the size differences are not too large, the bond lengths and angles in $Tm_5(BO_3)_2F_9$ are comparable to the values found in $Er_5(BO_3)_2F_9$ and $Yb_5(BO_3)_2F_9$. The Tm–O/F distances range from 219.5(4) to 292.7(4) pm, which fits well to the values of 220.9(3)–288.8(4) pm for Er–O/F in $Er_5(BO_3)_2F_9$ and 218.6(4)–294.2(5) pm for Yb–O/F in $Yb_5(BO_3)_2F_9$. The mean Tm–F distance is 237.5 pm, which is between the mean bond lengths of 238.1 and 236.9 pm in $Er_5(BO_3)_2F_9$ and $Yb_5(BO_3)_2F_9$, respectively. The B–O distances in $Tm_5(BO_3)_2F_9$ are in the range from 137.9(10) pm to 146.9(10) pm (Table 4), which is the same as found in $Er_5(BO_3)_2F_9$ (136.4(7)–146.1(8) pm) and $Yb_5(BO_3)_2F_9$ (138.7(10)–148.0(12) pm).

We also calculated the charge distribution of the atoms in $Tm_5(BO_3)_2F_9$ *via* bond valence sums (ΣV) using VALIST (*B*ond *V*alence *C*alculation and *L*isting) [19–21] and *via* the CHARDI (*char*ge *di*stribution in solids) concept (ΣQ) [22], verifying the formal valence states in the fluoride borate. Table 7 shows the formal ionic charges, received from the calculations, which correspond to the expected values.

IR Spectroscopy

An FT-IR absorbance spectrum of a single crystal of $Tm_5(BO_3)_2F_9$ was recorded with transmitted,

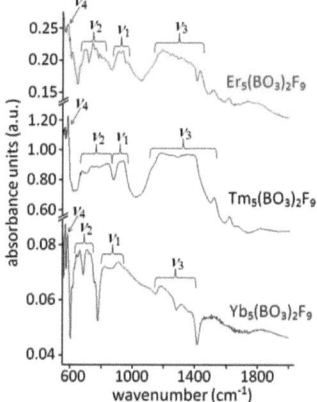

Fig. 3 (color online). FT-IR absorbance spectra of single crystals of $Tm_5(BO_3)_2F_9$, $Er_5(BO_3)_2F_9$, and $Yb_5(BO_3)_2F_9$ in the range of 550–2000 cm^{-1}.

polarized light on a BaF$_2$ plate, using a Bruker Vertex 70 spectrometer, attached to a Hyperion 3000 microscope, a MIR light source, and an LN-MCT detector. The spectral resolution was $\sim$4 cm^{-1}. The minimum-maximum normalization was realized by the OPUS 6.5 software (Bruker). In Fig. 3, the spectrum of Tm$_5$(BO$_3$)$_2$F$_9$ is compared to the spectra of RE_5(BO$_3$)$_2$F$_9$ (RE = Er, Yb). The absorption pattern of Tm$_5$(BO$_3$)$_2$F$_9$ resembles those of the isotypic compounds, showing the characteristic absorbances for triangular BO$_3$ groups [5, 29, 30]. Below 650 cm^{-1}, bands are caused by the in-plane bending (v_4) of the BO$_3$ groups. The strong absorptions, induced from the out-of-plane bending (v_2) of the trigonal group, appear as sharp bands in the range of 650–850 cm^{-1}. Due to the crystalline environment, the absorption bands between 900–1000 cm^{-1} can be assigned as symmetric stretching vibrations (v_1). The area of 1150–1500 cm^{-1} shows the asymmetric stretching vibrations (v_3) of the BO$_3$ groups. The already known single-crystal absorbance spectra of the isotypic phases show two bands in the region of 3400–3600 cm^{-1} (Fig. 4), which indicate hydroxyl groups. In the corresponding spectrum of the dried bulk material of the isotypic compound Yb$_5$(BO$_3$)$_2$F$_9$ [5], these bands were not present. The single crystals of all isotypic phases used for IR spectroscopy were exposed to the influence of air and humidity. In consequence, Tm$_5$(BO$_3$)$_2$F$_9$ shows a weak absorption pattern, which may be due to the substitution of fluoride by hydroxyl groups. This is a well known exchange, which apparently differs for the various fluoride borates [4, 5], corresponding to the hygroscopic properties. In apatites, the exchange of hydroxyl and fluoride ions was observed and studied in detail [31, 32].

Raman spectroscopy

In the range 100–1600 cm^{-1}, a Raman spectrum of a single crystal of Tm$_5$(BO$_3$)$_2$F$_9$ was measured without polarizers, using a Horiba LabRam HR-800 confocal micro-spectrometer and a 785 nm diode laser. The scattered light was dispersed by a grating with 1800 lines mm^{-1} and collected by a 1024 × 256 open electrode CCD detector. The laser focus and the power on the sample surface were $\sim$1 μm and $\sim$10 mW, respectively. The depth resolution was $\sim$4 μm and the spectral resolution $\sim$2 cm^{-1}. The wavenumber accuracy of $\sim$1 cm^{-1} was accomplished by regularly adjusting the zero-order position of the grating and checked by a neon calibration lamp. The third-order polynomial background subtraction, normalization, and band fitting by Gauss-Lorentz functions were done by the LABSPEC 5 software (Horiba).

The Raman spectra of the isotypic compounds RE_5(BO$_3$)$_2$F$_9$ (RE = Er, Tm, Yb) resemble each other

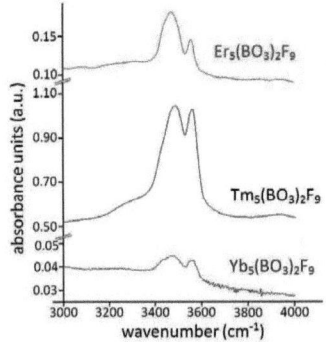

Fig. 4 (color online). FT-IR absorbance spectra of single crystals of Er$_5$(BO$_3$)$_2$F$_9$ (top), Tm$_5$(BO$_3$)$_2$F$_9$ (middle), and Yb$_5$(BO$_3$)$_2$F$_9$ (bottom) in the range of 3000–4000 cm^{-1}.

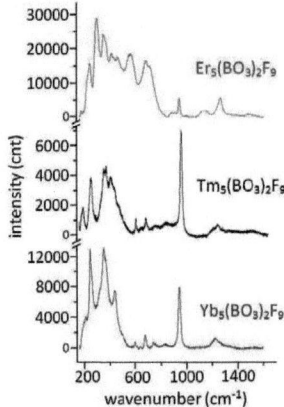

Fig. 5 (color online). Raman spectra of single crystals of Er$_5$(BO$_3$)$_2$F$_9$ (top), Tm$_5$(BO$_3$)$_2$F$_9$ (middle), and Yb$_5$(BO$_3$)$_2$F$_9$ (bottom) in the range of 150–1600 cm^{-1}.

(Fig. 5). The most intense bands < 500 cm^{-1} and the narrow, "isolated" band at ∼940 cm^{-1} occur in each spectrum. Above 1000 cm^{-1}, one broad band is noticed in the spectrum of Tm$_5$(BO$_3$)$_2$F$_9$. Bands around 900 cm^{-1} in borates are usually caused by stretching modes of BO$_4$ tetrahedra, whereas those of BO$_3$ groups are anticipated > 1100 cm^{-1} [4, 33–38]. However, in hydrated monoborates, very intense bands were observed at 882 and 501 cm^{-1} and assigned to symmetric stretching vibrations of the isolated BO$_3$ groups [35]. The unexpectedly high variation of B–O bond lengths inside the BO$_3$ groups (see above) might be the reason for the comparatively low wavenumber and the large wavenumber range of these bands. Bands < 500 cm^{-1} are evoked by the RE–O/ RE–F bond bending and stretching, as well as lattice vibrations.

Conclusion

With the synthesis of Tm$_5$(BO$_3$)$_2$F$_9$, the missing isotypic compound in the series RE_5(BO$_3$)$_2$F$_9$ (RE = Er, Tm, Yb) was found and characterized. To investigate the stability field of this structure type, additional experiments in the pressure range 3–7 GPa were performed with the neighboring, larger rare-earth cations Ho^{3+}, Dy^{3+}, and Tb^{3+}. Up to now, only structure types similar to that of the compounds RE_3(BO$_3$)$_2$F$_3$ (RE = Sm, Eu, Gd) [1, 2] and Gd$_2$(BO$_3$)F$_3$ [3] were identified.

Acknowledgement

We would like to thank Dr. G. Heymann for collecting the single-crystal data.

[1] G. Corbel, R. Retoux, M. Leblanc, *J. Solid State Chem.* 1998, *139*, 52.
[2] E. Antic-Fidancev, G. Corbel, N. Mercier, M. Leblanc, *J. Solid State Chem.* 2002, *153*, 270.
[3] H. Müller-Bunz, Th. Schleid, *Z. Anorg. Allg. Chem.* 2002, *628*, 2750.
[4] A. Haberer, R. Kaindl, J. Konzett, R. Glaum, H. Huppertz, *Z. Anorg. Allg. Chem.* 2010, *636*, 1526.
[5] A. Haberer, H. Huppertz, *J. Solid State Chem.* 2009, *182*, 888.
[6] A. Haberer, R. Kaindl, O. Oeckler, H. Huppertz, *J. Solid State Chem.* 2010, *183*, doi: 10.1016/j.jssc.2010.06.019.
[7] A. Haberer, R. Kaindl, H. Huppertz, *J. Solid State Chem.* 2010, *183*, 471.
[8] A. Haberer, R. Kaindl, H. Huppertz, *Solid State Sci.* 2010, *12*, 515.
[9] K. Kazmierczak, H. Höppe, *Eur. J. Inorg. Chem.* 2010, 2678.
[10] H. Huppertz, *Z. Kristallogr.* 2004, *219*, 330.
[11] D. Walker, M. A. Carpenter, C. M. Hitch, *Am. Mineral.* 1990, *75*, 1020.
[12] D. Walker, *Am. Mineral.* 1991, *76*, 1092.
[13] D. C. Rubie, *Phase Transitions* 1999, *68*, 431.
[14] N. Kawai, S. Endo, *Rev. Sci. Instrum.* 1970, *8*, 1178.
[15] Z. Otwinowski, W. Minor in *Methods in Enzymology*, Vol. 276, *Macromolecular Crystallography*, Part A (Eds.: C. W. Carter Jr., R. M. Sweet), Academic Press, New York, 1997, pp. 307.
[16] G. M. Sheldrick, SHELXL-97, Program for the Refinement of Crystal Structures, University of Göttingen, Göttingen (Germany) 1997.
[17] G. M. Sheldrick, *Acta Crystallogr.* 2008, *A64*, 112.
[18] R. D. Shannon, *Acta Crystallogr.* 1976, *A32*, 751.
[19] A. S. Wills, VALIST (version 4.0.0), University College London, London (UK) 1998–2008. Program available from www.ccp14.ac.uk.
[20] I. D. Brown, D. Altermatt, *Acta Crystallogr.* 1985, *B41*, 244.
[21] N. E. Brese, M. O'Keeffe, *Acta Crystallogr.* 1991, *B47*, 192.
[22] R. Hoppe, S. Voigt, H. Glaum, J. Kissel, H. P. Müller, K. J. Bernet, *J. Less-Common Met.* 1989, *156*, 105.
[23] R. Hoppe, *Angew. Chem.* 1966, *78*, 52; *Angew. Chem., Int. Ed. Engl.* 1966, *5*, 96.
[24] R. Hoppe, *Angew. Chem.* 1970, *82*, 7; *Angew. Chem., Int. Ed. Engl.* 1970, *9*, 25.
[25] R. Hübenthal, MAPLE (version 4.0), Program for the Calculation of Distances, Angles, Effective Coordination Numbers, Coordination Spheres, and Lattice Energies, University of Gießen, Gießen (Germany) 1993.
[26] W. H. Zachariasen, *Skr. Nor. Vidensk.-Akad.* 1928, 4.
[27] S. V. Berger, *Acta Crystallogr.* 1952, *5*, 389.
[28] A. Zalkin, D. H. Templeton, *J. Am. Chem. Soc.* 1953, *75*, 2453.
[29] J. P. Laperches, P. Tarte, *Spectrochim. Acta* 1966, *22*, 1201.
[30] G. Heymann, K. Beyer, H. Huppertz, *Z. Naturforsch.* 2004, *59b*, 1200.
[31] V. P. Orlovskii, S. P. Ionov, T. V. Belyaevskaya, S. M. Barinov, *Inorg. Mater.* 2002, *38*, 182.
[32] A. Knappwost, *Naturwissenschaften* 1959, *46*, 555.
[33] H. Emme, H. Huppertz, *Chem. Eur. J.* 2003, *9*, 3623.
[34] H. Huppertz, *J. Solid State Chem.* 2004, *177*, 3700.
[35] L. Jun, X. Shuping, G. Shiyang, *Spectrochim. Acta* 1995, *A51*, 519.
[36] G. Chadeyron, M. El-Ghozzi, R. Mahiou, A. Arbus, J. C. Cousseins, *J. Solid State Chem.* 1997, *128*, 261.
[37] J. C. Zhang, Y. H. Wang, X. Guo, *J. Lumin.* 2007, *122–123*, 980.
[38] G. Padmaja, P. Kistaiah, *J. Phys. Chem.* 2009, *A113*, 2397.

4.2.4 $Gd_4B_4O_{11}F_2$

Another rare-earth fluoride borate, obtained by high-pressure/high-temperature synthesis, is $Gd_4B_4O_{11}F_2$. Colorless crystals of this compound were synthesized at 7.5 GPa and 1100 °C (Figure 4.2-4). It not only exhibits a completely new crystal structure, but its structural motifs are unique in the whole borate chemistry. Here, the polyhedral B–O-clusters are described according to their linkage. Hawthorne *et al.* distinguish several fundamental building blocks (FBBs), that are known to be found in borates with specific arrangements of BO_3-groups and BO_4-tetrahedra [234-236]. In $Gd_4B_4O_{11}F_2$, the structural motif consists of two BO_3-groups (Δ) and two BO_4-tetrahedra (□) and can be characterized with the descriptor 2Δ2□:Δ□□Δ (after Burns *et al.* [237]). This FBB previously was not described in the literature.

Details on the crystal structure and characterization of $Gd_4B_4O_{11}F_2$ can be found in the following publication.

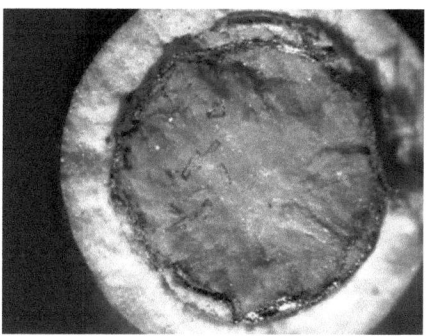

Figure 4.2-4
Crystalline sample of $Gd_4B_4O_{11}F_2$

Journal of Solid State Chemistry 183 (2010) 471–478

Contents lists available at ScienceDirect

Journal of Solid State Chemistry

journal homepage: www.elsevier.com/locate/jssc

$Gd_4B_4O_{11}F_2$: Synthesis and crystal structure of a rare-earth fluoride borate exhibiting a new "fundamental building block" in borate chemistry

Almut Haberer [a], Reinhard Kaindl [b], Hubert Huppertz [a],*

[a] *Institut für Allgemeine, Anorganische und Theoretische Chemie, Leopold-Franzens-Universität Innsbruck, Innrain 52a, A-6020 Innsbruck, Austria*
[b] *Institut für Mineralogie und Petrographie, Leopold-Franzens-Universität Innsbruck, Innrain 52, A-6020 Innsbruck, Austria*

ARTICLE INFO

Article history:
Received 3 July 2009
Received in revised form
24 November 2009
Accepted 7 December 2009
Dedicated to Dr. Klaus Römer on the
Occasion of His 70th Birthday
Available online 16 December 2009

Keywords:
High-pressure
Multianvil
Crystal structure
Fluoride borate

ABSTRACT

A new gadolinium fluoride borate $Gd_4B_4O_{11}F_2$ was yielded in a Walker-type multianvil apparatus at 7.5 GPa and 1100 °C. $Gd_4B_4O_{11}F_2$ crystallizes monoclinically in the space group C2/c with the lattice parameters $a=1361.3(3)$ pm, $b=464.2(2)$ pm, $c=1374.1(3)$ pm, and $\beta=91.32(3)°$ (Z=4). The crystal structure exhibits a structural motif not yet reported from borate chemistry: two BO_4-tetrahedra (□) and two BO_3-groups (△) are connected via common corners, leading to the fundamental building block $2\triangle 2\square:\triangle\square\square\triangle$. In the two crystallographically identical BO_4-tetrahedra, a distortion resulting in a very long B-O-bond is found.

© 2009 Elsevier Inc. All rights reserved.

1. Introduction

The possibility to obtain a large variety of new compounds under high-pressure/high-temperature conditions has been extensively studied for borates, for example the rare-earth borates $RE_4B_6O_{15}$ (RE=Dy, Ho) [1–3], α-$RE_2B_4O_9$ (RE=Sm–Ho) [4–6], β-$RE_2B_4O_9$ (RE=Gd [7], Dy [8]), $RE_3B_9O_{12}$ (RE=Er–Lu) [9], $Pr_4B_{10}O_{21}$ [10], and the meta-borates β-$RE(BO_2)_3$ (RE=Tb–Lu) [11–13], γ-$RE(BO_2)_3$ (RE=La–Nd) [14], and δ-$RE(BO_2)_3$ (RE=La, Ce) [15,16].

Until recently, the field of rare-earth fluoride borates was only represented by the compounds $RE_3(BO_3)_2F_3$ (RE=Sm, Eu, Gd) [17] and $Gd_2(BO_3)_2F$ [18], which were synthesized by heating a stoichiometric mixture of RE_2O_3, B_2O_3, and REF_3 under ambient pressure conditions. Under high-pressure/high-temperature conditions, the first ytterbium fluoride borate $Yb_5(BO_3)_2F_9$ could be added to the field by our group [19].

Borates, being glass formers in general, show an increased willingness to crystallize under pressure, which seems to hold true for fluoroborates, too. The chemistry of rare-earth fluoride borates under high-pressure/high-temperature conditions shows similarities to the chemistry of the borates concerning structural motifs. Trigonal-planar BO_3-groups as well as BO_4-tetrahedra can be identified and the pressure-induced transformation of

* Corresponding author. Fax: +435125072934.
E-mail address: hubert.huppertz@uibk.ac.at (H. Huppertz).

0022-4596/$ - see front matter © 2009 Elsevier Inc. All rights reserved.
doi:10.1016/j.jssc.2009.12.003

the former into the latter can be observed. The application of higher pressures on $Yb_5(BO_3)_2F_9$, which exhibits only BO_3-groups, yielded a structure built up solely from BO_4-tetrahedra [20]. The motif of edge-sharing tetrahedra could be realized under high pressure for the compounds $RE_4B_6O_{15}$ (RE=Dy, Ho) [1–3] and α-$RE_2B_4O_9$ (RE=Sm–Ho) [4–6]. Recently, HP-NiB_2O_4 [21] and β-FeB_2O_4 [22] were synthesized, in which each BO_4-tetrahedron shares a common edge with another tetrahedron. For fluoro- and fluoride borates, this structural feature is not known yet.

In this article, we present a new gadolinium fluoride borate $Gd_4B_4O_{11}F_2$, obtained by high-pressure/high-temperature synthesis. In the crystal structure of $Gd_4B_4O_{11}F_2$, BO_3-groups as well as BO_4-tetrahedra are found, connected via common corners. In borate chemistry, the polyhedra B-O-clusters are described according to their linkage. Hawthorne et al. distinguish several fundamental building blocks (FBBs), that are known to be found in borates, consisting of specific arrangements of BO_3-groups and BO_4-tetrahedra [23]. In $Gd_4B_4O_{11}F_2$, the polyhedra form a FBB not yet reported from borate chemistry. The structural motif consists of two BO_3-groups (△) and two BO_4-tetrahedra (□) and can be described with the fundamental building block $2\triangle 2\square:\triangle\square\square\triangle$ (after Burns et al. [24]). The two crystallographically identical tetrahedra in the FBB are distorted, showing a very long B-O-bond. A closer look reveals that the coordination sphere of the boron atom can be described as an intermediate state between a BO_3-group and a BO_4-tetrahedron. In the following, synthesis and structural details of the new compound $Gd_4B_4O_{11}F_2$ are reported.

4.2 Rare-Earth Fluoride and Fluorido Borates

2. Experimental section

2.1. Synthesis

The synthesis of the fluoride borate $Gd_6B_4O_{11}F_2$ took place under high-pressure/high-temperature conditions of 7.5 GPa and 1100 °C. Therefore, a mixture of Gd_2O_3 (Strem Chemicals, 99.99%), B_2O_3 (Strem Chemicals, 99.9+%), and GdF_3 (Strem Chemicals, 99.9%) at a molar ratio of 5:6:2 was ground and filled into a boron nitride crucible (Henze BNP GmbH, HeBoSint® S10, Kempten, Germany). This crucible was placed into the center of an 14/8-assembly, which was compressed by eight tungsten carbide cubes (TSM-10 Ceratizit, Reutte, Austria). The details of preparing the assembly can be found in Refs. [25–29]. Pressure was applied by a multianvil device based on a Walker-type module and a 1000 ton press (both devices from the company Voggenreiter, Mainleus, Germany). The sample was compressed up to 7.5 GPa in 3 h, then heated to 1100 °C in 15 min and kept there for 20 min. Afterwards, the sample was cooled down to 850 °C in 20 min, followed by natural cooling down to room temperature by switching off heating. The decompression required 9 h. The recovered experimental MgO-octahedron (pressure transmitting medium, Ceramic Substrates & Components Ltd., Newport, Isle of Wight, UK) was broken apart and the sample carefully separated from the surrounding boron nitride crucible, obtaining colorless, air- and water-resistant, crystal platelets of $Gd_6B_4O_{11}F_2$.

2.2. Crystal structure analysis

Sample characterization was performed by powder X-ray diffraction, carried out in transmission geometry on a flat sample of the reaction product, using a STOE STADI P powder diffractometer with MoKα_1 radiation (Ge monochromator, λ=71.073 pm). The powder pattern showed reflections of $Gd_6B_4O_{11}F_2$, as well as of α-$Gd_2B_4O_9$ [5] as a by-product of the synthesis. The experimental powder pattern tallies well with the theoretical patterns of the two compounds, simulated from single-crystal data. Indexing the reflections of the gadolinium fluoride borate, we got the parameters a=1362.4(4) pm, b=464.2(3) pm, and c=1374.8(5) pm, with β=91.18(3)° and a volume of 869.2(4) Å^3. This confirmed the lattice parameters, obtained from single-crystal X-ray diffraction (Table 1).

The intensity data of a single crystal of $Gd_6B_4O_{11}F_2$ were collected at room temperature by the use of a Kappa CCD diffractometer (Bruker AXS/Nonius, Karlsruhe), equipped with a Miracol Fiber Optics Collimator and a Nonius FR590 generator (graphite-monochromatized MoKα_1 radiation, λ=71.073 pm). A multi-scan absorption correction (Scalepack [30]) was applied to the intensity data. All relevant details of the data collection and evaluation are listed in Table 1.

Structure solution and parameter refinement (full-matrix least-squares against F^2) were successfully performed, using the SHELX-97 software suite [31,32] with anisotropic atomic displacement parameters for all atoms. According to the systematic

Table 1
Crystal data and structure refinement of $Gd_6B_4O_{11}F_2$.

Empirical formula	$Gd_6B_4O_{11}F_2$
Molar mass (g mol^{-1})	812.87
Crystal system	Monoclinic
Space group	C2/c (no.15)
Lattice parameters from powder data	
Powder diffractometer	Stoe Stadi P
Radiation	MoKα_1 (λ=71.073 pm)
a (pm)	1362.4(4)
b (pm)	464.2(3)
c (pm)	1374.8(5)
β (deg)	91.18(3)
Volume (Å^3)	869.2(4)
Single-crystal diffractometer	Bruker AXS/Nonius Kappa CCD
Radiation	MoKα_1 (λ=71.073 pm)
Single-crystal data	
a (pm)	1361.3(3)
b (pm)	464.2(2)
c (pm)	1374.1(3)
β (deg)	91.32(3)
Volume (Å^3)	868.1(3)
Formula units per cell	Z=4
Temperature (K)	293(2)
Calculated density (g cm^{-3})	6.781
Crystal size (mm^3)	0.03 × 0.02 × 0.02
Absorption coefficient (mm^{-1})	30.266
F(000)	1528
θ range (deg)	2.97 ≤ θ ≤ 30.00
Range in hkl	± 18, ± 6, ± 19
Total no. reflections	4570
Independent reflections	1271 (R_{int}=0.0310)
Reflections with $I > 2\sigma(I)$	1094 (R_σ=0.0262)
Data/parameters	1271/97
Absorption correction	Multi-scan
Goodness-of-fit (F^2)	1.114
Final R indices ($I > 2\sigma(I)$)	$R1$=0.0205
	$wR2$=0.0359
R indices (all data)	$R1$=0.0287
	$wR2$=0.0377
Largest differ. peak / deepest hole (e/Å^{-3})	0.93/−0.90

Table 2
Atomic coordinates and isotropic equivalent displacement parameters (U_{eq}/Å^2) for $Gd_6B_4O_{11}F_2$ (space group: C2/c).

Atom	Wyckoff site	x	y	z	U_{eq}
Gd1	8f	0.05835(2)	0.52008(5)	0.37047(2)	0.00601(7)
Gd2	8f	0.27992(2)	0.01782(5)	0.37101(2)	0.00577(7)
B1	8f	0.9072(4)	0.978(2)	0.2849(4)	0.0054(9)
B2	8f	0.0954(4)	0.957(2)	0.5235(4)	0.008(1)
O1	8f	0.9128(2)	0.8649(6)	0.3942(2)	0.0055(6)
O2	8f	0.1745(2)	0.8173(7)	0.2568(2)	0.0068(7)
O3	8f	0.0792(2)	0.6640(7)	0.5339(2)	0.0063(7)
O4	4e	0	0.8460(9)	¼	0.0047(9)
O5	8f	0.1145(3)	0.0621(7)	0.4334(3)	0.0089(7)
O6	8f	0.9015(3)	0.2807(7)	0.2765(2)	0.0070(7)
F1	8f	0.2312(2)	0.5248(6)	0.4236(2)	0.0149(6)

U_{eq} is defined as one-third of the trace of the orthogonalized U_{ij} tensor.

Table 3
Interatomic distances (pm) in $Gd_6B_4O_{11}F_2$, calculated with the single-crystal lattice parameters.

Gd1–O3a	235.4(3)	Gd2–O2a	230.0(3)	B1–O6	141.3(6)	B2–O5	136.1(6)
Gd1–O4	236.6(3)	Gd2–O2b	233.4(3)	B1–O2	144.7(6)	B2–O3	138.8(6)
Gd1–O6a	237.9(3)	Gd2–O6	239.5(3)	B1–O4	149.3(5)	B2–O1	140.7(6)
Gd1–O5a	241.3(3)	Gd2–O5	243.6(3)	B1–O1	159.0(6)		
Gd1–F1	244.8(3)	Gd2–O1	243.9(3)	⌀=148.8		⌀=138.5	
Gd1–O3b	246.5(3)	Gd2–O3	244.6(3)				
Gd1–O1	257.2(3)	Gd2–F1a	249.5(3)	F1–Gd1	244.8(3)		
Gd1–O2	263.8(3)	Gd2–F1b	255.4(3)	F1–Gd2a	249.5(3)		
Gd1–O6b	270.8(4)	Gd2–F1c	283.7(3)	F1–Gd2b	255.4(3)		
Gd1–O5b	276.3(3)			F1–Gd2c	283.7(3)		
⌀=251.0		⌀=247.1		⌀=258.4			

Table 4
Interatomic angles (deg) in $Gd_4B_4O_{11}F_2$, calculated with the single-crystal lattice parameters.

O1-B1-O2	102.9(3)	O1-B2-O3	118.5(4)	Gd1-F1-Gd2a	99.3(2)
O1-B1-O4	98.3(3)	O1-B2-O5	122.8(4)	Gd1-F1-Gd2b	100.5(2)
O1-B1-O6	114.1(4)	O3-B2-O5	118.6(4)	Gd1-F1-Gd2c	102.9(2)
O2-B1-O4	108.0(4)	∅ = 120.0		Gd2a-F1-Gd2b	133.7(2)
O2-B1-O6	116.1(4)			Gd2a-F1-Gd2c	103.7(2)
O4-B1-O6	115.3(4)			Gd2b-F1-Gd2c	111.9(2)
∅ = 109.1					∅ = 108.7

extinctions, the monoclinic space groups C2/c and Cc were derived. The structure solution in C2/c (no. 15) succeeded. The final difference Fourier syntheses did not reveal any significant residual peaks in all refinements. The positional parameters of the refinements, interatomic distances, and interatomic angles are listed in Tables 2–4. Further information of the crystal structure is available from the Fachinformationszentrum Karlsruhe (crysdata@fiz-karlsruhe.de), D-76344 Eggenstein-Leopoldshafen (Germany), quoting the registry no. CSD-420809.

2.3. Scanning electron microscopy

Scanning electron microscopy was performed on a JEOL JSM-6500F equipped with a field emission gun at an acceleration voltage of 15 kV. Samples were prepared by placing single crystals on adhesive conductive pads and subsequently coating them with a thin conductive carbon film. Each EDX spectrum (Oxford Instruments) was recorded with the analyzed area limited on one single crystal to avoid the influence of possible contaminating phases.

2.4. IR spectroscopy

FTIR absorption transmission spectra of the crystals were recorded on a BaF$_2$ plate with a Bruker Vertex 70 FT-IR spectrometer (resolution ~0.5 cm^{-1}) attached to a Hyperion 3000 microscope in a spectral range from 550 to 4000 cm^{-1}. A quadrangular diaphragm was set to a diameter of 100 μm for both spectrum and background measurement. For background correction, polynomial functions of second order were fitted to the spectra and subtracted. A rotatable polarizer was used to define a certain orientation of the electric field vector of the incoming IR light.

2.5. Raman spectroscopy

Confocal Raman spectra of single crystals were obtained with a HORIBA JOBIN YVON LabRam-HR 800 Raman micro-spectrometer. The sample was excited using the 633 nm emission line of a 17 mW He–Ne-laser and an OLYMPUS 100× objective (N.A. = 0.9). Size and power of the laser spot on the surface were approximately 1 μm and 5 mW. The spectral resolution, determined by measuring the Rayleigh line, was about 1.2 cm^{-1}. The dispersed light was collected by a 1024 × 256 open electrode CCD detector. Spectra were recorded without polarizer for the laser and the scattered Raman light. Raman bands were fitted by the built-in spectrometer software LabSpec to convoluted Gauss-Lorentz functions. Accuracy of Raman line shifts, calibrated by regular measuring the Rayleigh line, was in the order of 0.5 cm^{-1}.

3. Results and discussion

X-ray diffraction is a powerful tool for crystal structure solution and refinement. Nevertheless, in the chemistry of fluoride borates or fluoride minerals, it is nearly impossible to distinguish between fluoride ions and hydroxyl groups by means of electron density or bond lengths. Via crystal structure refinement, the composition $Gd_4B_4O_{11}F_2$ was determined. To assure the atom assignment in the structure, single crystals of our sample were subjected to elemental analysis. The crystals showed average atomic Gd:B:O:F compositions (%) of 14.7:20.5:53.2:11.6. Due to the light weight of boron, measurements have to be taken with caution, but still, these results confirm the presence of all elements and the composition obtained from the single crystal structure determination (calculated values (%) Gd:B:O:F: 19.0:19.0:52.5:9.5).

Due to the fact that a differentiation between fluoride anions and hydroxyl groups at a crystallographic site via structural refinement is impossible, IR-spectrocopic investigations were performed on single-crystals of the sample (Fig. 1). Looking closely at the region of 3100–3600 cm^{-1}, two small peaks could be detected that hardly overtopped the background noise. Absorption bands at those wavelengths can be assigned to OH-groups and typically reveal OH-containing borates [33]. A direction-dependent measurement clearly showed that the bands are not evoked by surface humidity of the crystals but indeed by internal OH-groups. On the basis of these IR measurements, the presence of traces of OH-groups in $Gd_4B_4O_{11}F_2$ has to be assumed. The problem of hydroxyl quantification is commonly known from halogen containing minerals and often the exact composition is merely estimated, as done for $Mn_5(PO_4)_3Cl_{0.9}(OH)_{0.1}$ [34]. The title compound could therefore also be described as "$Gd_4B_4O_{11}F_{2-x}(OH)_x$". Because of the diminutive fraction of what is known from water- or OH-containing borates) and the results of elemental analysis, the value of x has to be considered as minimal and is therefore neglected in the structural description.

Furthermore, we calculated the Madelung Part of Lattice Energy (MAPLE) [35–37] for $Gd_4B_4O_{11}F_2$ in order to compare it with the sum of the MAPLE values of the high-temperature modifications of Gd_2O_3 [38] and GdF_3 [39], and of the high-pressure modification of B_2O_3 (B_2O_3-II) [40]. The additive

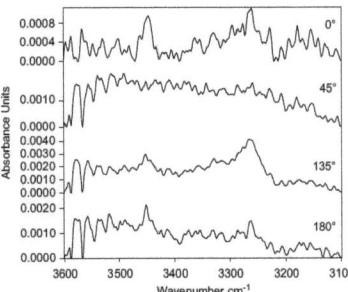

Fig. 1. Polarized IR absorption spectra of $Gd_4B_4O_{11}F_2$ in the region 3100–3600 cm^{-1}. The spectra were recorded at four different angles of the electric field vector of the IR-light relative to the crystal. The weak absorption bands around 3250 and 3450 cm^{-1} can be assigned to OH-groups.

4.2 Rare-Earth Fluoride and Fluorido Borates

potential of the MAPLE values allows the calculation of a hypothetical value for $Gd_4B_4O_{11}F_2$, starting from binary oxides and fluorides. As a result, we obtained a value of 72407 kJ/mol in comparison to 72481 kJ/mol (deviation: 0.1%), starting from the binary components [5/3 × Gd_2O_3 (14950 kJ/mol)+2 × B_2O_3–II (21938 kJ/mol)+2/3 × $GdF3$ (5532 kJ/mol)], which also indicates that the small amount of hydrogen groups can be neglected.

3.1. Crystal structure of $Gd_4B_4O_{11}F_2$

The structure of $Gd_4B_4O_{11}F_2$ consists of BO_3-groups, BO_4-tetrahedra, gadolinium cations, and fluoride anions (Fig. 2). The two BO_3-groups (△) and two BO_4-tetrahedra (□) are connected via common corners, forming a fundamental building block (FBB) 2△2□:△□□△, which has not yet been reported by Burns et al.

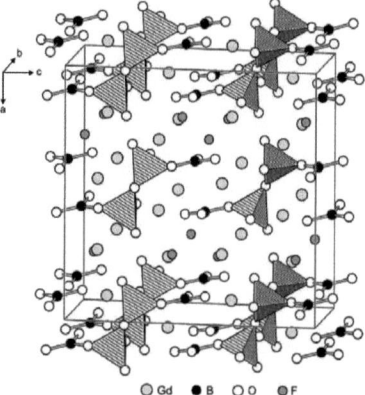

Fig. 2. Crystal structure of $Gd_4B_4O_{11}F_2$, showing the fundamental building block units consisting of BO_3-groups and BO_4-tetrahedra.

[24]. In 2002, Becker summarized the FBBs known from borate structures, mentioning four compounds with the motif 2△2□:△□□△ [41], namely the borates $BaBe_2(BO_3)_2$ [42], $SrCu_2(BO_3)_2$ [43], $Sr_{0.74}Ba_{0.26}Cu_2(BO_3)_2$, and $Sr_{0.66}Ca_{0.33}Cu_2(BO_3)_2$ [44]. However, in $BaBe_2(BO_3)_2$, the tetrahedra are not BO_4-tetrahedra, but BeO_4-tetrahedra, connected via common edges. For the structural motif of edge-sharing tetrahedra, our group established the new descriptor '⊠' in 2003, which extended the FBB concept [2]. Strictly speaking, the FBB descriptors according to Burns et al. are only applied to mere B–O-polyhedra. Nevertheless, if the FBB concept is applied to $BaBe_2(BO_3)_2$ after all, the correct FBB notation now should be 2△2□:△⊠△. Fig. 3 left shows the structure of $BaBe_2(BO_3)_2$ with interconnected building blocks, leading to a network structure.

In the compounds $SrCu_2(BO_3)_2$, $Sr_{0.74}Ba_{0.26}Cu_2(BO_3)_2$, and $Sr_{0.66}Ca_{0.33}Cu_2(BO_3)_2$, there are not even tetrahedra in the structural motif; the fourfold-coordinated copper ions show square-planar geometry. This is exemplified for $Sr_{0.74}Ba_{0.26}Cu_2(BO_3)_2$ in Fig. 3 (right). The so called FBBs build up two-dimensional sheets, connected via Sr, Ba, or Ca ions.

In $Gd_4B_4O_{11}F_2$, the isolated FBBs exclusively consist of B–O-polyhedra (Fig. 2). Their locations in the unit cell are displayed in Fig. 4 (view along [00$\bar{1}$]). To the best of our knowledge, this fluoride borate is the first compound exhibiting a real 2△2□:△□□△ motif according to Burns et al. [24] and is thus extending the borate fundamental building block concept by a new structural motif.

A closer look at the new FBB of $Gd_4B_4O_{11}F_2$ is given in Fig. 5. A twofold rotation axis is running through the tetrahedra-bridging oxygen O4, thus all threefold- and fourfold-coordinated boron atoms are crystallographically identical (Table 2). Inside the BO_3-groups, the B2–O-distances range from 136.1(6) to 140.7(6) pm (Table 3), which is typical for threefold-coordinated boron atoms, e.g. in borates with calcite structure ($AlBO_3$ (137.96(4) pm) [45], β-YBO_3 (137.8(4) pm) [46], and $FeBO_3$ (137.9(2) pm) [47]). The B2–O-angles sum up to 120° (Table 4), as would be expected from the trigonal-planar geometry.

Inside the BO_4-tetrahedra, the B1–O average bond lengths is 148.6 pm (Table 3) and thus fairly larger than the known average value of 147.6 pm for fourfold-coordinated boron atoms in borates [23]. The reason for this deviation can be seen in Fig. 6. The central boron atom B1 is dislocated from the center of the trigonal plane spanned up by O2, O6, and O4. In detail, the boron atom is shifted towards the oxygen atom O1 from the neighboring BO_3-group, indicating an intermediate state between a strongly distorted

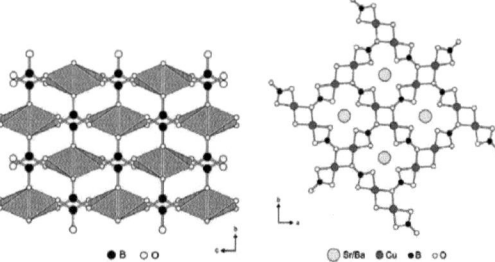

Fig. 3. Left: Structure of $BaBe_2(BO_3)_2$, showing edge-sharing BeO_4-tetrahedra, connected via BO_3-groups, resulting in a network structure. Right: Structure of $Sr_{0.74}Ba_{0.26}Cu_2(BO_3)_2$, built up from edge-sharing, square-planar CuO_4-units connected via BO_3-groups, resulting in a layered structure.

trigonal BO_3-group and a BO_4-tetrahedron. Thus, the tetrahedron could be seen as a former BO_3-group, in which the boron atom is drawn towards a fourth oxygen atom, resulting in the long B1–O1-bond of 159.0(6) pm displayed in Fig. 6. The non-planarity of BO_3-groups, due to the close proximity of another oxygen atom, is described in the literature [48]. Calculation of the coordination number of B1 with MAPLE (Madelung Part of Lattice Energy) [49–51] results in a value of 3.73, verifying the assumption of a transition state. The O–B–O angles in the BO_4-tetrahedra vary between 98.3(3)° and 116.1(4)° with a mean value of 109.1°, indicating a slight distortion.

Another point of interest in $Gd_4B_4O_{11}F_2$ are the BO_3-groups of two neighboring building blocks, facing each other at a relatively short B–B-distance of 269 pm (Figs. 7a and c). The distances of the boron atom to the next oxygen atom outside the coordination sphere account for 292.9 pm. Similar arrangements of BO_3-groups were also found in $M_3(BO_3)_2$ (M=Mg, Co, Ni), where the B–O-distances of one boron atom to the first oxygen atom of the second BO_3-group are < 260 pm [52]. In these compounds, a transition state between threefold and fourfold boron coordination is found, leading formally to two distorted BO_4-tetrahedra connected via common edges (Fig. 7b). While the B–B-distance in undistorted edge-sharing tetrahedra usually lies around 215 pm [1–6,21,22], the B–B-distance in $Mg_3(BO_3)_2$ measures 268 pm. A similar transition could be expected for $Gd_4B_4O_{11}F_2$, as exemplified in Figs. 7a and c. The application of higher pressures might transform the neighboring BO_3-groups into edge-sharing BO_4-tetrahedra, which are an extremely rare structural feature of high-pressure borates.

There are two crystallographically independent Gd^{3+} ions in the structure, which are nine-and tenfold coordinated by oxygen and fluorine (Fig. 8). Gd1 has got nine oxide ions and one fluoride ion in the coordination sphere, whereas Gd2 is surrounded by six oxide and three fluoride ions. As presented in Table 3, the average interatomic distance Gd2–O/F with 247.1 pm is in the same range as the average Gd–O/F distance of ninefold coordinated Gd^{3+} in $Gd_2[BO_3]F_3$ (242.7 pm [18]). The average distance Gd1–O/F of 251.1 pm is larger, as would be expected from a tenfold coordination.

The fluoride ion in $Gd_4B_4O_{11}F_2$ is coordinated by four gadolinium ions (Fig. 8). The bond lengths (Table 3) range between 244.8(3) and 283.7(3) pm, being in the same region as the bond lengths of the fourfold-coordinated fluoride ion in the high-temperature phase of GdF_3 (249.8–276.5 pm) [53]. The Gd–F angles sum up to 108.7° (Table 4), which is fairly close to the ideal tetrahedron angle.

The calculations of the charge distribution of the atoms in $Gd_4B_4O_{11}F_2$ via bond valence sums (ΣV) with ValList (bond valence calculation and listing) [54] and with the CHARDI (charge distribution in solids) concept (ΣQ) [55–57] confirm the formal valence states in the fluoride borate (Table 5).

The Raman spectrum of $Gd_4B_4O_{11}F_2$ is displayed in Fig. 9. A relatively large number of 38 bands from 113 to 1358 cm^{-1} could be detected (Table 6), consistent with the low symmetry of the structure. No Raman-active OH-stretching modes between 3100 and 3600 cm^{-1} (see above) were observed, which is probably caused by the low OH-content and high background in this spectral region.

Bands in the range between 800 and 1600 cm^{-1} are generally assigned to stretching vibrations of BO_4 tetrahedra and BO_3 groups [5,33,58–61]. The former ones are expected at wave numbers below, the latter ones above 1100 cm^{-1}. This is consistent with the strong band at 959 cm^{-1} and the two bands at 1244 and 1282 cm^{-1}. The large average bond length within the tetrahedra (147.6 pm, see above) results in the comparable low wave number of the vibrational mode. E.g. in $Co_6B_{22}O_{39} \cdot H_2O$ [62], which contains BO_4-tetrahedra with a mean B–O distance of 145.7 pm, the respective mode is observed at 1035 cm^{-1}. The variable B2–O-distances inside the BO_3-groups or their connection to BO_4-tetrahedra in fundamental building blocks might account for two instead of one stretching modes in this range as it was described for compounds containing isolated planar molecules (e.g. [58]).

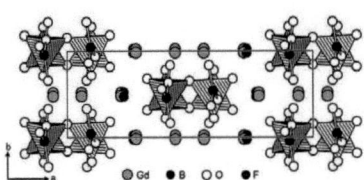

Fig. 4. Crystal structure of $Gd_4B_4O_{11}F_2$, viewing along [00$\bar{1}$].

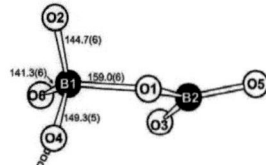

Fig. 6. Transition of B1 from a trigonal to a tetrahedral coordination, resulting in an elongated B1–O1-bond length (bond lengths in pm).

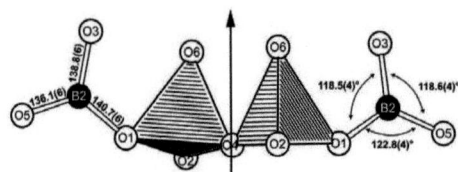

Fig. 5. Fundamental building block 2△☐:△☐☐△ in $Gd_4B_4O_{11}F_2$, showing a twofold rotation axis at O4 (bond lengths in pm).

4.2 Rare-Earth Fluoride and Fluorido Borates

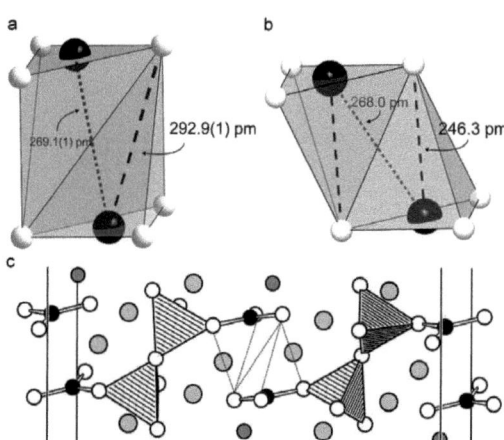

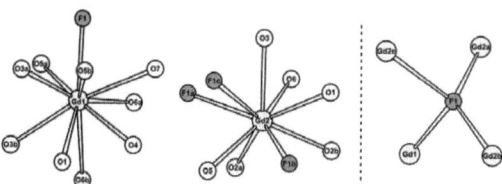

Fig. 7. (a) Possible transition of BO_3-groups into edge-sharing BO_4-tetrahedra in $Gd_4B_4O_{11}F_2$. (b) BO_3-groups in $Mg_3(BO_3)_2$ in an arrangement which could form edge-sharing BO_4-tetrahedra under elevated pressure conditions. (c) Neighboring BO_3-groups in $Gd_4B_4O_{11}F_2$ building blocks with implied edge-sharing BO_4-tetrahedra.

Fig. 8. Coordination spheres of the Gd^{3+} and F^- ions in $Gd_4B_4O_{11}F_2$.

Table 5
Charge distribution in $Gd_4B_4O_{11}F_2$, calculated with VaList (ΣV) and CHARDI (ΣQ).

	Gd1	Gd2	B1	B2		
ΣV	3.11	3.03	2.98	2.89		
ΣQ	2.97	3.04	2.97	3.02		
	O1	O2	O3	O4	O5	O6	F1
ΣV	−2.08	−2.04	−2.11	−2.32	−1.94	−1.91	−0.78
ΣQ	−1.92	−2.00	−2.10	−2.26	−1.95	−1.93	−0.97

After a wide spectral gap, numerous bands occur from 700 to 200 cm^{-1}, the most intense bands are observed at very low wave numbers below 200 cm^{-1}. These bands can be assigned to bending and stretching vibrations of various borate arrangements, Gd–O, F–O, and Gd–F bonds and complex lattice vibrations. In hydrated tetraborates [58], which contain a polyborate anion with two planar BO_3 triangles, two BO_4-tetrahedra, and OH-groups, very strong stretching vibrations of this unit between 543 and

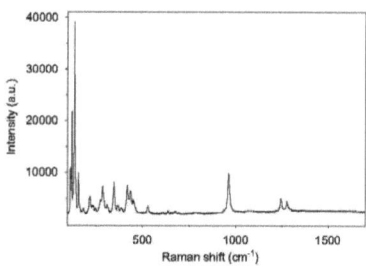

Fig. 9. Unpolarized micro-Raman spectrum of a single crystal of $Gd_4B_4O_{11}F_2$.

Table 6
Raman shifts of observed bands and tentative assignment of the spectrum of $Gd_4B_4O_{11}F_2$.

Raman shift (cm^{-1})	Assignment
113	
120	
134	
155	
158	
181	
212	
218	lattice, Gd–O, Gd–F, F–O
230	
245	
270	
285	
308	
345	
366	
383	
390	Bending (BO$_4$)
417	
434	
448	
455	
528	
609	
636	Pulse vibration (BO$_3$)–(BO$_4$)
660	
676	
693	
748	
785	
896	
932	Stretching (BO$_4$)
959	
1001	
1075	
1244	
1277	Stretching (BO$_3$)
1282	
1358	

583 cm^{-1} were observed, whereas bending vibrations of the tetrahedra are described around 391–461 cm^{-1}. The low wave number region was not studied by jun et al. [58]; we believe that the very strong bands of $Gd_4B_4O_{11}F_2$ below 200 cm^{-1} are dominantly related to the fourfold-coordinated fluoride and the Gd^{3+} ion. More reliable assignments of observed bands to vibrational modes would require quantum mechanical calculations of the structure, which are not available up to now.

4. Conclusions

In this article, we described the high-pressure synthesis and crystal structure of the new gadolinium fluoride borate $Gd_4B_4O_{11}F_2$. The structure is built up from a fundamental building block 2△2☐:△☐☐△, which has not yet been reported in borate chemistry. The tetrahedra in the structural motif are distorted, indicating a transition state between a distorted trigonal BO$_3$-group and a BO$_4$-tetrahedron. The possible transformation of neighboring BO$_3$-groups into edge-sharing tetrahedra is another point of interest and will be object of our further studies.

Acknowledgments

We thank Dr. Gunter Heymann for collecting the single-crystal data and Christian Minke (LMU München) for the scanning electron microscopy measurements. Special thanks go to Prof. Dr. W. Schnick (LMU München) for his continuous support of these investigations.

Appendix A. Supplementary material

Supplementary data associated with this article can be found in the online version at doi:10.1016/j.jssc.2009.12.003.

References

[1] H. Huppertz, B. von der Eltz, J. Am. Chem. Soc. 124 (2002) 9376.
[2] H. Huppertz, Z. Naturforsch. B 58 (2003) 278.
[3] H. Huppertz, H. Emme, J. Phys. Condens. Matter. 16 (2004) 1283.
[4] H. Emme, H. Huppertz, Z. Anorg. Allg. Chem. 628 (2002) 2165.
[5] H. Emme, H. Huppertz, Chem. Eur. J. 9 (2003) 3623.
[6] H. Emme, H. Huppertz, Acta Crystallogr. C 61 (2005) i29.
[7] H. Emme, H. Huppertz, Acta Crystallogr. C 61 (2005) i23.
[8] H. Huppertz, S. Altmannshofer, G. Heymann, J. Solid State Chem. 170 (2003) 320.
[9] H. Emme, M. Valldor, R. Pöttgen, H. Huppertz, Chem. Mater. 17 (2005) 2707.
[10] A. Haberer, G. Heymann, H. Huppertz, J. Solid State Chem. 180 (2007) 1595.
[11] T. Nikelski, Th. Schleid, Z. Anorg. Allg. Chem. 629 (2003) 1017.
[12] T. Nikelski, M.C. Schäfer, H. Huppertz, Th. Schleid, Z. Kristallogr. 233 (2008) 177.
[13] H. Emme, T. Nikelski, Th. Schleid, R. Pöttgen, M.H. Möller, H. Huppertz, Z. Naturforsch. B 59 (2004) 202.
[14] H. Emme, C. Despotopoulou, H. Huppertz, Z. Anorg. Allg. Chem. 630 (2004) 2450.
[15] G. Heymann, T. Soltner, H. Huppertz, Solid State Sci. 8 (2006) 821.
[16] A. Haberer, G. Heymann, H. Huppertz, Z. Naturforsch. B 62 (2007) 759.
[17] G. Corbel, R. Retoux, M. Leblanc, J. Solid State Chem. 139 (1998) 52.
[18] H. Müller-Bunz, Th. Schleid, Z. Anorg. Allg. Chem. 628 (2002) 2750.
[19] A. Haberer, H. Huppertz, J. Solid State Chem. 182 (2009) 888.
[20] A. Haberer, H. Huppertz, Angew. Chem., submitted.
[21] J.S. Knyrim, F. Roeßner, S. Jakob, D. Johrendt, I. Kinski, R. Glaum, H. Huppertz, Angew. Chem. 119 (2007) 9256;
J.S. Knyrim, F. Roeßner, S. Jakob, D. Johrendt, I. Kinski, R. Glaum, H. Huppertz, Angew. Chem. Int. Ed. 46 (2007) 9097.
[22] S.C. Neumair, R. Glaum, H. Huppertz, Z. Naturforsch. B64 (2009) 883.
[23] F.C. Hawthorne, P.C. Burns, J.D. Grice, Reviews in Mineralogy, vol. 33, Boron: Mineralogy, Petrology, and Geochemistry, vol. 33, Mineralogical Society of America, Washington, 1996 (Chapter 2).
[24] P.C. Burns, J.D. Grice, F.C. Hawthorne, Can. Miner. 33 (1995) 1131.
[25] D. Walker, M.A. Carpenter, C.M. Mitch, Am. Miner. 75 (1990) 1020.
[26] D. Walker, Am. Miner. 76 (1991) 1092.
[27] H. Huppertz, Z. Kristallogr. 219 (2004) 330.
[28] D.C. Rubie, Phase Transitions 68 (1999) 431.
[29] N. Kawai, S. Endo, Rev. Sci. Instrum. 8 (1970) 1178.
[30] Z. Otwinowski, W. Minor, Methods Enzymol. 276 (1997) 307.
[31] G.M. Sheldrick, SHELXS-97 and SHELXL-97, Program suite for the solution and refinement of crystal structures, University of Göttingen, Göttingen, Germany, 1997.
[32] G.M. Sheldrick, Acta Crystallogr. A 64 (2008) 112.
[33] H. Huppertz, J. Solid State Chem. 177 (2004) 3700.
[34] G. Engel, J. Pretzsch, V. Gramlich, W.H. Baur, Acta Crystallogr. B 31 (1975) 1854.
[35] R. Hoppe, Angew. Chem. 78 (1966) 52;
R. Hoppe, Angew. Chem. Int. Ed. 5 (1966) 96.
[36] R. Hoppe, Angew. Chem. 82 (1970) 7;
R. Hoppe, Angew. Chem. Int. Ed. 9 (1970) 25.
[37] R. Hübenthal, MAPLE-Program for the Calculation of MAPLE Values, Vers. 4, University of Gießen, Germany, 1993.
[38] O.J. Guentert, R.L. Mozzi, Acta Crystallogr. 11 (1958) 746.
[39] L.S. Garashina, B.P. Sobolev, V.B. Aleksandrov, Y.S. Vishnyakov, Sov. Phys. Crystallogr. 25 (1980) 171.
[40] C.T. Prewitt, R.D. Shannon, Acta Crystallogr. B 24 (1968) 869.
[41] P. Becker, Z. Kristallogr. 216 (2001) 523.
[42] K.I. Schaffers, D.A. Keszler, Inorg. Chem. 33 (1994) 1201.
[43] R.W. Smith, D.A. Keszler, J. Solid State Chem. 93 (1991) 430.
[44] R. Norrestam, S. Carlson, A. Sjödin, Acta Crystallogr. C 50 (1994) 1847.
[45] A. Vegas, Acta Crystallogr. B 33 (1977) 3607.
[46] H. Huppertz, Z. Naturforsch. B 56 (2001) 697.
[47] R. Diehl, Solid State Commun. 17 (1975) 743.
[48] E. Zobetz, Z. Kristallogr. 160 (1982) 81.

[49] R. Hoppe, Angew. Chem. 78 (1966) 52;
R. Hoppe, Angew. Chem. Int. Ed. 5 (1966) 96.
[50] R. Hoppe, Angew. Chem. 82 (1970) 7;
R. Hoppe, Angew. Chem. Int. Ed. 9 (1970) 25.
[51] R. Hübenthal, MAPLE Program for the Calculation of MAPLE Values, Vers. 4, University of Gießen, Gießen, Germany, 1993.
[52] H. Effenberger, F. Pertlik, Z. Kristallogr. 166 (1984) 129.
[53] R.E. Thoma, G.D. Brunton, Inorg. Chem. 5 (1966) 1937.
[54] A. S. Wills, VaList Version 3.0.13, University College London, UK, 1998–2008. Program available from ⟨http://www.ccp14.ac.ukwww.ccp14.ac.uk⟩.
[55] I.D. Brown, D. Altermatt, Acta Crystallogr. B 41 (1985) 244.
[56] N.E. Brese, M. O'Keeffe, Acta Crystallogr. B 47 (1991) 192.
[57] R. Hoppe, S. Voigt, H. Glaum, J. Kissel, H.P. Müller, K.J. Bernet, Less-Common Met. 156 (1989) 105.
[58] L. Jun, X. Shuping, G. Shiyang, Spectrochim. Acta A 51 (1995) 519.
[59] G. Chadeyron, M. El-Ghozzi, R. Mahiou, A. Arbus, J.C. Cousseins, J. Solid State Chem. 128 (1997) 261.
[60] J.C. Zhang, Y.H. Wang, X. Guo, J. Lumin. 122–123 (2007) 980.
[61] G. Padmaja, P. Kistaiah, J. Phys. Chem. A 113 (2009) 2397.
[62] S. C. Neumair, J. S. Knyrim, O. Oeckler, R. Glaum, R. Kaindl, R. Stalder, H. Huppertz, Chem. Eur. J., submitted.

4.2 Rare-Earth Fluoride and Fluorido Borates

4.2.5 $RE_4B_4O_{11}F_2$ (RE = Eu, Dy)

Further investigations on the formation range of $RE_4B_4O_{11}F_2$ soon led to two isotypic phases: $Eu_4B_4O_{11}F_2$ at 5 GPa and 900 °C and $Dy_4B_4O_{11}F_2$ at 8 GPa and 1000 °C. Colorless, air- and water-resistant crystals of both compounds could be obtained after the respective synthesis (Figure 4.2-3). This was achieved in cooperation with M. Enders during his Master's thesis in our working group. All relevant details on this phase can be found in the following publication.

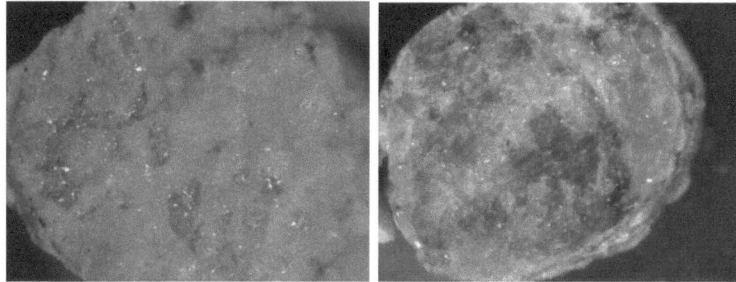

Figure 4.2-5
Crystalline samples of $Eu_4B_4O_{11}F_2$ (left) and $Dy_4B_4O_{11}F_2$ (right)

$RE_4B_4O_{11}F_2$ (RE = Eu, Dy): New Phases Isotypic to the Fluoride Borate $Gd_4B_4O_{11}F_2$

Almut Pitscheider, Michael Enders, and Hubert Huppertz

Institut für Allgemeine, Anorganische und Theoretische Chemie, Leopold-Franzens-Universität Innsbruck, Innrain 52a, 6020 Innsbruck, Austria

Reprint requests to H. Huppertz. E-mail: Hubert.Huppertz@uibk.ac.at

Z. Naturforsch. 2010, 65b, 1439–1444; received September 17, 2010

The rare-earth fluoride borates $RE_4B_4O_{11}F_2$ (RE = Eu, Dy) were synthesized in a Walker-type multianvil apparatus from the corresponding rare-earth oxides and fluorides, and boron oxide. $Eu_4B_4O_{11}F_2$ was obtained under high-pressure/high-temperature conditions of 5 GPa and 900 °C, and $Dy_4B_4O_{11}F_2$ at 8 GPa and 1000 °C. The single-crystal structure determinations revealed that both compounds are isotypic to $Gd_4B_4O_{11}F_2$, crystallizing in the space group $C2/c$ (Z = 4) with the parameters a = 1368.2(3), b = 465.4(1), c = 1376.6(3) pm, β = 91.2(1)°, V = 0.8765(3) nm^3, R_1 = 0.0232, and wR_2 = 0.0539 (all data) for $Eu_4B_4O_{11}F_2$ and a = 1349.5(3), b = 460.9(1), c = 1362.5(3) pm, β = 91.3(1)°, V = 0.8472(3) nm^3, R_1 = 0.0353, and wR_2 = 0.0729 (all data) for $Dy_4B_4O_{11}F_2$. These phases are entirely different from the recently discovered lanthanum fluoride borate $La_4B_4O_{11}F_2$, which exhibits the same constitution in another structure type with space group $P2_1/c$.

Key words: Rare Earth, Fluoride, Borate, High Pressure, Crystal Structure

Introduction

Considering the extensive research on oxoborates under high-pressure/high-temperature conditions in our group in the last decade, it was tempting to extend our investigations into fluoride borates. These compounds, especially "fluoroborate" glasses, have been the subject of much recent work, but crystalline fluorido and fluoride borates have not been that well studied, except for natural minerals with more than two cations. The difference between fluorido borates (old designation: fluoroborates) and fluoride borates lies in the coordination sphere of boron. While borates, that contain only isolated fluorine anions, are generally termed "fluoride borates", "fluorido borates" contain fluorine atoms covalently bound to the boron atoms. Due to the structural similarities of fluoride borates to oxoborates, it is more than likely that numerous new compounds and crystal structures with interesting physical properties can be obtained in the field of fluoride borates under pressure. A summary of the achievements reached so far can be found in [1].

For the composition $RE_4B_4O_{11}F_2$, two different structure types were obtained by high-pressure/high-temperature synthesis up to now. The recently presented compound $La_4B_4O_{11}F_2$ [2] crystallizes in the space group $P2_1/c$ with the lattice parameters a = 778.1(2), b = 3573.3(7), c = 765.7(2) pm, and β = 113.92(3)° (Z = 8). The crystal structure consists of BO_3 groups ($\triangle$), which are either isolated ($\triangle$), connected via common corners ($\triangle\triangle$), or connected via a BO_4 tetrahedron, forming a fundamental building block (FBB) $2\triangle\square$: $\triangle\square\triangle$.

The fluoride borate $Gd_4B_4O_{11}F_2$ discovered earlier shows the same atomic composition, except for a completely different crystal structure in the space group $C2/c$ [3]. In the crystal structure of $Gd_4B_4O_{11}F_2$, there are BO_3 groups and BO_4 tetrahedra, connected via common corners. The structural motif consists of two BO_3 groups ($\triangle$) and two BO_4 tetrahedra ($\square$), and can be described with the fundamental building block $2\triangle2\square$: $\triangle\square\square\triangle$ (after Burns et al. [4]), which represents a novelty in borate chemistry.

Here we report about two new compounds $RE_4B_4O_{11}F_2$ (RE = Eu, Dy), which are isotypic to $Gd_4B_4O_{11}F_2$. The syntheses and crystal structures of $RE_4B_4O_{11}F_2$ (RE = Eu, Dy) are presented in comparison to $Gd_4B_4O_{11}F_2$.

Experimental Section

Syntheses

For the syntheses of $RE_4B_4O_{11}F_2$ (RE = Eu, Dy), the reactions of the oxides RE_2O_3 (RE = Eu, Dy) and

4.2 Rare-Earth Fluoride and Fluorido Borates

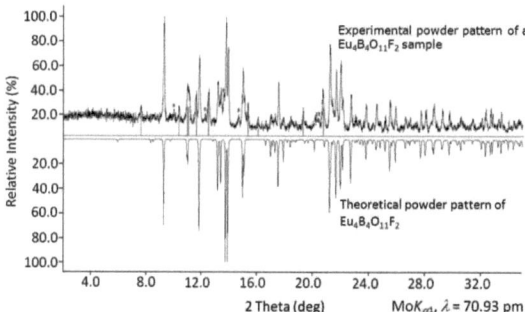

Fig. 1 (color online). Top: experimental powder pattern of $Eu_4B_4O_{11}F_2$; the reflections of EuF_3 and an unknown phase are indicated with lines and asterisks, respectively. Bottom: theoretical powder pattern of $Eu_4B_4O_{11}F_2$, based on single-crystal diffraction data.

B_2O_3 with the corresponding rare-earth fluoride REF_3 were performed under high-pressure/high-temperature conditions. For $Eu_4B_4O_{11}F_2$, the synthesis was carried out at 5 GPa and 900 °C, while $Dy_4B_4O_{11}F_2$ was obtained at 8 GPa and 1000 °C. Mixtures of Eu_2O_3 (Smart Elements, 99.99 %) or Dy_2O_3 (Strem Chemicals, 99.9 %), B_2O_3 (Strem Chemicals, 99.9+ %), and EuF_3 (Strem Chemicals, 99.9 %) or DyF_3 (Strem Chemicals, 99.9 %) with a molar ratio of 1 : 1 : 3 were ground inside of a glove box and filled into boron nitride crucibles (Henze BNP GmbH, HeBoSint® S100, Kempten, Germany). These crucibles were placed into the center of 18/11-assemblies, which were compressed by eight tungsten carbide cubes (TSM-10 Ceratizit, Reutte, Austria). The details of preparing the assemblies can be found in refs. [5–9]. Pressure was applied by a multianvil device, based on a Walker-type module and a 1000 ton press (both devices from the company Voggenreiter, Mainleus, Germany). For the syntheses of $Eu_4B_4O_{11}F_2$ / $Dy_4B_4O_{11}F_2$ the educts were compressed up to 5 / 8 GPa in 130 / 220 min, then heated to 900 / 1000 °C in 15 min, and kept there for 20 min. Afterwards, the temperature was lowered to 700 / 750 °C in 20 min, and finally to r.t. by switching off the heating. The decompression of the assemblies required 390 / 660 min. The recovered MgO octahedra (pressure-transmitting medium, Ceramic Substrates & Components Ltd., Newport, Isle of Wight, UK) were broken apart and the samples carefully separated from the surrounding boron nitride crucibles. Colorless, air- and water-resistant crystals of $Eu_4B_4O_{11}F_2$ and $Dy_4B_4O_{11}F_2$ were the products of the respective syntheses.

Since the syntheses were initially planned to give fluorine-rich borates an excess of rare-earth fluoride was always applied ($RE_2O_3 : B_2O_3 : REF_3 = 1 : 1 : 3$; the ideal ratio for the syntheses of the composition $RE_4B_4O_{11}F_2$ is 5 : 6 : 2). Notwithstanding, the products contained mainly the compounds $RE_4B_4O_{11}F_2$ (RE = Eu, Dy) besides remaining REF_3 and yet unidentified phases. Obviously, this composition and the associated $Gd_4B_4O_{11}F_2$ structure type seem to have a remarkable tendency towards formation under these extreme conditions.

Crystal structure analyses

$Eu_4B_4O_{11}F_2$ was identified by X-ray powder diffraction on a flat sample of the reaction product, using a Stoe Stadi P powder diffractometer with $MoK_{\alpha 1}$ radiation (transmission geometry, Ge monochromator, λ = 70.93 pm). Fig. 1 shows the powder pattern, displaying $Eu_4B_4O_{11}F_2$ as well as reflections of the excess educt EuF_3 (marked with lines in Fig. 1). Additional reflections, which could not be assigned to a known phase, are marked with asterisks. The experimental powder pattern (top) is in good agreement with the theoretical pattern (bottom), simulated from the single-crystal data. By indexing the reflections of the europium fluoride borate, the parameters a = 1368.2(3), b = 465.3(2), c = 1376.8(3) pm, β = 91.2(2)° with a volume of 0.8762(3) nm^3 were obtained. This validated the lattice parameters received from the single-crystal X-ray diffraction data (Table 1). The intensity data of single crystals of both $Eu_4B_4O_{11}F_2$ and $Dy_4B_4O_{11}F_2$ were collected at r.t. by use of a Kappa CCD diffractometer (Bruker AXS / Nonius, Karlsruhe), equipped with a Miracol fiber optics collimator and a Nonius FR590 generator (graphite-monochromatized MoK_{α} radiation, λ = 71.073 pm). Absorption corrections, based on multi-scans, were performed with SCALEPACK [10]. All significant details of the data collections and analyses are listed in Table 1. For the structural refinement, the positional parameters of the isotypic compound $Gd_4B_4O_{11}F_2$ were used as starting values [3]. The parameter refinements (full-matrix least-squares against F^2) were achieved by using the SHELX-97 software suite [11, 12]. All atoms were refined with anisotropic atomic displace-

Table 1. Crystal data and structure refinement of $RE_4B_4O_{11}F_2$ (RE = Eu, Dy) (standard deviations in parentheses).

Empirical formula	$Eu_4B_4O_{11}F_2$	$Dy_4B_4O_{11}F_2$
Molar mass, g mol^{-1}	865.08	907.24
Crystal system		monoclinic
Space group		$C2/c$
Powder diffractometer		Stoe Stadi P
Radiation; λ, pm		MoK$_{\alpha 1}$; 70.93
		(Ge(111) monochromator)
Powder data		
a, pm	1368.2(3)	
b, pm	465.3(2)	
c, pm	1376.8(3)	
β, deg	91.2(2)	
V, nm^3	0.8762(3)	
Single-crystal diffractometer		Nonius Kappa CCD
Radiation; λ, pm		MoK$_{\alpha}$; 71.073
		(graphite monochromator)
Single-crystal data		
a, pm	1368.2(3)	1349.5(3)
b, pm	465.4(1)	460.9(1)
c, pm	1376.6(3)	1362.5(3)
β, deg	91.2(1)	91.3(1)
V, nm^3	0.8765(3)	0.8472(3)
Formula units per cell	$Z = 4$	$Z = 4$
Calcd density, g cm^{-3}	6.56	7.11
$F(000)$, e	1512	1560
Temperature, K	293(2)	293(2)
Crystal size, mm^3	$0.04 \times 0.04 \times 0.07$	$0.03 \times 0.04 \times 0.05$
Absorption coeff, mm^{-1}	28.3	35.0
Absorption correction	multi-scan	multi-scan
θ range, deg	3.0–32.5	3.0–32.5
Range in hkl	$-18/+20, \pm 7, \pm 20$	$-20/+18, \pm 6, \pm 20$
Total no. of reflections	5011	5261
Independent refls / R_{int}	1576 / 0.0494	1533 / 0.0544
Refls with $I \geq 2\sigma(I) / R_\sigma$	1454 / 0.0374	1331 / 0.0394
Data / ref. parameters	1576 / 97	1533 / 97
Goodness-of-fit on F^2	1.156	1.098
Final indices R_1/wR_2 $[I \geq 2\sigma(I)]$	0.0210 / 0.0529	0.0288 / 0.0702
Indices R_1/wR_2 (all data)	0.0232 / 0.0539	0.0353 / 0.0729
Largest diff. peak / hole, $\times 10^{-6}$ e pm^{-3}	3.2 / −2.0	3.8 / −2.3

Table 2. Atomic coordinates and isotropic equivalent displacement parameters U_{eq} for $RE_4B_4O_{11}F_2$ (RE = Eu, Dy) (space group: $C2/c$) (standard deviations in parentheses). U_{eq} is defined as one third of the trace of the orthogonalized U_{ij} tensor.

Atom	W. position	x	y	z	U_{eq} (Å^2)
Eu1	$8f$	0.05887(2)	0.52031(4)	0.37058(2)	0.00489(8)
Eu2	$8f$	0.27958(2)	0.01747(4)	0.37066(2)	0.00456(8)
B1	$8f$	0.9070(3)	0.9774(8)	0.2857(3)	0.0052(7)
B2	$8f$	0.0953(3)	0.9557(8)	0.5251(3)	0.0052(6)
O1	$8f$	0.9122(2)	0.8673(5)	0.3937(2)	0.0053(4)
O2	$8f$	0.1736(2)	0.8139(5)	0.2577(2)	0.0061(4)
O3	$8f$	0.0790(2)	0.6653(5)	0.5351(2)	0.0063(4)
O4	$4e$	0	0.8476(7)	1/4	0.0046(6)
O5	$8f$	0.1146(2)	0.0609(6)	0.4348(2)	0.0078(5)
O6	$8f$	0.9012(2)	0.2796(5)	0.2751(2)	0.0064(4)
F1	$8f$	0.2308(2)	0.5244(5)	0.4248(2)	0.0114(5)
Dy1	$8f$	0.05887(2)	0.51733(5)	0.37032(2)	0.0063(2)
Dy2	$8f$	0.28029(2)	0.01473(5)	0.37012(2)	0.0058(2)
B1	$8f$	0.9065(5)	0.974(2)	0.2867(5)	0.006(2)
B2	$8f$	0.0963(5)	0.958(2)	0.5226(5)	0.009(2)
O1	$8f$	0.9122(6)	0.8625(8)	0.3952(3)	0.0055(6)
O2	$8f$	0.1761(2)	0.8147(8)	0.2576(3)	0.0066(7)
O3	$8f$	0.0786(3)	0.6664(8)	0.5335(3)	0.0069(7)
O4	$4e$	0	0.841(2)	1/4	0.0059(9)
O5	$8f$	0.1156(3)	0.634(8)	0.4322(3)	0.0078(7)
O6	$8f$	0.9015(3)	0.2795(8)	0.2763(3)	0.0074(7)
F1	$8f$	0.2312(3)	0.5272(7)	0.4244(3)	0.0127(7)

Results and Discussion

The structures of $RE_4B_4O_{11}F_2$ (RE = Eu, Dy) contain BO_3 groups and BO_4 tetrahedra, connected via common corners (Fig. 2). The main structural motif

ment parameters. The final difference Fourier syntheses did not reveal any significant residual peaks in all refinements. The positional parameters of the atom refinements, interatomic distances, and interatomic angles are listed in the Tables 2–4.

Further details of the crystal structure investigation may be obtained from Fachinformationszentrum Karlsruhe, 76344 Eggenstein-Leopoldshafen, Germany (fax: +49-7247-808-666; e-mail: crysdata@fiz-karlsruhe.de, http://www.fiz-informationsdienste.de/en/DB/icsd/depot_anforderung.html) on quoting the deposition number CSD-422159 ($Eu_4B_4O_{11}F_2$) and CSD-422160 ($Dy_4B_4O_{11}F_2$).

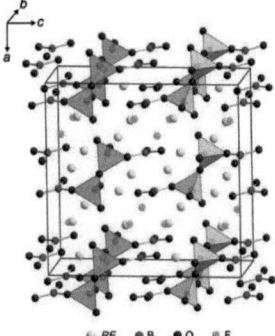

Fig. 2 (color online). Crystal structure of $RE_4B_4O_{11}F_2$ (RE = Eu, Gd, Dy), showing the fundamental building block 2△2☐: △☐☐△.

4.2 Rare-Earth Fluoride and Fluorido Borates

Table 3. Interatomic distances (pm) in $RE_4B_4O_{11}F_2$ (RE = Eu, Dy) (space group: $C2/c$), calculated with the single-crystal lattice parameters (standard deviations in parentheses).

Eu1–O6a	237.1(2)	Eu2–O2a	230.7(2)	B1–O6	141.6(4)		
Eu1–O3a	237.4(2)	Eu2–O2b	234.2(2)	B1–O2	145.8(5)		
Eu1–O4	238.1(2)	Eu2–O6	241.1(4)	B1–O4	150.1(4)		
Eu1–O5a	243.0(3)	Eu2–O5	244.9(3)	B1–O1	157.3(5)		
Eu1–F1	245.3(3)	Eu2–O1	245.4(2)	av. = 148.7			
Eu1–O3b	246.8(2)	Eu2–O3	246.0(2)				
Eu1–O1	260.1(2)	Eu2–F1a	250.8(2)	B2–O5	136.7(5)		
Eu1–O2	261.6(3)	Eu2–F1b	256.7(2)	B2–O3	137.7(4)		
Eu1–O6b	274.2(2)	Eu2–F1c	282.9(3)	B2–O1	139.4(5)		
Eu1–O5b	276.9(3)	av. = 248.1		av. = 137.9			
av. = 252.1							
F1–Eu1	245.3(3)						
F1–Eu2a	250.8(2)						
F1–Eu2b	256.7(2)						
F1–Eu2c	282.9(3)						
av. = 258.9							
Dy1–O3a	233.6(4)	Dy2–O2a	225.3(4)	B1–O6	141.5(7)		
Dy1–O4	234.2(4)	Dy2–O2b	230.9(4)	B1–O2	145.6(8)		
Dy1–O6a	235.1(4)	Dy2–O6	236.3(4)	B1–O4	149.9(7)		
Dy1–O5a	237.5(4)	Dy2–O5	240.6(4)	B1–O1	156.6(8)		
Dy1–F1	242.3(4)	Dy2–O1	241.3(4)	av. = 148.4			
Dy1–O3b	244.7(4)	Dy2–O3	243.5(4)				
Dy1–O1	256.8(4)	Dy2–F1a	246.1(3)	B2–O5	135.7(8)		
Dy1–O2	261.8(4)	Dy2–F1b	256.7(3)	B2–O3	137.1(7)		
Dy1–O6b	268.8(4)	Dy2–F1c	281.4(5)	B2–O1	140.0(8)		
Dy1–O5b	275.7(4)	av. = 244.7		av. = 137.6			
av. = 249.1							
F1–Dy1	242.3(4)						
F1–Dy2a	246.1(3)						
F1–Dy2b	256.7(3)						
F1–Dy2c	281.4(5)						
av. = 256.6							

Table 4. Interatomic angles (deg) in $RE_4B_4O_{11}F_2$ (RE = Eu, Dy) (space group: $C2/c$), calculated with the single-crystal lattice parameters (standard deviations in parentheses).

O6–B1–O2	115.9(3)	O5–B2–O3	118.5(3)	Eu1–F1–Eu2a	100.2(1)
O6–B1–O4	114.3(3)	O5–B2–O1	122.4(3)	Eu1–F1–Eu2b	99.1(1)
O2–B1–O4	107.2(3)	O3–B2–O1	119.0(3)	Eu1–F1–Eu2c	103.6(1)
O6–B1–O1	115.0(3)	av. = 120.0		Eu2a–F1–Eu2b	133.0(2)
O2–B1–O1	103.7(3)			Eu2a–F1–Eu2c	112.2(1)
O4–B1–O1	99.0(3)			Eu2b–F1–Eu2c	104.1(1)
av. = 109.2				av. = 108.7	
O6–B1–O2	115.3(5)	O5–B2–O3	119.2(6)	Dy1–F1–Dy2a	100.9(2)
O6–B1–O4	114.4(5)	O5–B2–O1	122.3(5)	Dy1–F1–Dy2b	98.4(2)
O2–B1–O4	107.3(4)	O3–B2–O1	118.4(6)	Dy1–F1–Dy2c	103.1(2)
O6–B1–O1	115.0(5)	av. = 120.0		Dy2a–F1–Dy2b	132.8(3)
O2–B1–O1	104.1(4)			Dy2a–F1–Dy2c	112.5(2)
O4–B1–O1	99.0(4)			Dy2b–F1–Dy2c	104.1(2)
av. = 109.2				av. = 108.6	

consists of two BO$_3$ groups (△) and two BO$_4$ tetrahedra (□), which can be described with the fundamental building block 2△2□: △□□△ (after Burns et al. [4]), a novelty in borate chemistry. For a detailed depiction of the structure, the reader is referred to the description of the isotypic compound Gd$_4$B$_4$O$_{11}$F$_2$ [3]. In this paper, a comparison of the three isotypic compounds $RE_4B_4O_{11}F_2$ (RE = Eu, Gd, Dy) is given.

Fig. 3 shows a comparison of the values of the lattice parameters of Eu$_4$B$_4$O$_{11}$F$_2$, Gd$_4$B$_4$O$_{11}$F$_2$ [3], and Dy$_4$B$_4$O$_{11}$F$_2$. The exact values are given in Table 5. The difference of the lattice parameters corresponds to the decreasing ionic radii (lanthanoid contraction) of the rare-earth cations. The values for the ionic radius of ninefold coordinated cations are taken from reference [13]: Eu^{3+} (126.0 pm), Gd^{3+} (124.7 pm), and Dy^{3+} (122.3 pm). Because the size differences

A. Pitscheider et al. · Rare-earth Fluoride Borates $RE_4B_4O_{11}F_2$ (RE = Eu, Dy)

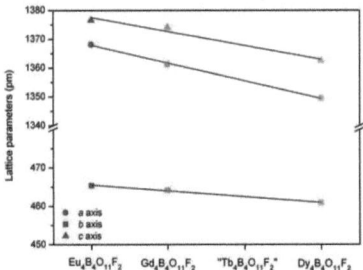

Fig. 3 (color online). Visualization of the progression of the lattice parameters (pm) of $RE_4B_4O_{11}F_2$ (RE = Eu, Gd, Dy) with the decrease due to the lanthanoid contraction.

Table 5. Comparison of the single-crystal lattice parameters and volumes of $RE_4B_4O_{11}F_2$ (RE = Eu, Gd, Dy) (standard deviations in parentheses).

Compound	a (pm)	b (pm)	c (pm)	β (deg)	V (nm³)
Eu₄B₄O₁₁F₂	1368.2(3)	465.4(1)	1376.6(3)	91.2(1)	0.8765(3)
Gd₄B₄O₁₁F₂	1361.3(3)	464.2(2)	1374.1(3)	91.3(1)	0.8681(3)
Dy₄B₄O₁₁F₂	1349.5(3)	460.9(1)	1362.5(3)	91.3(1)	0.8472(3)

Table 6. Charge distribution in $RE_4B_4O_{11}F_2$ (RE = Eu, Dy), calculated with VALIST (ΣV) [17] and the CHARDI concept (ΣQ) [18–20].

	Dy1	Dy2	B1	B2	Eu1	Eu2	B1	B2
ΣV	2.80	2.79	2.98	2.96	3.11	3.01	2.96	2.94
ΣQ	2.98	3.06	2.96	3.00	2.98	3.04	2.97	3.01
	O1	O2	O3	O4	O1	O2	O3	O4
ΣV	−2.06	−1.93	−2.01	−2.21	−2.12	−2.04	−2.13	−1.84
ΣQ	−2.00	−1.98	−2.12	−2.25	−2.00	−2.00	−2.10	−2.22
	O5	O6	F1		O5	O6	F1	
ΣV	−1.87	−1.81	−0.76		−1.91	−1.90	−0.97	
ΣQ	−1.90	−1.94	−0.94		−1.90	−1.92	−0.98	

are not too large, the bond lengths and angles of $RE_4B_4O_{11}F_2$ (RE = Eu, Dy) are comparable to the values found in $Gd_4B_4O_{11}F_2$ [3]. The Eu–O/F distances are in the range 230.7(2)–282.9(3) pm and the Dy–O/F distances within 225.3(4)–281.4(5) pm. This fits well to the values of 230.0(3)–283.7(3) pm for Gd–O/F in $Gd_4B_4O_{11}F_2$. The mean Gd–F distance is 258.4 pm, which lies between the mean RE–F bond lengths of 258.9 and 256.6 pm in $Eu_4B_4O_{11}F_2$ and $Dy_4B_4O_{11}F_2$, respectively. In the crystal structure of $Gd_4B_4O_{11}F_2$, a distorted tetrahedron was found and interpreted as a BO_3 group, in which the boron atom is drawn towards a fourth oxygen atom, resulting in a long B–O bond [3]. For $Gd_4B_4O_{11}F_2$, this long B1–O1 bond measures 159.0(6) pm. The corresponding B–O distances in $Eu_4B_4O_{11}F_2$ and $Dy_4B_4O_{11}F_2$ are slightly shorter with values of 157.3(5) and 156.6(8) pm, respectively. Inside the BO_3 groups of these structures, average B–O distances of 137.9 and 137.6 pm are found for $RE_4B_4O_{11}F_2$ (RE = Eu, Dy) – in perfect agreement with the literature.

We also calculated the charge distribution of the atoms in $RE_4B_4O_{11}F_2$ (RE = Eu, Dy) via bond valence sums (ΣV), using VALIST (Bond Valence Calculation and Listing) [14] and the CHARDI (charge distribution in solids) concept (ΣQ) [15–17], verifying the formal valence states in the fluoride borates. Table 6 shows the formal ionic charges, received from the calculations, which correspond to the expected values.

Additionally, we calculated the MAPLE values (Madelung Part of Lattice Energy according to Hoppe [18–20]) of $RE_4B_4O_{11}F_2$ (RE = Eu, Dy), which were checked against the data of the binary compounds. We obtained a value of 72337 kJ mol^{-1} for $Eu_4B_4O_{11}F_2$, to be compared with 72620 kJ mol^{-1} (deviation: 0.4 %), based on the binary components [5/3× Eu_2O_3 (14991 kJ mol^{-1} [21]) +2× B_2O_3-I (21938 kJ mol^{-1} [22]) +2/3× EuF_3 (5797 kJ mol^{-1} [23])]. For $Dy_4B_4O_{11}F_2$, the resulting value is 72799 kJ mol^{-1}, compared to 73179 kJ mol^{-1} (deviation: 0.5 %) based on the binary components [5/3× Dy_2O_3 (14991 kJ mol^{-1} [24]) +2× B_2O_3-I (21938 kJ mol^{-1} [22]) +2/3× DyF_3 (5797 kJ mol^{-1} [23])].

Conclusion

With the synthesis of $RE_4B_4O_{11}F_2$ (RE = Eu, Dy), the range of compounds with the composition $RE_4B_4O_{11}F_2$ has been extended and explored in more detail. The existence of the isotypic compound "$Tb_4B_4O_{11}F_2$" still missing in the series is very likely; further studies on this phase and the possible formation of $RE_4B_4O_{11}F_2$ with RE = Ho–Lu are planned. Additional experiments with the larger rare-earth cations RE = Ce–Nd and Sm will be performed to investigate the transformation into the structure type of $La_4B_4O_{11}F_2$ and any polymorphs.

Acknowledgement

We would like to thank Dr. G. Heymann for collecting the single-crystal data.

[1] H. Huppertz, *Chem. Commun.* **2011**, DOI:10.1039/C0CC02715D.
[2] A. Haberer, R. Kaindl, O. Oeckler, H. Huppertz, *J. Solid State Chem.* **2010**, *183*, 1970.
[3] A. Haberer, R. Kaindl, H. Huppertz, *J. Solid State Chem.* **2010**, *183*, 471.
[4] P. C. Burns, J. D. Grice, F. C. Hawthorne, *Can. Mineral.* **1995**, *33*, 1131.
[5] D. Walker, M. A. Carpenter, C. M. Hitch, *Am. Mineral.* **1990**, *75*, 1020.
[6] D. Walker, *Am. Mineral.* **1991**, *76*, 1092.
[7] H. Huppertz, *Z. Kristallogr.* **2004**, *219*, 330.
[8] D. C. Rubie, *Phase Transitions* **1999**, *68*, 431.
[9] N. Kawai, S. Endo, *Rev. Sci. Instrum.* **1970**, *8*, 1178.
[10] SCALEPACK. Z. Otwinowski, W. Minor in *Methods in Enzymology*, Vol. 276, *Macromolecular Crystallography*, Part A (Eds.: C. W. Carter Jr., R. M. Sweet), Academic Press, New York, 1997, pp. 307.
[11] G. M. Sheldrick, SHELXS/L-97, Programs for Crystal Structure Determination, University of Göttingen, Göttingen (Germany) 1997.
[12] G. M. Sheldrick, *Acta Crystallogr.* **2008**, *A64*, 112.
[13] R. D. Shannon, *Acta Crystallogr.* **1976**, *A32*, 751.
[14] A. S. Wills, VALIST (version 4.0.0), University College London, London (U. K.) **1998**–**2008**; programm available from www.ccp14.ac.uk.
[15] I. D. Brown, D. Altermatt, *Acta Crystallogr.* **1985**, *B41*, 244.
[16] N. E. Brese, M. O'Keeffe, *Acta Crystallogr.* **1991**, *B47*, 192.
[17] R. Hoppe, S. Voigt, H. Glaum, J. Kissel, H. P. Müller, K. J. Bernet, *J. Less-Common Met.* **1989**, *156*, 105.
[18] R. Hoppe, *Angew. Chem.* **1966**, *78*, 52; *Angew. Chem., Int. Ed. Engl.* **1966**, *5*, 96.
[19] R. Hoppe, *Angew. Chem.* **1970**, *82*, 7; *Angew. Chem., Int. Ed. Engl.* **1970**, *9*, 25.
[20] R. Hübenthal, M. Serafin, R. Hoppe, MAPLE (version 4.0), Program for the Calculation of Distances, Angles, Effective Coordination Numbers, Coordination Spheres, and Lattice Energies, University of Gießen, Gießen (Germany) 1993.
[21] D. H. Templeton, C. H. Dauben, *J. Am. Chem. Soc.* **1954**, *76*, 5237.
[22] C. T. Prewitt, R. D. Shannon, *Acta Crystallogr.* **1968**, *B24*, 869.
[23] A. Zalkin, D. H. Templeton, *J. Am. Chem. Soc.* **1953**, *75*, 2453.
[24] P. Karen, *J. Solid State Chem.* **2004**, *177*, 281.

4.2.6 La$_4$B$_4$O$_{11}$F$_2$

For the composition $RE_4B_4O_{11}F_2$, another structure type was obtained by high-pressure / high-temperature synthesis. The lanthanum fluoride borate La$_4$B$_4$O$_{11}$F$_2$ was synthesized at 6 GPa and 1300 °C, yielding a colorless crystalline sample (Figure 4.2-6). While the compound Gd$_4$B$_4$O$_{11}$F$_2$, presented above, crystallizes in the space group $C2/c$, La$_4$B$_4$O$_{11}$F$_2$ shows the same atomic composition, but exhibits a completely different crystal structure in the space group $P2_1/c$. Nevertheless, the crystal structure of La$_4$B$_4$O$_{11}$F$_2$ also exhibits a new fundamental building block (FBB) in borate chemistry, namely a 2 △◻: △◻△ unit, not yet reported by Burns *et al.* [237].

Another interesting feature of the crystal structure is the very long b lattice parameter of La$_4$B$_4$O$_{11}$F$_2$, which is about five times longer than the other cell edges. A wave-like modulation of the cations and the B-O-polyhedra with a formal wavelength $\lambda = b$ is observed. In the following publication, these results are presented in detail.

Figure 4.2-6
Crystalline sample of La$_4$B$_4$O$_{11}$F$_2$

Journal of Solid State Chemistry 183 (2010) 1970–1979

Contents lists available at ScienceDirect

Journal of Solid State Chemistry

journal homepage: www.elsevier.com/locate/jssc

A new structure type of $RE_4B_4O_{11}F_2$: High-pressure synthesis and crystal structure of $La_4B_4O_{11}F_2$

Almut Haberer [a], Reinhard Kaindl [b], Oliver Oeckler [c], Hubert Huppertz [a,*]

[a] *Institut für Allgemeine, Anorganische und Theoretische Chemie, Leopold-Franzens-Universität Innsbruck, Innrain 52a, A-6020 Innsbruck, Austria*
[b] *Institut für Mineralogie und Petrographie, Leopold-Franzens-Universität Innsbruck, Innrain 52, A-6020 Innsbruck, Austria*
[c] *Department Chemie, Ludwig-Maximilians-Universität München, Butenandtstrasse 5–13, D-81377 München, Germany*

ARTICLE INFO

Article history:
Received 5 May 2010
Received in revised form
28 June 2010
Accepted 28 June 2010
Available online 7 July 2010

Keywords:
High-pressure
Multianvil
Crystal structure
Fluoride borate

ABSTRACT

The first lanthanum fluoride borate $La_4B_4O_{11}F_2$ was obtained in a Walker-type multianvil apparatus at 6 GPa and 1300 °C. $La_4B_4O_{11}F_2$ crystallizes in the monoclinic space group $P2_1/c$ with the lattice parameters $a=778.1(2)$ pm, $b=3573.3(7)$ pm, $c=765.7(2)$ pm, $\beta=113.92(3)°$ ($Z=8$), and represents a new structure type in the class of compounds with the composition $RE_4B_4O_{11}F_2$. The crystal structure contains BO_4-tetrahedra interconnected with two BO_3-groups via common vertices, B_2O_5-pyroborate units, and isolated BO_3-groups. The structure shows a wave-like modulation along the b-axis. The crystal structure and properties of $La_4B_4O_{11}F_2$ are discussed and compared to $Gd_4B_4O_{11}F_2$.

© 2010 Elsevier Inc. All rights reserved.

1. Introduction

Borates have been extensively examined under high-pressure/high-temperature conditions and a large variety of new compounds could be obtained, for example the rare-earth borates $RE_4B_6O_{15}$ ($RE=$ Dy, Ho) [1–3], α-$RE_2B_4O_9$ ($RE=$ Sm–Ho) [4–6], β-$RE_2B_4O_9$ ($RE=$ Gd [7], Dy [8]), $RE_3B_5O_{12}$ ($RE=$ Er–Lu) [9], and $Pr_4B_{10}O_{21}$ [10]. Borates, being glass formers in general, show an increased tendency to crystallize under pressure, which seems to be valid for fluorine-containing borates, too. Before we started our high-pressure/high-temperature research, rare-earth fluoride borates were only represented by the compounds $RE_3(BO_3)_2F_3$ ($RE=$ Sm, Eu, Gd) [11,12] and $Gd_2(BO_3)_3F_3$ [13], synthesized by heating stoichiometric mixtures of RE_2O_3, B_2O_3, and REF_3 under ambient pressure. Recently, the field of rare-earth fluoride borates could be extended with the high-pressure phases $Yb_5(BO_3)_2F_5$ [14], $Pr_4B_3O_{10}F$ [15], $Gd_4B_4O_{11}F_2$ [16], and the ambient pressure phase $Eu_6(BO_3)_3F$ [17].

Concerning structural motifs, the chemistry of rare-earth fluoride borates under high-pressure/high-temperature conditions shows similarities to that of oxoborates in general. Trigonal-planar BO_3-groups as well as BO_4-tetrahedra were identified and the pressure-induced transformation of the former into the latter ones was observed. The motif of edge-sharing tetrahedra could be realized under high pressure for the compounds $RE_4B_6O_{15}$ ($RE=$ Dy, Ho) [1–3] and α-$RE_2B_4O_9$ ($RE=$ Sm–Ho) [4–6]. Recently, HP-NiB_2O_4 [18] and β-FeB_2O_4 [19] were synthesized, in which each BO_4-tetrahedron shares a common edge with another one. In fluorido and fluoride borates, this structural feature could not be observed till date.

For the composition $RE_4B_4O_{11}F_2$, two different structure types were obtained by high-pressure/high-temperature synthesis. While the recently presented compound $Gd_4B_4O_{11}F_2$ crystallizes in the space group $C2/c$ [16], the new compound $La_4B_4O_{11}F_2$ shows the same atomic composition, but exhibits a completely different crystal structure in the space group $P2_1/c$. In the crystal structure of $Gd_4B_4O_{11}F_2$, there are BO_3-groups and BO_4-tetrahedra, connected via common corners. The structural motif consists of two BO_3-groups (Δ) and two BO_4-tetrahedra ($\square$), and can be described with the fundamental building block $2\Delta 2\square:\Delta\square\square\Delta$ (after Burns et al. [20]), which is a novelty in borate chemistry. In $La_4B_4O_{11}F_2$, the building blocks $\Delta\square\Delta$ and $\Delta\Delta$ are present, along with isolated BO_3-groups. In the following, the synthesis and structural details of the new compound $La_4B_4O_{11}F_2$ are reported.

2. Experimental section

2.1. Synthesis

The reaction of the oxides La_2O_3 and B_2O_3 with LaF_3 took place under high-pressure/high-temperature conditions of 6 GPa

---

* Corresponding author.
 E-mail address: hubert.huppertz@uibk.ac.at (H. Huppertz).

0022-4596/$ - see front matter © 2010 Elsevier Inc. All rights reserved.
doi:10.1016/j.jssc.2010.06.019

4.2 Rare-Earth Fluoride and Fluorido Borates

and 1300 °C, leading to the fluoride borate $La_4B_4O_{11}F_2$ (Eq. (1)):

$$5\ La_2O_3 + 6\ B_2O_3 + 2\ LaF_3 \xrightarrow{6\ GPa,\ 1300\ °C} 3\ La_4B_4O_{11}F_2 \quad (1)$$

A mixture of La_2O_3 (Strem Chemicals, 99.9%), B_2O_3 (Strem Chemicals, 99.9+%), and LaF_3 (Strem Chemicals, 99.9%) at a molar ratio of 5:6:2 (Eq. (1)) was ground and filled into a boron nitride crucible (Henze BNP GmbH, HeBoSint® S100, Kempten, Germany). This crucible was placed into the center of an 14/8-assembly, which was compressed by eight tungsten carbide cubes (TSM-10 Ceratizit, Reutte, Austria). The details of preparing the assembly can be found in Refs. [21–25]. Pressure was applied by a multianvil device based on a Walker-type module and a 1000 ton press (both devices from the company Voggenreiter, Mainleus, Germany). The sample was compressed up to 6 GPa in 3 h, then heated to 1300 °C in 15 min and kept there for 10 min. Afterwards, the sample was cooled down to 800 °C in 10 min, and cooled down to room temperature by switching off the heating. The decompression of the assembly required 9 h. The recovered MgO-octahedron (pressure transmitting medium, Ceramic Substrates & Components Ltd., Newport, Isle of Wight, UK) was broken apart and the sample was carefully separated from the surrounding boron nitride crucible, and colorless, air- and water-resistant crystal platelets of $La_4B_4O_{11}F_2$ were obtained.

2.2. Crystal structure analysis

The sample characterization was performed by powder X-ray diffraction, carried out in transmission geometry on a flat sample of the reaction product, using a STOE STADI P powder diffractometer with MoKα_1 radiation (Ge monochromator, $\lambda = 70.93$ pm). The powder pattern showed reflections of $La_4B_4O_{11}F_2$ as well as of γ-$La(BO_2)_3$ [26] as a by-product of the synthesis. The experimental powder pattern tallies well with the superimposed theoretical patterns of the two compounds simulated from single-crystal data. Indexing the reflections of the lanthanum fluoride borate, we got the lattice parameters $a = 779.6(6)$ pm, $b = 3573.4(17)$ pm, and $c = 765.8(5)$ pm, with $\beta = 113.94(7)°$ and a unit-cell volume of $1949.9(15)$ Å^3. This confirmed the lattice parameters, taken from the single-crystal X-ray diffraction study (Table 1).

The intensity data of a single crystal of $La_4B_4O_{11}F_2$ were collected at room temperature using a Nonius Kappa-CCD diffractometer with graded multilayer X-ray optics (MoKα radiation, $\lambda = 71.073$ pm). Semiempirical scaling complemented by a spherical absorption correction [27,28] was applied to the intensity data. All relevant details of the data collection and evaluation are listed in Table 1.

According to the systematic extinctions, the monoclinic space group $P2_1/c$ was derived. Structure solution and parameter refinement (full-matrix least-squares against F^2) were performed using the SHELX-97 software suite [29,30] with anisotropic atomic displacement parameters for all atoms. The residual peaks after the final difference Fourier syntheses (Table 1) are located very close to the lanthanum cations. Fluorine and oxygen atoms are difficult to distinguish by means of electron density and bond lengths. Therefore, an assignment of the atom types was based on the charge distribution calculations with VaList (*vide infra*) and charge neutrality. The positional parameters of the refinements, interatomic distances, and interatomic angles are listed in Tables 2–4. Further information of the crystal structure is available from the Fachinformationszentrum Karlsruhe (crysdata@fiz-karlsruhe.de), D-76344 Eggenstein-Leopoldshafen (Germany), quoting the Registry no. CSD-421688.

Table 1
Crystal data and structure refinement of $La_4B_4O_{11}F_2$.

Empirical formula	$La_4B_4O_{11}F_2$
Molar mass (g mol^{-1})	812.88
Crystal system	Monoclinic
Space group	$P2_1/c$ (No. 14)
Lattice parameters from powder data	
Powder diffractometer	Stoe StadiP
Radiation	MoKα_1 ($\lambda = 70.93$ pm)
a (pm)	779.6(6)
b (pm)	3573.4(17)
c (pm)	765.8(5)
β (deg.)	113.94(7)
Volume (Å^3)	1949.9(15)
Single-crystal diffractometer	Nonius Kappa-CCD
Radiation	MoKα ($\lambda = 71.073$ pm)
Single-crystal data	
a (pm)	778.1(2)
b (pm)	3573.3(7)
c (pm)	765.7(2)
β (deg.)	113.92(3)
Volume (Å^3)	1946.3(7)
Formula units per cell	$Z = 8$
Temperature (K)	293(2)
Calculated density (g cm^{-3})	5.548
Crystal size (mm^3)	0.04 × 0.02 × 0.02
Absorption coefficient (mm^{-1})	17.3
$F(000)$	2832
θ range (deg.)	$3.3 \leq \theta \leq 45.4$
Range in $h\ k\ l$	± 15, ± 69, −14/15
Total no. reflections	64178
Independent reflections	15796 ($R_{int} = 0.0532$, $R_\sigma = 0.0488$)
Reflections with $I > 2\sigma(I)$	12135
Data/parameters	15796/379
Absorption correction	Semiempirical
Transm. ratio (min/max)	0.667/0.708
Goodness-of-fit (F^2)	1.024
Final R indices ($I > 2\sigma(I)$)	$R1 = 0.0353$
	$wR2 = 0.0650$
R indices (all data)	$R1 = 0.0577$
	$wR2 = 0.0718$
Largest differ. peak, deepest hole (e/Å^{-3})	4.78/−4.38

2.3. Scanning electron microscopy

Scanning electron microscopy was performed on a JEOL JSM-6500F equipped with a field emission gun at an acceleration voltage of 15 kV. The samples were prepared by placing single crystals on adhesive conductive pads and subsequently coating them with a thin conductive carbon film. Each EDX spectrum (Oxford Instruments) was recorded within a limited area on one single crystal, to avoid the influence of possible contaminating phases.

2.4. IR spectroscopy

FTIR-ATR (Attenuated Total Reflection) spectra of the crystals were recorded with a Bruker Vertex 70 FT-IR spectrometer (resolution ~ 0.5 cm^{-1}), attached to a Hyperion 3000 microscope in a spectral range from 600 to 4000 cm^{-1}. A frustrum shaped germanium ATR-crystal with a tip diameter of 100 μm was pressed on the surface of the crystals with a power of 5 N, which caused them to crush into pieces of μm-size. 64 scans for sample and background were acquired. Beside spectra correction for atmospheric influences, an enhanced ATR-correction [31], using the OPUS 6.5 software, was performed. A mean refraction index of the sample of 1.6 was assumed for the ATR-correction. Background correction and peak fitting followed via polynomial and folded Gaussian–Lorentzian functions.

Table 2
Atomic coordinates (all Wyckoff sites 4e) and isotropic equivalent displacement parameters (U_{eq}/Å^2) for La$_4$B$_4$O$_{11}$F$_2$ (space group: $P2_1/c$). U_{eq} is defined as one-third of the trace of the orthogonalized U_{ij} tensor.

Atom	x	y	z	U_{eq}
La1	0.48200(2)	0.234243(5)	0.25646(2)	0.00730(3)
La2	0.86556(2)	0.117015(5)	0.89156(2)	0.00820(3)
La3	0.95956(2)	0.227106(5)	0.74423(2)	0.00921(3)
La4	0.85704(2)	0.138323(5)	0.38092(2)	0.00670(3)
La5	0.36110(2)	0.142117(5)	0.87716(2)	0.00756(3)
La6	0.38039(3)	0.098889(6)	0.38476(3)	0.01150(3)
La7	0.50962(2)	0.016079(5)	0.75693(2)	0.00665(3)
La8	0.99082(2)	0.017185(5)	0.23637(2)	0.00719(3)
B1	0.6053(5)	0.1772(2)	0.6157(4)	0.0080(5)
B2	0.1340(5)	0.0636(2)	0.6169(5)	0.0126(6)
B3	0.2228(4)	0.0588(1)	0.9872(5)	0.0082(5)
B4	0.7638(5)	0.0550(2)	0.5326(5)	0.0085(5)
B5	0.1364(5)	0.1721(2)	0.1780(6)	0.0140(6)
B6	0.6731(6)	0.0695(2)	0.1400(7)	0.0191(7)
B7	0.1625(6)	0.2941(2)	0.9853(6)	0.0184(7)
B8	0.7590(6)	0.1990(2)	0.0783(7)	0.0196(8)
O1	0.6413(3)	0.03099(7)	0.1076(3)	0.0091(3)
O2	0.7167(3)	0.16148(7)	0.0277(3)	0.0107(4)
O3	0.2222(3)	0.04418(7)	0.8153(3)	0.0090(3)
O4	0.5641(4)	0.21490(7)	0.5829(3)	0.0128(4)
O5	0.7887(3)	0.16743(7)	0.6471(3)	0.0100(4)
O6	0.1869(3)	0.09572(7)	0.0025(4)	0.0127(4)
O7	0.1214(3)	0.17874(8)	0.5938(4)	0.0163(5)
O8	0.7443(3)	0.09222(7)	0.5613(3)	0.0125(4)
O9	0.1674(3)	0.25722(8)	0.0479(4)	0.0140(4)
O10	0.1014(3)	0.13927(8)	0.2468(3)	0.0132(4)
O11	0.1172(4)	0.17505(8)	0.9898(4)	0.0164(5)
O12	0.8620(4)	0.08154(8)	0.1746(4)	0.0196(5)
O13	0.9324(3)	0.04124(7)	0.5315(3)	0.0119(4)
O14	0.1279(4)	0.10204(8)	0.6182(4)	0.0183(5)
O15	0.2312(3)	0.04787(8)	0.5059(3)	0.0119(4)
O16	0.4851(3)	0.15034(8)	0.6240(3)	0.0134(4)
O17	0.7890(3)	0.22647(7)	0.9714(3)	0.0108(4)
O18	0.2517(3)	0.03512(8)	0.1360(3)	0.0138(4)
O19	0.6161(4)	0.03160(8)	0.5004(3)	0.0155(5)
O20	0.5589(4)	0.09462(8)	0.1704(4)	0.0154(4)
O21	0.1794(4)	0.20327(8)	0.3039(4)	0.0172(5)
O22	0.7885(4)	0.20890(8)	0.2775(5)	0.0244(6)
F1	0.8728(3)	0.03625(7)	0.8853(3)	0.0147(4)
F2	0.5482(3)	0.08495(7)	0.8028(4)	0.0251(5)
F3	0.4851(3)	0.16134(7)	0.2555(3)	0.0175(4)
F4	0.4119(5)	0.28744(8)	0.4254(4)	0.0282(6)

2.5. Raman spectroscopy

Confocal Raman spectra of single crystals were obtained with a HORIBA JOBIN YVON LabRam-HR 800 Raman micro-spectrometer. The sample was excited by the 532 nm emission line of a 30 mW Nd–YAG-laser under an OLYMPUS 100× objective (N.A.=0.9). The size and power of the laser spot on the surface were approximately 1 μm and 5 mW. The scattered light was dispersed by a grating with 1800 lines/mm and collected by a 1024 × 256 open electrode CCD detector. The spectral resolution, determined by measuring the Rayleigh line, was about 1.4 cm^{-1}. The polynomial and convoluted Gauss–Lorentz functions were applied for background correction and band fitting. The wavenumber accuracy of about 0.5 cm^{-1} was achieved by adjusting the zero-order position of the grating and regularly checked by a Neon spectral calibration lamp.

3. Results and discussion

3.1. Crystal structure of La$_4$B$_4$O$_{11}$F$_2$

At first glance, the crystal structure of La$_4$B$_4$O$_{11}$F$_2$ is fairly complicated with a large unit cell (1946.3(7) Å^3) and a total of 42 independent atomic positions. A detailed discussion of the structure mainly based on the different borate fundamental building blocks (FBBs) is presented in the following.

La$_4$B$_4$O$_{11}$F$_2$ consists of BO$_3$-groups (Δ), BO$_4$-tetrahedra (□), lanthanum cations, and fluoride anions (Fig. 1). The BO$_3$-groups are either isolated (Δ), connected via common corners (ΔΔ), or connected via a BO$_4$-tetrahedron, forming a fundamental building block (FBB) 2ΔO:ΔΔ, not yet reported by Burns et al. [20].

With 3573.3(7) pm, the b lattice parameter of La$_4$B$_4$O$_{11}$F$_2$ is about five times longer than the other cell edges. Looking at the structure along [001], a wave-like modulation of the cations and the B–O-polyhedra with a formal wavelength $\lambda = b$ is observed (Fig. 2). In contrast to the well known Vernier phases [32–35], no separation into mismatching substructures was found in La$_4$B$_4$O$_{11}$F$_2$. The three discrete fundamental building blocks Δ, ΔΔ, and ΔΠΔ are arranged along each wave in a sheet-like manner, as depicted in Fig. 3 (view along [010]), with the "sheets" expanding in the bc-plane. Nevertheless, it should be emphasized that La$_4$B$_4$O$_{11}$F$_2$ is not a layered structure, because the FBBs are not interconnected. There is a set of three different FBB arrangements in each wave, as indicated with dashed lines in Fig. 4. Each set is multiplied by inversion centers, twofold screw axes parallel to the b-axis, and c glide planes perpendicular to b. In arrangement 1, there are the building block ΔΠΔ and the isolated BO$_3$-group of B6 (Fig. 5 top). A detailed view of arrangement 2 is given in Fig. 5 (middle), showing B5 and B7 in form of a pyroborate group. Two further isolated BO$_3$-groups (B1 and B8, Fig. 5 bottom) are found in arrangement 3, where the planes of the BO$_3$-groups are orthogonally orientated to each other.

In the isolated BO$_3$-groups, the B–O-distances range from 134.9(5) to 149.0(6) pm (B1, B6, and B8 in Table 3). This is slightly larger than the values typical for threefold-coordinated boron atoms, e.g. in borates with calcite structure (AlBO$_3$ (137.96(4) pm) [36], β-YbBO$_3$ (137.8(4) pm) [37]), but similarly enlarged bond lengths for BO$_3$-groups are found as well, e.g. in BaB$_8$O$_{13}$ (153.5 pm) [38]. The reason for the bond stretching in La$_4$B$_4$O$_{11}$F$_2$ lies in the coordination of the oxygen ions. All isolated BO$_3$-groups are surrounded by seven lanthanum cations, coordinating their oxygen atoms, as displayed in Fig. 6 for the BO$_3$-group of B8. This leads to a close packing of La^{3+}O$_n$-polyhedra, which involves stretching of the BO$_3$-groups and thus enlarge the B–O-bond lengths. The strain of the BO$_3$-groups can also be seen in the interatomic angles, which slightly deviate from the ideal value of 120° (Table 4).

The corner-sharing BO$_3$-groups of the pyroborate building block ΔΔ show B–O-distances from 139.1(5) to 145.1(5) pm (B5 and B7 in Table 3). Here, the bond lengths between the boron atoms and the connecting oxygen atom are enlarged, resulting in varying interatomic angles (Table 4). This is known from other pyroborate compounds as well, e.g. from Eu$_2$B$_2$O$_5$ [39].

The BO$_3$-groups of the FBB ΔΠΔ show regular average bond-lengths of 137.7 and 138.0 pm, and uniform angles of 120° (B3 and B4 in Tables 3 and 4). Inside the BO$_4$-tetrahedron, the bond lengths range from 137.3(5) to 164.2(5) pm. The boron atom is displaced from the center of the tetrahedron towards the non-bridging oxygen atoms.

There are eight crystallographically independent La^{3+} ions in the structure, which are nine-, ten-, and elevenfold coordinated by oxygen and fluorine (Fig. 7 and Table 3). The average interatomic distances of the ninefold coordinated La–O/F are 254.1, 253.3, and 255.6 pm and thus in the same range as the average La–O distance of ninefold coordinated La^{3+} cations in LaB$_5$O$_9$ (261.6 pm [40]). The average La–O distances of tenfold coordinated La^{3+} cations are larger (263.0, 265.1, 261.8, and 272.3 pm) than would be expected from a tenfold coordination. For La2, the tenfold coordination should rather be described as a 9+1 coordination,

4.2 Rare-Earth Fluoride and Fluorido Borates

Table 3
Interatomic distances (pm) in $La_4B_4O_{11}F_2$, calculated with the single-crystal lattice parameters.

La1-O4a	241.7(2)	La2-O6	241.5(2)	La3-O5	246.3(2)	La4-O10	250.1(2)
La1-O9	245.9(3)	La2-O2	243.6(2)	La3-O9	247.7(3)	La4-O7	250.1(3)
La1-O4b	248.3(2)	La2-O8	247.7(3)	La3-O11	257.5(3)	La4-O14	251.4(3)
La1-O22	249.4(3)	La2-O5	249.0(2)	La3-O17a	257.5(2)	La4-O8	252.5(2)
La1-O17	268.3(3)	La2-O12	252.1(3)	La3-O17b	258.1(2)	La4-O5	252.9(2)
La1-O21	275.6(3)	La2-O10	271.1(3)	La3-O7	266.0(3)	La4-O12	258.1(3)
La1-F4a	248.1(3)	La2-O11	274.0(3)	La3-O9	267.6(3)	La4-O2	260.7(2)
La1-F4b	249.0(2)	La2-O16	307.6(3)	La3-O22	270.9(3)	La4-O22	263.2(3)
La1-F3	260.5(3)	La2-F2	255.2(2)	La3-O4	284.7(3)	La4-O20	271.8(3)
		La2-F1	288.7(2)	La3-O21	294.8(3)	La4-F3	277.9(2)
La5-O16	250.9(2)	La6-O10	246.0(3)	La7-O19a	246.0(3)	La8-O15	241.5(2)
La5-O14	252.2(3)	La6-O16	248.8(3)	La7-O19b	248.6(3)	La8-O12	247.6(3)
La5-O6	256.1(3)	La6-O15	253.1(2)	La7-O18	249.6(3)	La8-O18	252.7(2)
La5-O7	257.3(3)	La6-O20	254.9(2)	La7-O1a	250.4(2)	La8-O1	253.7(2)
La5-O2	262.3(2)	La6-O8	260.7(3)	La7-O15	251.0(3)	La8-O13a	262.4(2)
La5-O11	265.7(3)	La6-O6	269.7(3)	La7-O1b	251.3(2)	La8-O13b	264.8(3)
La5-O20	273.7(3)	La6-O18	287.7(3)	La7-O3	265.0(2)	La8-O3	268.0(2)
La5-F4	255.1(3)	La6-O19	293.5(3)	La7-F2	248.7(3)	La8-F1a	253.8(2)
La5-F2	270.0(3)	La6-O14	314.8(3)	La7-F1	268.7(2)	La8-F1b	255.6(2)
La5-F3	274.4(2)	La6-F3	269.8(2)				
		La6-F2	296.8(3)				
B1-O16	136.0(4)	B2-O14	137.3(5)	B3-O18	136.3(4)	B4-O19	136.1(4)
B1-O4	138.4(4)	B2-O15	146.0(4)	B3-O6	136.3(4)	B4-O8	136.6(4)
B1-O5	139.2(4)	B2-O3	155.5(4)	B3-O3	141.4(4)	B4-O13	140.5(4)
	∅=137.9	B2-O13	164.2(5)		∅=138.0		∅=137.7
			∅=150.8				
B5-O10	139.1(5)	B6-O20	134.9(5)	B7-O9	139.6(6)	B8-O17	135.8(5)
B5-O11	139.1(5)	B6-O1	140.0(5)	B7-O7	139.7(5)	B8-O2	139.7(5)
B5-O21	142.1(5)	B6-O12	145.0(5)	B7-O21	145.1(5)	B8-O22	149.0(6)
	∅=140.1		∅=140.0		∅=141.5		∅=141.5
F1-La8a	253.8(2)	F2-La7	248.7(3)	F3-La1	260.5(3)	F4-La1a	248.1(3)
F1-La8b	255.6(2)	F2-La2	255.2(2)	F3-La6	269.8(2)	F4-La1b	249.0(2)
F1-La7	268.7(2)	F2-La5	270.0(3)	F3-La5	274.4(2)	F4-La5	255.1(3)
F1-La2	288.7(2)	F2-La6	296.8(3)	F3-La4	277.9(2)		

Table 4
Interatomic angles (deg.) in $La_4B_4O_{11}F_2$, calculated with the single-crystal lattice parameters.

O16-B1-O4	125.6(3)	O14-B2-O15	114.7(3)	O18-B3-O6	120.0(3)	O19-B4-O8	119.1(3)
O16-B1-O5	119.4(3)	O14-B2-O3	116.3(3)	O18-B3-O3	119.1(3)	O19-B4-O13	120.4(3)
O4-B1-O5	115.0(3)	O15-B2-O3	105.6(3)	O3-B3-O6	120.9(3)	O13-B4-O8	120.5(3)
	∅=120.0	O14-B2-O13	117.4(3)		∅=120.0		∅=120.0
		O15-B2-O13	102.5(3)				
		O3-B2-O13	98.2(2)				
			∅=109.1				
O10-B5-O11	120.8(3)	O20-B6-O1	127.1(4)	O9-B7-O7	115.6(3)	O17-B8-O2	127.2(4)
O10-B5-O21	116.3(3)	O20-B6-O12	117.4(4)	O9-B7-O21	112.9(4)	O17-B8-O22	116.9(4)
O21-B5-O11	122.7(4)	O12-B6-O1	114.7(3)	O21-B7-O7	131.0(4)	O22-B8-O2	115.7(3)
	∅=119.9		∅=119.7		∅=119.8		∅=119.9
La8a-F1-La8b	99.20(8)	La7-F2-La2	122.0(2)	La1-F3-La6	145.25(9)	La1a-F4-La1b	106.8(2)
La8a-F1-La7	101.43(8)	La7-F2-La5	137.2(2)	La1-F3-La5	104.69(8)	La1a-F4-La5	137.2(2)
La8b-F1-La7	100.92(8)	La2-F2-La5	98.66(9)	La5-F3-La6	98.69(8)	La1b-F4-La5	114.3(1)
La8a-F1-La2	140.53(9)	La7-F2-La6	92.22(9)	La1-F3-La4	107.72(8)		∅=119.4
La8b-F1-La2	104.48(7)	La6-F2-La2	99.62(9)	La4-F3-La6	93.18(8)		
La7-F1-La2	104.46(7)	La5-F2-La6	93.52(8)	La5-F3-La4	99.33(8)		
	∅=108.5		∅=107.2		∅=108.1		

as the largest La2–ligand distance is 307.6 pm (La2–O16) and thus about 19 pm larger than the second largest (La2–F1). A similar case is found for elevenfold-coordinated La6, better described as a 10+1 coordination; here, a significant cut in the distance histogram can be observed as well (La6–O14: 314.8(3) pm, Table 3).

The fluoride ions in $La_4B_4O_{11}F_2$ are coordinated by either three or four lanthanum ions (Fig. 8). The bond lengths of the fourfold-coordinated fluoride ions range between 248.7(3) and 296.8(3) pm (Table 3), according with the values for similarly coordinated ions in LaF_3 (247.1–307.0 pm) [41]. The average La–F angles are 108.5°, 107.2°, and 108.1° (Table 4), and thus fairly close to the ideal tetrahedral angle. The bond lengths of the threefold-coordinated fluoride ion with 248.1(3) to 255.1(3) pm are shorter than expected, and the interatomic angles sum up to 119.4.

The calculations of the charge distribution of the atoms in $La_4B_4O_{11}F_2$ via bond valence sums (ΣV) with ValList (bond valence calculation and listing) [42] were performed and confirm the formal valence states in the fluoride borate (Table 5). Slight deviations, for example for O22, occur frequently when calculating bond valence sums with the bond-length/bond-strength concept and are evoked by larger-than-average bond lengths of the corresponding atoms.

To assure the atom assignment in the structure, single crystals of our sample were subjected to elemental analysis via SEM/EDX experiments. The crystals showed average atomic La:B:O:F compositions (%) of 16:16:53:15. Due to the light weight of boron, measurements have to be taken with caution, but still, these results confirm the presence of all elements and the composition, obtained from the single crystal structure determination (calculated values (%) La:B:O:F: 19.0:19.0:52.5:9.5).

Furthermore, we calculated the Madelung Part of Lattice Energy (MAPLE) [43–45] for $La_4B_4O_{11}F_2$ in order to compare it with the sum of the MAPLE values of the ambient-temperature modification of La_2O_3 [46], LaF_3 [47], and of the high-pressure modification of B_2O_3 (B_2O_3-II) [48]. The additive potential of the MAPLE values allows the calculation of a hypothetical value for $La_4B_4O_{11}F_2$, starting from binary oxides and fluorides. As result, we obtained a value of 70766 kJ/mol in comparison to 71136 kJ/mol (deviation: 0.5%), starting from the binary components

[5/3 × La_2O_3 (14234 kJ/mol)+2 × B_2O_3-II (21938 kJ/mol)+2/3 × LaF_3 (5306 kJ/mol)].

3.2. FTIR spectroscopy

In Figs. 9 and 10, the FTIR-ATR spectra of $La_4B_4O_{11}F_2$ are displayed. The wavenumbers of ATR-bands are given in Table 6. In the range between 600 and 1600 cm^{-1}, four main groups of

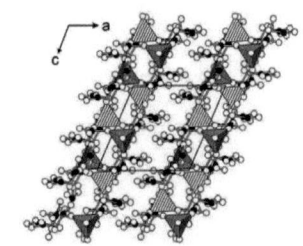

Fig. 3. Sheet-like arrangement of fundamental building blocks in $La_4B_4O_{11}F_2$, viewing along [0$\bar{1}$0].

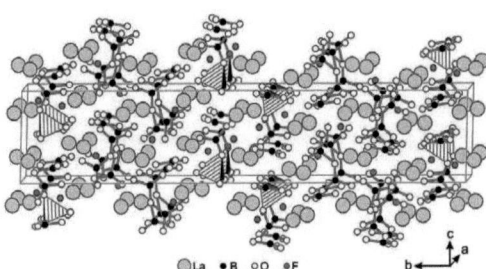

Fig. 1. Crystal structure of $La_4B_4O_{11}F_2$, consisting of La^{3+} and F^- ions, BO_3-groups, and BO_4-tetrahedra.

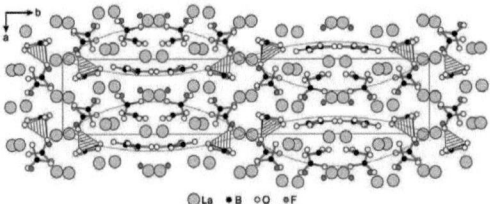

Fig. 2. Wave-like modulations in the crystal structure of $La_4B_4O_{11}F_2$, showing a wavelength of $\lambda = b$.

4.2 Rare-Earth Fluoride and Fluorido Borates

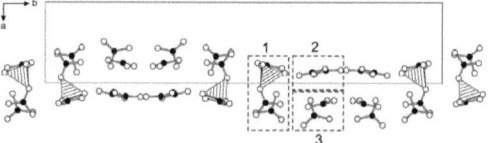

Fig. 4. Location of the FBBs in the crystal structure of $La_4B_6O_{11}F_2$. Each wave is built up from a set of three FBB arrangements (dashed lines).

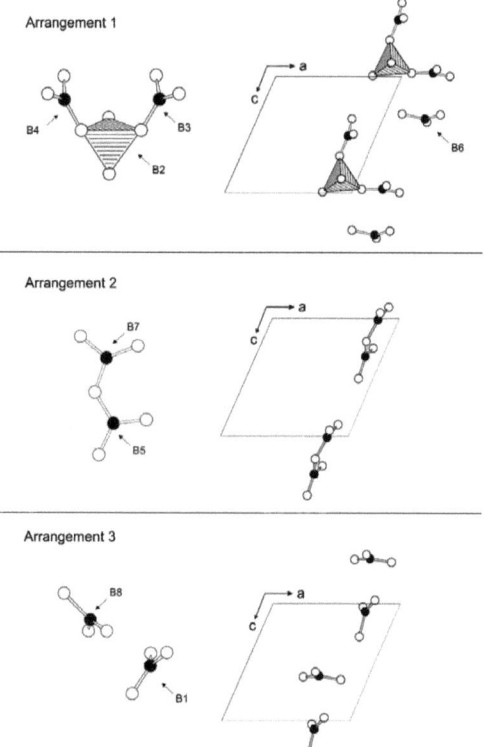

Fig. 5. Detailed view of the FBB arrangements in $La_4B_6O_{11}F_2$. Top: arrangement 1, showing the FBB $\triangle\square\triangle$ (left). Between each FBB, an isolated BO_3-group is positioned (right). Middle: detailed view of arrangement 2, showing the corner-sharing BO_3-groups of the FBB $\triangle\triangle$. Bottom: detailed view of arrangement 3, showing two isolated BO_3-groups in an orthogonal orientation.

bands could be distinguished: bands around 700 cm^{-1} are typical for in-plane and out-of-plane bending vibrations of BO$_3$-groups; however, in rare-earth metaborates, bands at 725 and 662 cm^{-1} were taken as indication for both three- and fourfold coordinated boron [16,49–51]. Between 900 and 1100 cm^{-1}, stretching vibrations of tetrahedrally coordinated boron atoms are expected. The strong bands in the range of 1200–1500 cm^{-1} are indicative of trigonal borate groups. From 3300 to 3550 cm^{-1}, several weak bands could be detected, suggesting a substitution of trace amounts of hydroxyl groups for fluoride.

3.3. Raman spectroscopy

The Raman spectrum and the corresponding wavenumbers of bands of a La$_4$B$_4$O$_{11}$F$_2$ single crystal are displayed in Fig. 11 and Table 7. In total, 40 bands in the range of 100–1600 cm^{-1} could be detected. In contrast to FTIR, no bands were observed between 3000 and 4000 cm^{-1}, which is probably related to the comparatively low sensitivity of Raman spectroscopy for hydroxyl vibrational modes. The Raman spectrum will be compared to the polymorph Gd$_4$B$_4$O$_{11}$F$_2$; the band assignments were made in analogy to structurally and chemically similar borate compounds [5,52–56].

In agreement with the ATR-spectra, the Raman spectrum of La$_4$B$_4$O$_{11}$F$_2$ showed four groups of bands; the three most intense bands occurred around 300 cm^{-1}. In Gd$_4$B$_4$O$_{11}$F$_2$, which contains corner-sharing BO$_3$-groups and BO$_4$-tetrahedra, intense bands were observed below 200 cm^{-1} and assigned to cation-oxygen and cation-fluorine bonds, as well as complex lattice vibrations. It has been speculated that these bands are dominantly related to the fluoride ions, which are all fourfold-coordinated by Gd^{3+} ions in Gd$_4$B$_4$O$_{11}$F$_2$. In La$_4$B$_4$O$_{11}$F$_2$, the fluoride ions are both three- and fourfold coordinated and the range of bond lengths (248.7–307.0 pm) is larger than in Gd$_4$B$_4$O$_{11}$F$_2$ (244.8–283.7 pm). This might account for the band differences in this area.

For La$_4$B$_4$O$_{11}$F$_2$, the area of 400–800 cm^{-1} is characterized by 13 weaker bands. The bending and the pulse vibrations of BO$_3$-groups and BO$_4$-tetrahedra are expected to be the main vibrational modes contributing to these bands.

Bands between 800 and 1100 cm^{-1} are most frequently assigned to stretching vibrations of the BO$_4$-tetrahedra. Three intense bands at 825, 916, and 960 cm^{-1} are observed here in contrast to Gd$_4$B$_4$O$_{11}$F$_2$, which shows only one intense band at 959 cm^{-1}. Bands in the range from 1100 to 1600 cm^{-1} are attributed to stretching vibrations of the BO$_3$-groups. For La$_4$B$_4$O$_{11}$F$_2$, at least 6 intense bands were detected compared to only two for Gd$_4$B$_4$O$_{11}$F$_2$. The most probable explanation for these differences are the different building blocks of the two

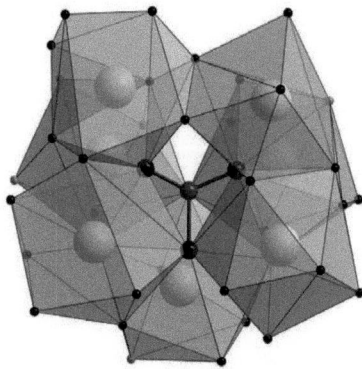

Fig. 6. Surrounding of the isolated BO$_3$-group (B8) in La$_4$B$_4$O$_{11}$F$_2$.

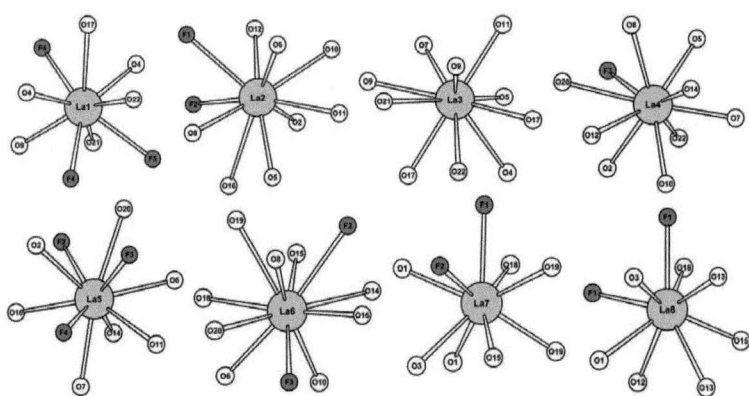

Fig. 7. Coordination spheres of the La^{3+} ions in La$_4$B$_4$O$_{11}$F$_2$.

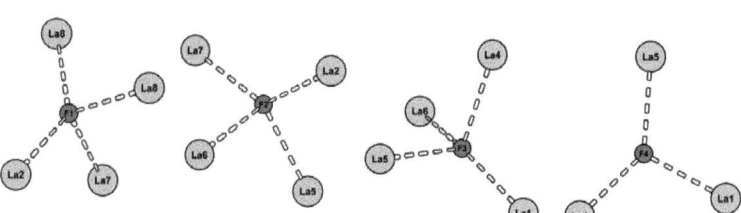

Fig. 8. Coordination of the fluoride ions in $La_4B_4O_{11}F_2$.

Table 5
Bond valence sums (ΣV) in $La_4B_4O_{11}F_2$, calculated with ValList [42].

	La1	La2	La3	La4	La5	La6	La7	La8
ΣV	3.19	3.19	2.94	3.29	2.89	2.80	3.28	3.11
	B1	B2	B3	B4	B5	B6	B7	B8
ΣV	2.94	2.87	2.93	2.95	2.85	2.79	2.67	2.69
	O1	O2	O3	O4	O5	O6	O7	O8
ΣV	2.10	2.03	2.03	2.07	2.20	2.13	1.95	2.15
	O9	O10	O11	O12	O13	O14	O15	O16
ΣV	−2.09	−2.14	−1.77	−1.97	−1.96	−1.81	−2.09	−1.95
	O17	O18	O19	O20	O21	O22		
ΣV	−1.95	−1.97	−2.01	−1.87	−2.01	−1.67		
	F1	F2	F3	F4				
ΣV	−0.87	−0.89	−0.75	−0.95				

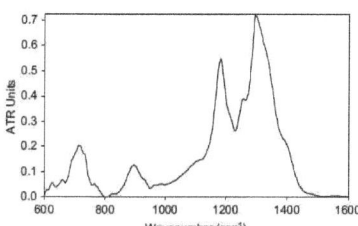

Fig. 9. FTIR-ATR spectrum of $La_4B_4O_{11}F_2$ in the range 600–1600 cm^{-1}.

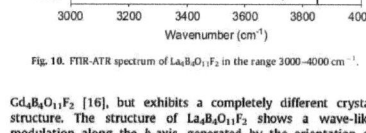

Fig. 10. FTIR-ATR spectrum of $La_4B_4O_{11}F_2$ in the range 3000–4000 cm^{-1}.

compounds and the larger range of B–O bond distances in the BO$_3$-groups ($La_4B_4O_{11}F_2$: 134.9–149.0 pm; $Gd_4B_4O_{11}F_2$: 136.1–140.7 pm) and the BO$_4$-tetrahedra ($La_4B_4O_{11}F_2$: 137.3–164.2 pm; $Gd_4B_4O_{11}F_2$: 141.3–159.0 pm).

4. Conclusions

In this article, we described the high-pressure synthesis and crystal structure of the new lanthanum fluoride borate $La_4B_4O_{11}F_2$. It shows the same composition as the fluoride borate $Gd_4B_4O_{11}F_2$ [16], but exhibits a completely different crystal structure. The structure of $La_4B_4O_{11}F_2$ shows a wave-like modulation along the b-axis, generated by the orientation of trigonal BO$_3$-groups and BO$_4$-tetrahedra. Elemental analysis confirmed the presence and quantity of fluoride ions in the structure. Single-crystal IR-/Raman spectroscopy complete the structural characterization of the new fluoride borate.

For the future, experiments on the formation range of the compounds $RE_4B_4O_{11}F_2$ shall follow. It will be of great interest to see, which crystal structure is favored by the rare-earth ions of intermediate size and whether it will be possible to obtain both structure types as polymorphs for one and the same cation.

Table 6
Wavenumbers and possible assignment of FTIR-absorption bands in the spectrum of $La_4B_4O_{11}F_2$.

Band	assignment	Band	assignment
629	$\delta(BO_3)$	895	
		932	
658	$\delta(BO_3)$, $\delta(BO_4)$	974	
		1093	$\nu(BO_4)$
		1161	
700	$\delta(BO_3)$	1182	
		1217	
729	$\delta(BO_3)$, $\delta(BO_4)$	1253	
		1294	$\nu(BO_3)$
		1325	
770	$\delta(BO_3)$	1401	
		3302	
		3435	
		3506	$\nu(OH)$
		3548	

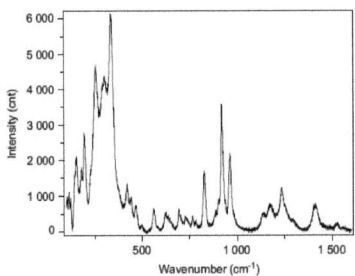

Fig. 11. Raman spectrum of $La_4B_4O_{11}F_2$ in the range 100–1600 cm^{-1}.

Table 7
Wavenumbers and possible assignment of Raman bands in the spectrum of $La_4B_4O_{11}F_2$.

Band	assignment	Band	assignment
104		699	bending (BO_4)
115		728	
125	Lattice	733	pulse vibration (BO_3)
145		764	
153	La-O, F-O	779	
179	bending, stretching (BO_3)	825	
197		834	
228		899	
254		916	stretching (BO_4)
299		923	
338		960	
373		973	
421		1134	
442		1173	
467	bending (BO_4)	1233	
497		1257	
562	pulse vibration (BO_3)	1297	stretching (BO_3)
622		1411	
637		1494	
692		1525	

Acknowledgments

We thank Dr. Gunter Heymann and Dr. Peter Mayer (LMU München) for collecting the single-crystal data. Special thanks go to Prof. Dr. W. Schnick (LMU München) for his continuous support of these investigations and to Prof. Dr. E. Bertel (LFU Insbruck) for useful discussions.

Appendix A. Supporting Materials

Supplementary data associated with this article can be found in the online version at doi:10.1016/j.jssc.2010.06.019.

References

[1] H. Huppertz, B. von der Eltz, J. Am. Chem. Soc. 124 (2002) 9376.
[2] H. Huppertz, Z. Naturforsch. B 58 (2003) 278.
[3] H. Huppertz, H. Emme, J. Phys.: Condens. Matter 16 (2004) 1283.
[4] H. Emme, H. Huppertz, Z. Anorg. Allg. Chem. 628 (2002) 2165.
[5] H. Emme, H. Huppertz, Chem. Eur. J. 9 (2003) 3623.
[6] H. Emme, H. Huppertz, Acta Crystallogr. C 61 (2005) i29.
[7] H. Emme, H. Huppertz, Acta Crystallogr. C 61 (2005) i23.
[8] H. Huppertz, S. Altmannshofer, G. Heymann, J. Solid State Chem. 170 (2003) 320.
[9] H. Emme, M. Valldor, R. Pöttgen, H. Huppertz, Chem. Mater. 17 (2005) 2707.
[10] A. Haberer, G. Heymann, H. Huppertz, J. Solid State Chem. 180 (2007) 1595.
[11] G. Corbel, R. Retoux, M. Leblanc, J. Solid State Chem. 139 (1998) 52.
[12] E. Antic-Fidancev, G. Corbel, N. Mercier, M. Leblanc, J. Solid State Chem. 153 (2000) 270.
[13] H. Müller-Bunz, Th. Schleid, Z. Anorg. Allg. Chem. 628 (2002) 2750.
[14] A. Haberer, H. Huppertz, J. Solid State Chem. 182 (2009) 888.
[15] A. Haberer, R. Kaindl, H. Huppertz, Solid State Sci. 12 (2010) 515.
[16] A. Haberer, R. Kaindl, H. Huppertz, J. Solid State Chem. 183 (2010) 471.
[17] K. Kazmierczak, H. A. Höppe, Eur. J. Inorg. Chem. (2010) DOI:10.1002/ejic.201000105.
[18] J. S. Knyrim, F. Roeßner, S. Jakob, D. Johrendt, I. Kinski, R. Glaum, H. Huppertz, Angew. Chem. 119 (2007) 9256; Angew. Chem. Int. Ed. 46 (2007) 9097.
[19] S.C. Neumair, R. Glaum, H. Huppertz, Z. Naturforsch. B 64 (2009) 883.
[20] P.C. Burns, J.D. Grice, F.C. Hawthorne, Can. Mineral. 33 (1995) 1131.
[21] D. Walker, M.A. Carpenter, C.M. Hitch, Am. Mineral. 75 (1990) 1020.
[22] D. Walker, Am. Mineral. 76 (1991) 1092.
[23] H. Huppertz, Z. Kristallogr. 219 (2004) 330.
[24] D.C. Rubie, Phase Transitions 68 (1999) 431.
[25] N. Kawai, S. Endo, Rev. Sci. Instrum. 8 (1970) 1178.
[26] H. Emme, C. Despotopoulou, H. Huppertz, Z. Anorg. Allg. Chem. 630 (2004) 2450.
[27] L.J. Farrugia, J. Appl. Cryst. 32 (1999) 837.
[28] Z. Otwinowski, W. Minor, Methods Enzymol 276 (1997) 307.
[29] G.M. Sheldrick, SHELXS-97 and SHELXL-97, Program suite for the solution and refinement of crystal structures, University of Göttingen, Göttingen, Germany, 1997.
[30] G.M. Sheldrick, Acta Crystallogr. A 64 (2008) 112.
[31] F.M. Mirabella Jr. (ed.), Internal Reflection Spectroscopy, Theory and Applications, Marcel Dekker, Inc. (1992) p. 276.
[32] B.G. Hyde, A.N. Bagshaw, S. Andersson, M. O'Keeffe, Ann. Rev. Mater. Sci. 4 (1974) 43.
[33] D.J.M. Bevan, J. Mohyla, B.F. Hoskins, R.J. Steen, Eur. J. Solid State Inorg. Chem. 27 (1990) 451.
[34] J.P. Laval, A. Taoudi, A. Abaouz, B. Frit, J. Solid State Chem. 119 (1995) 125.
[35] H. Müller-Bunz, O. Janka, Th. Schleid, Z. Anorg. Allg. Chem. 633 (2007) 37.
[36] A. Vegas, Acta Crystallogr. B 33 (1977) 3607.
[37] H. Huppertz, Z. Naturforsch. B 56 (2001) 697.
[38] J. Krogh-Moe, M. Ihara, Acta Crystallogr. B 25 (1969) 2153.
[39] K. Machida, H. Hata, K. Okuno, G. Adachi, J. Shiokawa, J. Inorg. Nucl. Chem. 41 (1979) 1425.
[40] L. Li, X. Jin, G. Li, Y. Wang, F. Liao, G. Yao, J. Lin, Chem. Mater. 15 (2003) 2253.
[41] M. Mansmann, Z. Kristallogr. 122 (1965) 375.
[42] A. S. Wills, VaList Version 3.0.13, University College London, UK, 1998–2008. Program available from www.ccp14.ac.uk.
[43] R. Hoppe, Angew. Chem. 78 (1966) 52; Angew. Chem. Int. Ed. 5 (1966) 96.
[44] R. Hoppe, Angew. Chem. 82 (1970) 7; Angew. Chem. Int. Ed. 9 (1970) 25.
[45] R. Hübenthal, MAPLE - Program for the Calculation of MAPLE Values, Vers. 4, University of Gießen, Gießen, Germany, 1993.
[46] N. Hirosaki, S. Ogata, C. Kocer, J. Alloys Compd. 351 (2003) 31.
[47] E. Staritzky, D.B. Asprey, Anal.Chem. 29 (1957) 856.

[48] C.T. Prewitt, R.D. Shannon, Acta Crystallogr. B 24 (1968) 869.
[49] J.P. Laperches, P. Tarte, Spectrochim. Acta. 22 (1966) 1201.
[50] G. Heymann, K. Beyer, H. Huppertz, Z. Naturforsch. B 59 (2004) 1200.
[51] C.E. Weir, R.A. Schroeder, J. Res. Nat. Bur. Stands 68A (1964) 465.
[52] L. Jun, X. Shuping, G. Shiyang, Spectrochim. Acta A 51 (1995) 519.
[53] G. Chadeyron, M. El-Ghozzi, R. Mahiou, A. Arbus, J.C. Cousseins, J. Solid State Chem. 128 (1997) 261.
[54] H. Huppertz, J. Solid State Chem. 177 (2004) 3700.
[55] C. Zhang, Y.H. Wang, X. Guo, J. Lumin. 122–123 (2007) 980.
[56] G. Padmaja, P. Kistaiah, J. Phys. Chem. A 113 (2009) 2397.

4.2 Rare-Earth Fluoride and Fluorido Borates

4.2.7 Pr$_4$B$_3$O$_{10}$F

With Pr$_4$B$_3$O$_{10}$F, the first praseodymium fluoride borate could be obtained at 3 GPa and 800 °C. Its crystal structure shows an interesting feature: an intermediate state between a BO$_3$-group and a BO$_4$-tetrahedron. This result gives insight in the pressure transformation of the boron coordination sphere, which is a typical issue in high-pressure research on borates. The green crystals of Pr$_4$B$_3$O$_{10}$F (Figure 4.2-7) were further investigated by means of IR and Raman spectroscopy, elemental analysis, and temperature-dependent X-ray spectroscopy. The results can be found in the following publication.

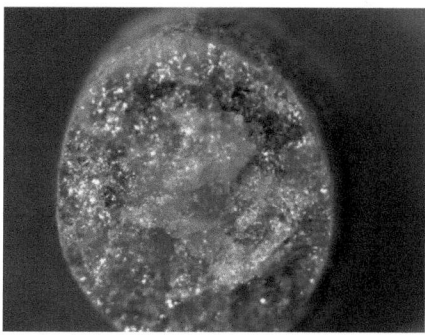

Figure 4.2-7
Crystalline sample of Pr$_4$B$_3$O$_{10}$F

146 4 Experimental Part

Solid State Sciences 12 (2010) 515-521

Contents lists available at ScienceDirect

Solid State Sciences

journal homepage: www.elsevier.com/locate/ssscie

The first praseodymium fluoride borate $Pr_4B_3O_{10}F$ – High-pressure synthesis and characterization

Almut Haberer[a], Reinhard Kaindl[b], Hubert Huppertz[a],*

[a] Institut für Allgemeine, Anorganische und Theoretische Chemie, Leopold-Franzens-Universität Innsbruck, Innrain 52a, A-6020 Innsbruck, Austria
[b] Institut für Mineralogie und Petrographie, Leopold-Franzens-Universität Innsbruck, Innrain 52, A-6020 Innsbruck, Austria

ARTICLE INFO

Article history:
Received 29 November 2009
Received in revised form
13 December 2009
Accepted 17 December 2009
Available online 28 December 2009

Keywords:
High-pressure phase
Fluoride borate
Crystal structure

ABSTRACT

The first praseodymium fluoride borate $Pr_4B_3O_{10}F$ was synthesized under high-pressure/high-temperature conditions in a Walker-type multianvil apparatus at 3 GPa and 800 °C, starting from stoichiometric mixtures of Pr_6O_{11}, B_2O_3, and PrF_3. $Pr_4B_3O_{10}F$ crystallizes in space group $P\bar{1}$ with $Z = 2$ and lattice parameters $a = 662.8(5)$ pm, $b = 882.5(4)$ pm, and $c = 894.1(5)$ pm, with $\alpha = 106.77(4)°$, $\beta = 108.67(3)°$, $\gamma = 104.92(4)°$ ($R1 = 0.0235$ and $wR2 = 0.0537$ (all data)). The crystal structure is built up exclusively from isolated BO_3-groups. One of the boron atoms is located on a split position, leading to a BO_3-group, that can be orientated in the crystal structure in two different ways. This can be interpreted as a special case of an intermediate state between a BO_3-group and a BO_4-tetrahedron. The fluoride position is occupied to a small percentage by OH^- ions.

© 2009 Elsevier Masson SAS. All rights reserved.

1. Introduction

The possibility to obtain a large variety of new compounds under high-pressure/high-temperature conditions has been extensively examined in borates, for example the rare-earth borates $RE_4B_6O_{15}$ (RE = Dy, Ho) [1–3], α-$RE_2B_4O_9$ (RE = Sm–Ho) [4–6], β-$RE_2B_4O_9$ (RE = Gd [7], Dy [8]), $RE_3B_5O_{12}$ (RE = Er–Lu) [9], $Pr_4B_{10}O_{21}$ [10], and the meta-borates β-$RE(BO_2)_3$ (RE = Tb–Lu) [11–13], γ-$RE(BO_2)_3$ (RE = La–Nd) [14], and δ-$RE(BO_2)_3$ (RE = La, Ce) [15,16].

Borates, being glass formers in general, show an increased willingness to crystallize under pressure, which seems to be valid for fluoroborates, too. Concerning structural motifs, the chemistry of rare-earth fluoride borates under high-pressure/high-temperature conditions generally shows similarities to the chemistry of oxoborates. There are trigonal-planar BO_3-groups and BO_4-tetrahedra, whose pressure-induced transformation – the former into the latter ones – can be observed. That leads to crystal structures, where intermediate states between these two structural features appear, e.g. in $M_6B_{22}O_{39} \cdot H_2O$ (M = Fe, Co) [17].

Until recently, rare-earth fluoride borates were only represented by the compounds $RE_3(BO_3)_2F_3$ (RE = Sm, Eu, Gd) [18] and $Gd_2(BO_3)F_3$ [19], synthesized by heating a stoichiometric mixture of RE_2O_3, B_2O_3, and REF_3 under ambient pressure conditions. Under high-pressure/high-temperature conditions, the compounds $Gd_4B_4O_{11}F_2$ and $Yb_5(BO_3)_2F_9$ could be added to this field by our group [20,21].

While the other rare-earth fluoride borates, above mentioned, have only BO_3-groups, $Gd_4B_4O_{11}F_2$ exhibits a BO_4-group as well. Due to a strong distortion, this BO_4-tetrahedron can also be regarded as an intermediate between a trigonal-planar and a fully tetrahedral coordination sphere [20]. The application of higher pressure on $Yb_5(BO_3)_2F_9$, showing only BO_3-groups, yielded a structure built up solely of BO_4-tetrahedra [22].

In this article, we report the first praseodymium fluoride borate $Pr_4B_3O_{10}F$, obtained by high-pressure/high-temperature synthesis. In the crystal structure of $Pr_4B_3O_{10}F$, only isolated BO_3-groups are found. One of the boron atoms is located on a split position, leading to a BO_3-group, that can be orientated in the crystal structure in two different ways. This can also be interpreted as a special case of an intermediate state between a BO_3-group and a BO_4-tetrahedron. In the following, we present the synthesis, crystal structure, thermal behaviour, and spectroscopic properties of the new high-pressure compound.

2. Experimental section

2.1. Synthesis

To synthesize the compound $Pr_4B_3O_{10}F$, high-pressure/high-temperature conditions of 3 GPa and 800 °C were applied, starting from Pr_6O_{11} (Strem Chemicals, 99.9%), B_2O_3 (Strem Chemicals, 99.9+%), and PrF_3 (Strem Chemicals, 99.9%) at equal molar ratios.

---

* Corresponding author. Fax: +43 5125072934.
E-mail address: hubert.huppertz@uibk.ac.at (H. Huppertz).

1293-2558/$ – see front matter © 2009 Elsevier Masson SAS. All rights reserved.
doi:10.1016/j.solidstatesciences.2009.12.017

4.2 Rare-Earth Fluoride and Fluorido Borates

The starting materials were ground up and filled into a boron nitride crucible (Henze BNP GmbH, HeboSint® S100, Kempten, Germany). The compounds were compressed and heated via a multianvil assembly, based on a Walker-type module and a 1000 ton press (both devices from the company Voggenreiter, Mainleus, Germany). Eight tungsten carbide cubes (TSM-20, Ceratizit, Reutte, Austria) compressed the boron nitride crucible, which was positioned inside the centre of an 18/11 assembly. A detailed description of the assembly can be found in reference [23].

The 18/11 assembly was compressed up to 3 GPa in 1.25 h and heated to 800 °C (cylindrical graphite furnace) in the following 10 min, kept there for 60 min, and cooled down to 600 °C in 20 min at constant pressure. After natural cooling down to room temperature by switching off heating, a decompression period of 3.75 h was required. Then the recovered octahedral pressure medium (MgO, Ceramic Substrates & Components Ltd., Newport, Isle of Wight, UK) was broken apart and the sample carefully separated from the surrounding graphite and boron nitride. Green air- and water-resistant crystalline platelets of $Pr_4B_3O_{10}F$ emerged.

2.2. Crystal structure analysis

The sample characterization was performed by powder X-ray diffraction, carried out in transmission geometry on a flat sample of the reaction product, using a STOE STADI P powder diffractometer with $MoK_{\alpha 1}$ radiation (Ge monochromator, $\lambda = 71.073$ pm). The powder pattern showed reflections of $Pr_4B_3O_{10}F$ and of PrOF [24] as a by-product of the synthesis (Fig. 1). The experimental powder pattern tallied well with the theoretical pattern of $Pr_4B_3O_{10}F$, simulated from single-crystal data. The diffraction pattern was indexed with the program ITO [25] on the basis of a triclinic unit cell. The lattice parameters were calculated from least-squares fits of the powder data. The correct indexing of the pattern of $Pr_4B_3O_{10}F$ was confirmed by intensity calculations, taking the atomic positions from the structure refinement [26]. Indexing the reflections of the praseodymium fluoride borate, we got the parameters $a = 662.8(5)$ pm, $b = 882.5(4)$ pm, and $c = 894.1(5)$ pm, with $\alpha = 106.77(4)°$, $\beta = 108.67(3)°$, $\gamma = 104.92(4)°$, and a volume of 437.7(3) Å^3. This confirmed the lattice parameters, obtained from single-crystal X-ray diffraction (Table 1).

Single crystal intensity data of $Pr_4B_3O_{10}F$ were measured with an Enraf-Nonius Kappa CCD with graphite monochromatized MoK_{α} ($\lambda = 71.073$ pm) radiation. Afterwards, a multi-scan absorption correction was applied to the data (Scalepack [27]). Structure solution and parameter refinement (full-matrix least-squares against F^2) were successfully performed, using the SHELX-97 software suite [28,29] with anisotropic atomic displacement parameters for all atoms. The partial replacement of fluorine by OH$^-$ was not accounted for in the structure solution, since a differentiation, based on X-ray diffraction, is not possible. According to the missing extinctions, the triclinic space groups $P1$ and $P\bar{1}$ were derived. The structure solution in $P\bar{1}$ (No. 2) succeeded. The final difference

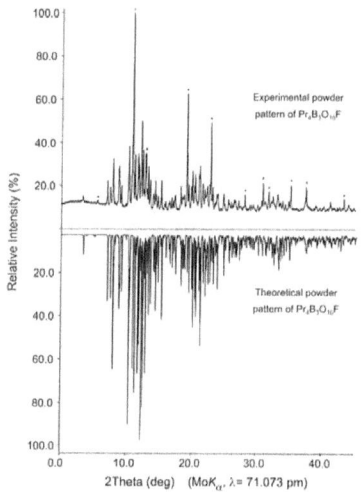

Fig. 1. Top: Experimental powder pattern of $Pr_4B_3O_{10}F$; reflections of PrOF are indicated with asterisks. Bottom: theoretical powder pattern of $Pr_4B_3O_{10}F$, based on single-crystal diffraction data.

Table 1
Crystal data and structure refinement of $Pr_4B_3O_{10}F$.

Empirical Formula	$Pr_4B_3O_{10}F$
Molar mass (g mol^{-1})	775.07
Crystal system	triclinic
Space group	$P\bar{1}$ (No. 2)
Powder diffractometer	Stoe StadiP
Radiation	$MoK_{\alpha 1}$ ($\lambda = 71.073$ pm)
Lattice parameters from powder data	
a (pm)	662.8(5)
b (pm)	882.5(4)
c (pm)	894.1(5)
α (deg)	106.77(4)
β (deg)	108.67(3)
γ (deg)	104.92(4)
Volume (Å^3)	437.7(3)
Single-crystal data	
Single-crystal diffractometer	Bruker AXS/Nonius Kappa CCD
Radiation	$MoK_{\alpha 1}$ ($\lambda = 71.073$ pm)
a (pm)	662.03(1)
b (pm)	881.06(2)
c (pm)	892.35(2)
α (deg)	106.93(1)
β (deg)	108.66(1)
γ (deg)	104.82(1)
Volume (Å^3)	435.4(1)
Formula units per cell	$Z = 2$
Temperature (K)	293(2)
Calculated density (g cm^{-3})	5.915
Crystal size (mm^3)	0.03 × 0.03 × 0.03
Absorption coefficient (mm^{-1})	22.05
$F(000)$	680
θ range (deg)	$3.4 \leq \theta \leq 35.0$
Range in h k l	±10, ±14, ±14
Total no. reflections	13 243
Independent reflections	3815 ($R_{int} = 0.0253$)
Reflections with $I > 2\sigma(I)$	3711 ($R_{\sigma} = 0.0195$)
Data/parameters	3815/174
Absorption correction	Multi-scan [27]
Goodness-of-fit (F^2)	1.055
Final R indices ($I > 2\sigma(I)$)	$R1 = 0.0228$
	$wR2 = 0.0532$
R indices (all data)	$R1 = 0.0235$
	$wR2 = 0.0537$
Largest differ. peak / deepest hole (e/Å^{-3})	4.09/−3.62

Table 2
Atomic coordinates and isotropic equivalent displacement parameters U_{eq} (Å^2) for Pr$_4$B$_3$O$_{10}$F (space group: $P\bar{1}$). U_{eq} is defined as one third of the trace of the orthogonalized U_{ij} tensor. Wyckoff sites for all atoms: 2i.

Atom	x	y	z	U_{eq}	Site occupancy
Pr1	0.31984 (3)	0.53371 (2)	0.64833 (2)	0.00887 (4)	1.00
Pr2	0.32846 (3)	0.16323 (2)	0.81599 (2)	0.00969 (4)	1.00
Pr3	0.99177 (3)	0.74999 (3)	0.85469 (2)	0.01458 (5)	1.00
Pr4	0.23774 (4)	0.76856 (3)	0.36476 (3)	0.01704 (5)	1.00
B1	0.7402 (6)	0.4987 (4)	0.9851 (5)	0.0094 (5)	1.00
B2	0.5704 (7)	0.9265 (5)	0.8353 (5)	0.0144 (6)	1.00
B3a	0.747 (2)	0.856 (2)	0.587 (2)	0.010 (3)	0.42 (3)
B3b	0.814 (3)	0.863 (2)	0.537 (2)	0.044 (5)	0.58 (3)
O1	0.1104 (5)	0.2400 (4)	0.5951 (4)	0.0186 (5)	1.00
O2	0.9126 (5)	0.6608 (3)	0.0771(4)	0.0141(4)	1.00
O3	0.6705 (5)	0.0966 (4)	0.8623 (4)	0.0186 (5)	1.00
O4	0.2999 (6)	0.1481 (5)	0.0802 (4)	0.0237 (7)	1.00
O5	0.6699 (5)	0.3999 (3)	0.0664 (3)	0.0158 (5)	1.00
O6	0.6306 (5)	0.4340 (4)	0.8068 (3)	0.0145 (5)	1.00
O7	0.3363 (5)	0.8399 (4)	0.7350 (6)	0.0310 (9)	1.00
O8	0.6495 (5)	0.9262 (3)	0.4669 (3)	0.0135 (4)	1.00
O9	0.6129 (6)	0.6816 (5)	0.5659 (4)	0.0236 (6)	1.00
O10	0.9609 (5)	0.9526 (3)	0.7163 (3)	0.0142 (4)	1.00
F1	0.0737 (4)	0.4977 (3)	0.3687 (3)	0.0177 (4)	1.00

Fourier syntheses did not reveal any significant residual peaks in all refinements. The positional parameters of the refinements, anisotropic displacement parameters, interatomic distances, and interatomic angles are listed in the Tables 2–5. Further information of the crystal structure is available from the Fachinformationszentrum Karlsruhe (crysdata@fiz-karlsruhe.de), D-76344 Eggenstein-Leopoldshafen (Germany), quoting the Registry No. CSD-421268.

2.3. Temperature programmed X-ray powder diffraction experiments

In situ X-ray powder diffraction experiments were done on a STOE STADI P powder diffractometer [MoK$_{\alpha 1}$ radiation ($\lambda = 71.073$ pm)] with a computer controlled STOE furnace: The sample was enclosed in a silica capillary and heated from room temperature to 500 °C in 100 °C steps, and from 500 °C to 1100 °C in 50 °C steps. The heating rate was set to 40 °C/min. Afterwards, the sample was cooled down to 500 °C in 50 °C steps, and from 500 °C to room temperature in 100 °C steps (cooling rate: 50 °C/min). After each heating step, a diffraction pattern was recorded over the angular range 6° ≤ 2θ ≤ 30°.

2.4. Scanning electron microscopy

Scanning electron microscopy was performed on a JEOL JSM-6500F, equipped with a field emission gun at an acceleration voltage of 15 kV. The samples were prepared by placing single crystals on adhesive conductive pads and subsequently coating them with a thin conductive carbon film. Each EDX spectrum (Oxford Instruments) was recorded with the analyzed area limited on one single crystal to avoid the influence of possible contaminating phases.

2.5. IR spectroscopy

The infrared spectrum of Pr$_4$B$_3$O$_{10}$F was recorded on a Nicolet 5700 FT-IR spectrometer, scanning a range from 400 to 4000 cm^{-1}. Before measuring, the sample was thoroughly dried under high vacuum for several days.

2.6. Raman spectroscopy

Confocal Raman spectra of single crystals were obtained with a HORIBA JOBIN YVON LabRam-HR 800 Raman micro-spectrometer. The sample was excited by the 633 nm emission line of a 17 mW He–Ne-laser and an OLYMPUS 100× objective (N.A. = 0.9). The size and power of the laser spot on the surface were approximately 1 µm and 5 mW. The spectral resolution, determined by measuring the Rayleigh line, was about 1.2 cm^{-1}. The dispersed light was collected by a 1024 × 256 open electrode CCD detector. The spectra were recorded without polarizer for the laser and the scattered Raman light. The Raman bands were fitted by the built-in spectrometer software LabSpec to convoluted Gauss-Lorentz functions. The accuracy of Raman line shifts, calibrated by regular measuring the Rayleigh line, was in the order of 0.5 cm^{-1}.

3. Results and discussion

The crystal structure of Pr$_4$B$_3$O$_{10}$F consists of isolated BO$_3$-groups, praseodymium cations, and fluoride anions (Fig. 2). The fluoride position is occupied to a small percentage by OH$^-$ ions, which were detected via IR spectroscopy (vide infra). The problem of mixed halogenide-hydroxyl sites is commonly known, e.g. in apatite-structured compounds like Mn$_5$(PO$_4$)$_3$Cl$_{0.9}$(OH)$_{0.1}$ [30] or the rare-earth fluoride borate Gd$_4$B$_4$O$_{11}$F$_2$ [20]. The title compound

Table 3
Anisotropic equivalent displacement parameters (U_{ij}/Å^2) for Pr$_4$B$_3$O$_{10}$F (space group: $P\bar{1}$).

Atom	U_{11}	U_{22}	U_{33}	U_{12}	U_{13}	U_{23}
Pr1	0.01053 (8)	0.00921 (7)	0.00690 (7)	0.00428 (5)	0.00366 (5)	0.00295 (5)
Pr2	0.00817 (7)	0.00967 (7)	0.00954 (7)	0.00275 (5)	0.00327 (6)	0.00283 (5)
Pr3	0.01199 (8)	0.0239 (2)	0.01037 (8)	0.00962 (7)	0.00533 (6)	0.00732 (7)
Pr4	0.01256 (9)	0.01178 (8)	0.01655 (9)	−0.00123 (6)	0.00452 (7)	−0.00064 (6)
B1	0.009 (2)	0.008 (2)	0.008 (2)	0.002 (1)	0.002 (2)	0.003 (2)
B2	0.009 (2)	0.015 (2)	0.017 (2)	0.006 (2)	0.002 (2)	0.004 (2)
B3a	0.009 (4)	0.004 (4)	0.011 (5)	−0.002 (2)	0.000 (3)	0.004 (3)
B3b	0.058 (9)	0.065 (8)	0.008 (3)	0.055 (7)	0.003 (4)	−0.001 (4)
O1	0.011 (2)	0.014 (2)	0.024 (2)	0.0006 (9)	−0.002 (2)	0.012 (2)
O2	0.013 (2)	0.010 (2)	0.016 (2)	0.0010 (8)	0.0040 (9)	0.0048 (8)
O3	0.011 (2)	0.013 (2)	0.023 (2)	0.0017 (9)	0.004 (2)	0.001 (2)
O4	0.021 (2)	0.048 (2)	0.026 (2)	0.025 (2)	0.017 (2)	0.029 (2)
O5	0.021 (2)	0.011 (1)	0.009 (9)	−0.0024 (9)	0.0061 (9)	0.0032 (8)
O6	0.020 (2)	0.018 (2)	0.0075 (9)	0.012 (2)	0.0048 (9)	0.0046 (8)
O7	0.010 (2)	0.014 (2)	0.054 (2)	0.002 (2)	−0.003 (2)	0.016 (2)
O8	0.014 (2)	0.011 (2)	0.013 (2)	0.0072 (8)	0.0034 (9)	0.0026 (8)
O9	0.027 (2)	0.028 (2)	0.016 (2)	0.010 (2)	0.013 (2)	0.006 (2)
O10	0.009 (2)	0.013 (2)	0.013 (2)	0.0001 (8)	0.0014 (8)	0.0012 (8)
F1	0.015 (2)	0.019 (2)	0.016 (2)	0.0040 (8)	0.0051 (8)	0.0062 (8)

4.2 Rare-Earth Fluoride and Fluorido Borates

Table 4
Interatomic distances (pm) in $Pr_4B_3O_{10}F$, calculated on the basis of single crystal lattice parameters.

Pr1 – F1a	237.8 (2)	Pr2 – O10	234.5 (3)	Pr3 – F1	231.4 (3)	Pr4 – F1	237.1 (3)
Pr1 – O5	241.5 (3)	Pr2 – O3	242.2 (3)	Pr3 – O3	246.1 (3)	Pr4 – O1	242.9 (3)
Pr1 – O1	242.3 (3)	Pr2 – O1	242.5 (3)	Pr3 – O10	246.2 (3)	Pr4 – O8a	243.5 (3)
Pr1 – O9a	248.6 (3)	Pr2 – O4	246.4 (3)	Pr3 – O4	248.4 (3)	Pr4 – O8b	243.9 (3)
Pr1 – F1b	249.8 (3)	Pr2 – O5	246.9 (3)	Pr3 – O2	248.9 (3)	Pr4 – O2	247.5 (3)
Pr1 – O8b	251.3 (3)	Pr2 – O8	248.0 (3)	Pr3 – O9	273.5 (4)	Pr4 – O6	251.8(3)
Pr1 – O7	254.3 (3)	Pr2 – O2	269.1 (3)	Pr3 – O7	284.6 (4)	Pr4 – O3	277.7 (3)
Pr1 – O6	260.6 (3)	Pr2 – O6	273.5 (3)	Pr3 – O5	288.1 (3)	Pr4 – O9	298.3(3)
Pr1 – O4	293.2 (4)	Pr2 – O7	274.6 (3)	Pr3 – O6	298.3 (3)	Pr4 – O7	298.9 (5)
	∅ = 253.3		∅ = 253.1		∅ = 262.8		∅ = 260.2
B1 – O5	137.0 (4)	B2 – O7	137.7 (5)	B3a – O10	135.4 (9)	B3b – O8	140.3(9)
B1 – O6	137.8 (4)	B2 – O4	137.9 (5)	B3a – O8	144.4 (9)	B3b – O10	141.7 (9)
B1 – O2	138.0 (4)	B2 – O3	138.6 (5)	B3a – O9	148.8 (9)	B3b – O1	156 (2)
	∅ = 137.6		∅ = 138.1		∅ = 142.9		∅ = 146.0
F1 – Pr3	231.4 (3)						
F1 – Pr1a	237.8 (2)						
F1 – Pr4	237.1 (3)						
F1 – Pr1b	249.8 (3)						
	∅ = 239.0						

could therefore also be described as "$Pr_4B_3O_{10}F_{1-x}(OH)_x$". The OH$^-$ bands are small, and the elemental analysis showed a high fluorine percentage in the sample, so that the amount of OH$^-$ seems to be marginal, even though an exact quantification was not possible. For simplification, the position will thus be called F1 in the text, figures, and tables, and whenever fluoride (F1) is mentioned, it should be kept in mind that this is also OH$^-$ to a small percentage.

The boron positions of the two connected BO$_3$-groups (marked with arrows in Fig. 2) are not simultaneously occupied; we have rather a split position, leading to a statistically 50% occupation (the possibility of weak reflections, indicating a superstructure, could safely be ruled out). A closer look at this structural feature is given in Fig. 3. As depicted on the left, the atom B3 is either located at position B3a or at position B3b. This results in the formation of two possible BO$_3$-groups, both equally present in the structure. The anisotropic refinement of both boron positions was possible and resulted in fairly large displacement ellipsoids, which are directed towards the center of the four surrounding oxygen atoms. When B3 is located at position B3a, it is coordinated by the oxygen atoms O8, O9, and O10 (Fig. 3 right), and the atom O1 is isolated. On the other side, the occupation of position B3b leads to a BO$_3$-group with O1, O8, and O10 (Fig. 3 center), while O9 is isolated. Since B3 is placed on a split position with about 50% occupancy for each position, we have a statistical distribution in the structure. Nevertheless, we want to emphasize that all oxygen positions are fully occupied, regardless of the orientation of the BO$_3$-group (Table 2).

While position B3a is only located slightly off the plane of its trigonal coordination sphere, B3b shows a deviation of around 20 pm. It was also possible to refine the structure without a split position, resulting in slightly worse R-values, a very large atomic displacement parameter for atom B3, and a strongly distorted tetrahedral coordination sphere. We thus interpret the splitted

arrangement as another example for an intermediate state between a BO$_3$-group and a BO$_4$-tetrahedron. While for position B3a, the distance to the fourth oxygen atom O1 still measures 220 pm, position B3b is already moving towards O9 (Fig. 3 left), the fourth oxygen atom in close proximity (194 pm). If the B–O9-distance decreases only marginally, the split position will be neutralized by the formation of a BO$_4$-tetrahedron.

Looking at the B–O-distances in the other two BO$_3$-groups of $Pr_4B_3O_{10}F$, we find average bond lengths of 137.6 and 138.1 pm for the B1 and B2 coordination spheres (Table 4). These are typical values for BO$_3$-groups, as for example in the praseodymium borate $Pr_4B_{10}O_{21}$ [10]. Also the interatomic angles in these BO$_3$-groups sum up to 120.0 and 119.9°, confirming the trigonal-planar geometry (Table 5). For the BO$_3$-groups of the split position, we obtain average bond lengths of 142.9 and 146.0 pm for B3a–O and B3b–O-distances, respectively (Table 4 and Fig. 3). These values are fairly large and more in the range of distances, known for BO$_4$-tetrahedra. Especially the distances B3a–O9 and B3b–O1 are exceptionally large. The reason for this can be found in the orientation of the

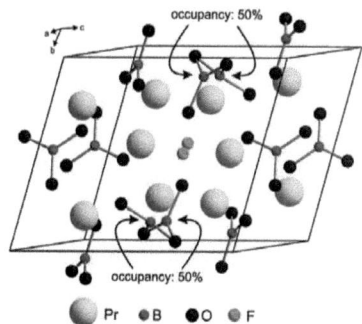

Fig. 2. Crystal structure of $Pr_4B_3O_{10}F$. The boron positions, marked with an arrow, are statistically occupied to 50%.

Table 5
Interatomic angles (deg) in $Pr_4B_3O_{10}F$, calculated on the basis of single crystal lattice parameters.

O6–B1–O5	118.5 (3)	O7–B2–O4	121.1 (4)	O10–B3a–O8	118.3 (6)
O6–B1–O2	119.9 (3)	O7–B2–O3	118.4 (3)	O10–B3a–O9	121.5 (6)
O5–B1–O2	121.7 (3)	O4–B2–O3	120.2 (4)	O9–B3a–O8	120.2 (6)
	∅ = 120.0		∅ = 119.9		∅ = 120.0
O10–B3b–O8	116.9 (7)	Pr3–F1–Pr1a	103.88 (9)	Pr3–F1–Pr1b	114.4 (2)
O10–B3b–O1	122.5 (7)	Pr3–F1–Pr4	119.3 (2)	Pr1b–F1–Pr4	102.54 (9)
O1–B3b–O8	115.0 (8)	Pr1a–F1–Pr4	106.6 (1)	Pr1a–F1–Pr1b	109.8 (2)
	∅ = 118.1				∅ = 109.4

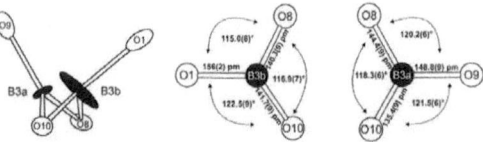

Fig. 3. View of the split position for atom B3 (left) with thermal displacement ellipsoids (probability: 50%); bond lengths and angles for the two possible BO$_3$-groups (right).

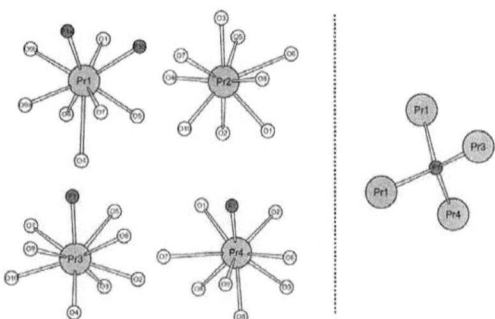

Fig. 4. Coordination spheres of the Pr^{3+} and F$^-$ ions in Pr$_4$B$_7$O$_{10}$F.

boron atoms towards the fourth neighbouring oxygen atom in order to obtain a fourfold coordination, that is not yet possible due to oversized B–O-distances. While the interatomic angles between the oxygen atoms around B3a show no irregularity, the average angle around B3b is slightly decreased to 118.1°, due to the off-planar position of 20.2 pm, mentioned before. We also estimated bond valence sums for the split boron positions. Here, very low values for the threefold-coordinated boron atoms (B3a: 2.6, B3b: 2.4) were derived, reflecting the need for further coordinating atoms. An inclusion of a fourth oxygen atom to the coordination sphere of the boron atoms improves the values (B3a: 2.7, B3b: 2.6), but due to the large distances, a full compensation is not yet possible.

In Pr$_4$B$_3$O$_{10}$F, four crystallographically independent praseodymium cations can be differentiated. They are all coordinated ninefold, either by oxide anions or by fluoride ions (Fig. 4). While Pr3 and Pr4 have one fluoride ion in their coordination sphere and Pr1 two, Pr2 is only coordinated by oxygen ions. The average Pr–O/F-distances are consistent with 253.3, 253.1, 262.8, and 260.2 pm for Pr1, Pr2, Pr3, and Pr4, respectively (Table 4). These values are similar to the Pr–O-distances, found in Pr$_4$B$_{10}$O$_{21}$ [10]. The single fluoride ion in Pr$_4$B$_3$O$_{10}$F is coordinated fourfold by

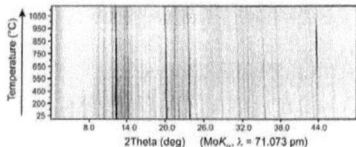

Fig. 5. Temperature-programmed X-ray powder patterns, showing the temperature stability of Pr$_4$B$_3$O$_{10}$F.

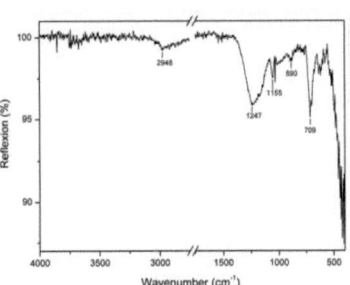

Fig. 6. IR spectrum of Pr$_4$B$_3$O$_{10}$F.

4.2 Rare-Earth Fluoride and Fluorido Borates

Table 6
Frequencies of absorption bands and tentative assignment of the IR-spectra of $Pr_4B_3O_{10}F$.

Bands (cm^{-1})	Assignment
417	bending (BO$_3$)
623	symmetric pulse vibration (BO$_3$)
709	out-of-plane bending (BO$_3$)
890	
997	
1033	symmetric stretching (BO$_3$)
1057	
1155	
1247	asymmetric stretching (BO$_3$)
2948	OH stretching

praseodymium ions, showing bond lengths between 231.4(3) and 249.8(3) pm. The average interatomic angle of 109.4° is the ideal tetrahedral angle.

Temperature programmed X-ray powder diffraction investigations were performed to study the temperature stability of $Pr_4B_3O_{10}F$. Fig. 5 shows that the compound is stable up to a temperature of 1100 °C (heating rate: 40 °C/min). All the high-pressure phases, we synthesized so far, showed decomposition or transformation into another phase up to 1100 °C, due to their metastable character. We therefore conclude that $Pr_4B_3O_{10}F$ is not a metastable, but a thermodynamically stable compound in the area of ambient pressure conditions up to 3 GPa. Since the applied pressure during the synthesis was low, it seems likely that $Pr_4B_3O_{10}F$ could also be obtained via normal pressure solid state synthesis. This is in agreement with the crystal structure, that exhibits only isolated BO$_3$-groups. The occurrence of the split position as a possible transition state towards a BO$_4$-tetrahedron (a structural feature, obtained under ambient and higher pressures) indicates that a metastable high-pressure phase might form under slightly higher pressures. This will be an object of our future studies.

The analysis of the elemental constitution of $Pr_4B_3O_{10}F$ single crystals showed average atomic Pr/B/O/F compositions (%) of 17.6 : 17.7 : 57.6 : 7.1. Due to the light weight, boron measurements have to be taken cautiously, nevertheless, these results confirm the presence of the elements and are in agreement with the compositions, obtained from single crystal X-ray studies (the calculated values Pr/B/O/F are 22.0 : 16.0 : 55.5 : 5.5). Especially, the high percentage of fluorine, measured in the crystals, clearly confirms the atom assignment, made on the basis of single crystal data, and indicates that only traces of OH$^-$ ions are substituting fluoride atoms.

Fig. 6 shows the IR measurement, performed on the sample. The complete spectral region between 400 and 4000 cm^{-1} was measured. The absorptions between 1200 and 1400 cm^{-1}, between 600 and 800 cm^{-1}, and below 500 cm^{-1} characterize triangular BO$_3$-groups as in λ-LaBO$_3$ [31], H–LaBO$_3$ [32], or EuB$_2$O$_4$ [33]. The absorption area between 1200 and 1400 cm^{-1} is slightly broadened towards lower wavelengths. This might be evoked by the transition state towards a tetrahedral borate group, that typically shows absorption peaks between 790 and 1150 cm^{-1}. In the region of 3000-3500 cm^{-1}, a small peak was detected. Peaks at those wavelengths can be assigned to OH-groups and typically reveal water-containing borates. On the basis of this IR measurement, we have to assume the presence of a small amount of OH-groups in $Pr_4B_3O_{10}F$. A detailed assignment of the IR bands is made in Table 6.

In Fig. 7 and Table 7, the Raman spectrum and the frequencies of the determined bands of a single crystal of $Pr_4B_3O_{10}F$ are given. The most intense bands above 500 cm^{-1} are located at 619, 925, and 1208 cm^{-1}. In hydrated borates with isolated planar BO$_3$ molecules, strong symmetric stretching modes of the BO$_3$-group at 501, 882, and 1171 cm^{-1} were reported by Jun et al. [34]. The same authors observed also Raman bands in the spectra of triborates, which consist of one planar BO$_3$. triangle and two BO$_4$- tetrahedra. They assigned the Raman frequencies of a band between 613 and 630 cm^{-1} to a symmetric pulse vibration of triborate anions and a strong Raman band at 855 cm^{-1} to symmetric stretching modes of tetra-coordinated boron. For alkali borate glasses, Raman bands at a high wavenumber of ~1345 cm^{-1} were assigned to B$_2$O$^-$ triangles, linked to other borate triangular units [35].

In summary, the Raman spectroscopic investigations confirm the presence of triangular BO$_3$-groups. The most intense band at 925 cm^{-1} is most probably related to symmetric stretching vibrations of this group; its relatively high frequency can be explained by the exceptionally large B-O distances, obtained in the X-ray diffraction study. The band at 619 cm^{-1} is caused by the symmetric pulse vibrations of the BO$_3$-groups. The high frequency bands above 1100 cm^{-1} cannot be assigned unequivocally. They might result from the vibrations of the triangular units and their

Table 7
Raman shifts of observed bands and tentative assignment of the spectrum of $Pr_4B_3O_{10}F$.

Band (cm^{-1})	Assignment
115	
141	
166	
197	
232	
276	lattice
338	Pr–O
375	bending (BO$_3$)
423	
468	
517	
582	
619	symmetric pulse vibration (BO$_3$)
750	
767	
920	
925	symmetric stretching (BO$_3$)
934	
1082	
1208	
1241	symmetric stretching (BO$_3$)? overtone?
1509	

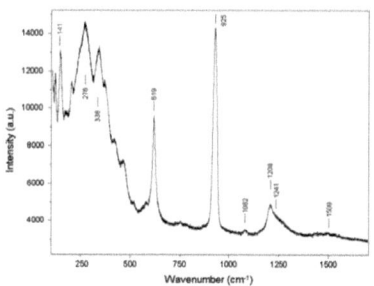

Fig. 7. Raman spectrum of $Pr_4B_3O_{10}F$.

orientation in two different ways in the structure, from the transition state towards a BO_4-tetrahedron, or represent simple overtones of the 619 cm^{-1} mode.

4. Conclusion

In this article, we described the high-pressure synthesis and the crystal structure of the first praseodymium fluoride borate $Pr_4B_3O_{10}F$. The crystal structure is built up exclusively from isolated BO_3-groups. One of the boron atoms is located on a split position, leading to a BO_3-group, that can be orientated in the crystal structure in two different ways, which is interpreted as a special case of an intermediate state between a BO_3-group and a BO_4-tetrahedron. Temperature programmed X-ray powder diffraction investigations indicate that $Pr_4B_3O_{10}F$ is a thermodynamically stable phase, which might also be obtainable under ambient pressure conditions. The synthesis of a metastable high-pressure phase, in which the split position resulted in a tetrahedrally coordinated boron position, will be part of further investigations.

Acknowledgements

We thank Dr. Gunter Heymann and Dr. Peter Mayer for their collecting the single-crystal data, Thomas Miller for the temperature-programmed in situ X-ray diffraction experiments, and Christian Minke for the electron microscopy measurements. Special thanks go to PD Dr. Oliver Oeckler for crystallographic advice and assistance.

References

[1] H. Huppertz, B. von der Eltz, J. Am. Chem. Soc. 124 (2002) 9376.
[2] H. Huppertz, Z. Naturforsch. B58 (2003) 278.
[3] H. Huppertz, H. Emme, J. Phys. Condens. Matter 16 (2004) 1283.
[4] H. Emme, H. Huppertz, Z. Anorg. Allg. Chem. 628 (2002) 2165.
[5] H. Emme, H. Huppertz, Chem. Eur. J. 9 (2003) 3623.
[6] H. Emme, H. Huppertz, Acta Crystallogr. C61 (2005) i29.
[7] H. Emme, H. Huppertz, Acta Crystallogr. C61 (2005) i23.
[8] H. Huppertz, S. Altmannshofer, G. Heymann, J. Solid State Chem. 170 (2003) 320.
[9] H. Emme, M. Valldor, R. Pöttgen, H. Huppertz, Chem. Mater. 17 (2005) 2707.
[10] A. Haberer, G. Heymann, H. Huppertz, J. Solid State Chem. 180 (2007) 1595.
[11] T. Nikelski, Th. Schleid, Z. Anorg. Allg. Chem. 629 (2003) 1017.
[12] T. Nikelski, M.C. Schäfer, H. Huppertz, Th. Schleid, Z. Kristallogr. 233 (2008) 177.
[13] H. Emme, T. Nikelski, Th. Schleid, R. Pöttgen, M.H. Möller, H. Huppertz, Z. Naturforsch 59b (2004) 202.
[14] H. Emme, C. Despotopoulou, H. Huppertz, Z. Anorg. Allg. Chem. 630 (2004) 2450.
[15] G. Heymann, T. Soltner, H. Huppertz, Solid State Sci. 8 (2006) 821.
[16] A. Haberer, G. Heymann, H. Huppertz, Z. Naturforsch. B62 (2007) 759.
[17] S.C. Neumair, J.S. Knyrim, O. Oeckler, R. Glaum, R. Kaindl, R. Stalder, H. Huppertz, Chem. Eur. J. submitted for publication.
[18] G. Corbel, R. Retoux, M. Leblanc, J. Solid State Chem. 139 (1998) 52.
[19] H. Müller-Bunz, Th. Schleid, Z. Anorg. Allg. Chem. 628 (2002) 2750.
[20] A. Haberer, H. Huppertz, J. Solid State Chem. 183 (2010), doi:10.1016/j.jssc.2009.12.003.
[21] A. Haberer, H. Huppertz, J. Solid State Chem. 182 (2009) 888.
[22] A. Haberer, H. Huppertz, unpublished results.
[23] H. Huppertz, Z. Kristallogr. 219 (2004) 330.
[24] N.C. Baenziger, J.R. Holden, G.E. Knudson, A.I. Popov, in: Golden Book of Phase Transitions, vol. 1, Wroclaw, 2002, pp. 1–123.
[25] J.W. Visser, J. Appl. Crystallogr. 2 (1969) 89.
[26] WinX^{POW} Software; STOE & CIE GmbH, Darmstadt, Germany, 1998.
[27] Z. Otwinowski, W. Minor, Methods Enzymol. 276 (1997) 307.
[28] G.M. Sheldrick, SHELXS-97 and SHELXL-97, Program Suite for the Solution and Refinement of Crystal Structures. University of Göttingen, Göttingen, Germany, 1997.
[29] G.M. Sheldrick, Acta Crystallogr. A64 (2008) 112.
[30] G. Engel, J. Pretzsch, V. Gramlich, W.H. Baur, Acta Crystallogr. 831 (1975) 1854.
[31] J.P. Laperches, P. Tarte, Spectrochim. Acta 22 (1966) 1201.
[32] R. Böhlhoff, U. Bambauer, W. Hoffmann, Z. Kristallogr. 133 (1971) 386.
[33] H. Machida, H. Hata, K. Okuno, G. Adachi, J. Shiokawa, J. Inorg. Nucl. Chem. 41 (1979) 1425.
[34] L. Jun, X. Shuping, G. Shiyang, Spectrochim. Acta A51 (1995) 519.
[35] G. Padmaja, P. Kistaiah, J. Phys. Chem. A113 (2009) 2397.

4.2.8 "$RE_5(BO_{3.66}F_{0.34})_3F$" ($RE$ = Gd, Yb)

Since the earlier described compounds $RE_5(BO_3)_2F_9$ (RE = Er - Yb) contain trigonal BO_3-groups only (Sections 4.2.1 - 4.2.3), it was tempting to induce a transformation into BO_4-groups by application of higher pressure. Consequently, we were able to synthesize the first apatite-structured fluorido borates "$RE_5(BO_{3.66}F_{0.34})_3F$" ($RE$ = Gd, Yb) from similar reaction mixtures at 10 GPa and 1000 °C (Figure 4.2-8). The sum formulas of these compounds are given in parentheses because until the end of this thesis, the exact distribution of fluorine and oxygen could not be verified (see below).

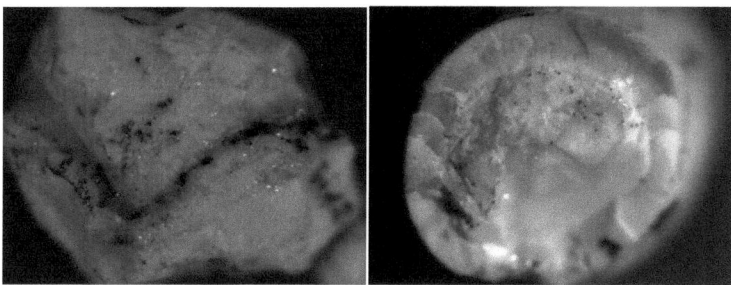

Figure 4.2-8
Crystalline samples of "$Gd_5(BO_{3.66}F_{0.34})_3F$" (left) and "$Yb_5(BO_{3.66}F_{0.34})_3F$" (right)

The syntheses of these phases were challenging. The starting materials were stoichiometric mixtures of gadolinium oxide (Strem Chemicals, 99.99%) or ytterbium oxide (Smart Elements, 99.99%), gadolinium fluoride (Strem Chemicals, 99.9%) or ytterbium fluoride (Strem Chemicals, 99.9%), and B_2O_3 (Strem Chemicals, 99.9+%). For both compounds, the syntheses were carried out in 18/11 multianvil assemblies. The mixtures were compressed within 240 min to 10 GPa and heated to 1000 °C in the following 15 min. After holding this temperature for 15 min, the samples were cooled down to 800 °C in 20 minutes, followed by quenching to room temperature after switching off heating.

The first syntheses yielded "$RE_5(BO_{3.66}F_{0.34})_3F$" ($RE$ = Gd, Yb) as the only reaction products, which were interspersed with the crucible material BN in both cases. Due to the extreme hardness of cubic boron nitride, the sample could barely be used for analytical purposes. Numerous other syntheses yielded the desired phases but the phase-pure syntheses from the beginning could never be reproduced. Several side phases, e.g. the previously mentioned compounds $Yb_5(BO_3)_2F_9$ and $Gd_4B_4O_{11}F_2$, made the characterization of the phases very difficult.

The isotypic crystal structures of "$RE_5(BO_{3.66}F_{0.34})_3F$" ($RE$ = Gd, Yb) were solved and refined on the basis of single-crystal data (space group: $P6_3/m$). All significant details of the data collections and analyses are listed in Table 4.2-2. The key issue in the characterization is the verification of the assignment of the anionic positions to either oxygen or fluorine. The distinction between fluorine and oxygen by means of X-ray diffraction experiments is not trivial, so microprobe measurements and solid-state-NMR investigations were carried out in addition. Unfortunately, no definite results could be obtained until the end of this thesis. The sum formulas given in parentheses resulted from several considerations described below, while a final prove is still missing.

There is no doubt about the overall atomic positions, as displayed in Table 4.2-3 and Figure 4.2-9, resulting in an apatite-type crystal structure. A large variety of compositions with the apatite structure type $M_5(AO_4)_3X$ is known [238]. The tetrahedrally coordinated atom A can either be a transition metal (V, Mn, Cr) or an element of the 14th or 15th group of the periodic table (Si, Ge, P, or As) [239]. With "$RE_5(BO_{3.66}F_{0.34})_3F$" ($RE$ = Gd, Yb), we synthesized the first two apatite-type compounds, in which the atomic position A is completely occupied by fourfold coordinated boron atoms. The structural diversity of apatites is thus extended towards an element of the 13th group as a full substitute of the original phosphorus atom.

4.2 Rare-Earth Fluoride and Fluorido Borates

Table 4.2-2
Crystal data and structure refinement of "$RE_5(BO_{3.66}F_{0.34})_3F$" ($RE$ = Gd, Yb) (standard deviations in parentheses).

Empirical formula	"$Gd_5(BO_{3.66}F_{0.34})_3F$"	"$Yb_5(BO_{3.66}F_{0.34})_3F$"
Molar mass, g mol^{-1}	2065.4	2223.26
Crystal system	hexagonal	
Space group	$P6_3/m$	
Powder diffractometer	Stoe Stadi P	
Radiation	Mo$K_{\alpha 1}$ (λ = 70.93 pm)	
Single-crystal diffractometer	Nonius Kappa CCD	
Radiation	MoK_α (λ = 71.073 pm)	
Single-crystal data		
a, pm	900.2(2)	879.4(2)
c, pm	696.9(2)	661.8(2)
V, nm^3	0.489(2)	0.443(2)
Formula units per cell	$Z = 2$	
Calculated density, g cm^{-3}	7.011	8.329
Crystal size, mm^3	0.03 × 0.02 × 0.02	0.07 × 0.06 × 0.05
Temperature, K	293(2)	293(2)
Absorption coefficient, mm^{-1}	33.52	52.33
$F(000)$, e	882	942
θ range, deg	2.6 - 30.0	2.7 - 29.0
Range in hkl	±12, 12/-10, ±9	±11, ±11, ±8
Total no. of reflections	4833	4208
Independent reflections	512 (R_{int} = 0.0534, R_σ = 0.0207)	421 (R_{int} = 0.0521, R_σ = 0.0272)
Reflections with $I \geq 2\sigma(I)$	460	355
Data / ref. Parameters	512 / 40	421 / 41
Absorption correction	semiempirical	numerical
Goodness-of-fit on F^2	1.003	1.047
Final R indices [$I \geq 2\sigma(I)$]	$R1$ = 0.0279	$R1$ = 0.0279
	$wR2$ = 0.0617	$wR2$ = 0.0695
R indices (all data)	$R1$ = 0.0335	$R1$ = 0.0340
	$wR2$ = 0.0647	$wR2$ = 0.0710
Largest diff. peak and hole, e·Å^{-3}	2.5 / -2.2	1.9 / -1.9

Table 4.2-3

Atomic coordinates and isotropic equivalent displacement parameters (U_{eq} / Å^2) for "$RE_5(BO_{3.66}F_{0.34})_3F$" ($RE$ = Gd, Yb) (standard deviations in parentheses). U_{eq} is defined as one third of the trace of the orthogonalized U_{ij} tensor.

Atom	Wyckoff-position	x	y	z	U_{eq}
Gd1	6h	0.25273(7)	0.23803(7)	1/4	0.0194(2)
Gd2	4f	1/3	2/3	0.00409(9)	0.0143(2)
B	6h	0.025(2)	0.409(2)	1/4	0.058(8)
O1	6h	0.121(2)	0.590(2)	1/4	0.017(2)
O2	6h	0.486(2)	0.155(2)	1/4	0.016(2)
O3/F3	12i	0.0969(9)	0.3492(9)	0.063(2)	0.032(2)
F1	2a	0	0	1/4	0.46(9)
Yb1	6h	0.25336(7)	0.24172(6)	1/4	0.0179(3)
Yb2	4f	1/3	2/3	0.0028(2)	0.0179(3)
B	6h	0.024(2)	0.399(2)	1/4	0.039(5)
O1	6h	0.119(2)	0.588(2)	1/4	0.018(2)
O2	6h	0.484(2)	0.163(2)	1/4	0.018(2)
O3/F3	12i	0.0931(9)	0.344(2)	0.063(2)	0.031(2)
F1	4e	0	0	0.198(7)	0.05(2)

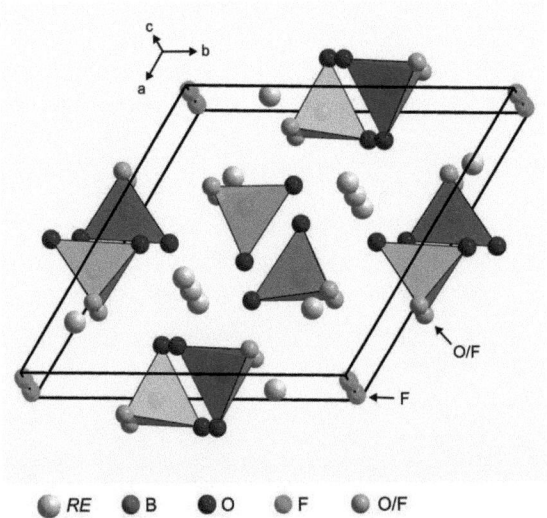

Figure 4.2-9
Apatite structure of "$RE_5(BO_{3.66}F_{0.34})_3F$" ($RE$ = Gd, Yb)

4.2 Rare-Earth Fluoride and Fluorido Borates

According to our current knowledge, each boron atom in "$RE_5(BO_{3.66}F_{0.34})_3F$" ($RE$ = Gd, Yb) is fourfold coordinated by two oxygen atoms O1 and O2 and by two mixed positions, that can be either oxygen or fluorine (O3/F3). There are several indications towards this fluorine-oxygen distribution. A free refinement of the occupancy of these positions led to a ratio O/F of about 1.5, confirming the mixed occupancy of this crystallographic site. Furthermore, a low charge distribution value of -1.67 is calculated with VALIST [134], supporting the O3/F3 assignment. In order to obtain electroneutrality, a ratio of O : F = 0.83 : 0.17 was assigned to these positions in the two compounds, leading to the overall sum formula "$RE_5(BO_{3.66}F_{0.34})_3F$".

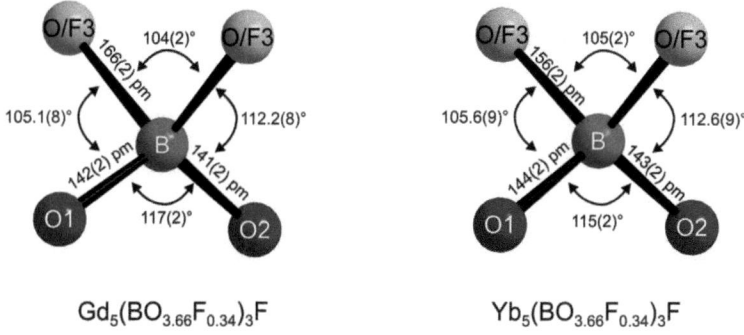

Figure 4.2-10
Bond lengths and angles of the $BO_{3.66}F_{0.34}$-tetrahedra in "$Gd_5(BO_{3.66}F_{0.34})_3F$" (left) and "$Yb_5(BO_{3.66}F_{0.34})_3F$" (right).

In Figure 4.2-10, the $BO_{3.66}F_{0.34}$-tetrahedra of both structures are compared. The B–O-bonds B–O1 (142(2) (RE = Gd) and 144(2) pm (RE = Yb)), and B–O2 (141(2) and 143(2) pm, respectively) are slightly shorter than the average of 147.6 pm in tetrahedral BO_4-groups of borates [235, 240]. The B–O3/F3-distances (166(2) and 156(2) pm) are longer and in a range known for distorted BO_4-tetrahedra, e.g. in $LiBa_2B_{10}O_{16}(OH)_3$ (160(2) pm) [241] or β-MB_4O_7 (M = Mn, Ni, Cu) (157.6(2) pm) [242]. The tetrahedral distortion appears also in the interatomic O–B–O-angles, leading to values between 104 and 117° for

"Gd$_5$(BO$_{3.66}$F$_{0.34}$)$_3$F" and angles between 105 and 115° for "Yb$_5$(BO$_{3.66}$F$_{0.34}$)$_3$F". Nevertheless, the average tetrahedral angles are 109.6 and 110.0°, respectively, and thus fairly close to the ideal tetrahedral angle.

In "RE_5(BO$_{3.66}$F$_{0.34}$)$_3$F" (RE = Gd, Yb), there are two independent rare-earth atom sites. RE1 shows a sevenfold coordination in a distorted pentagonal-bipyramidal geometry, involving two oxygen atoms, four mixed oxygen/fluorine positions, and one fluoride ion with bond lengths of 221 – 256 pm (Gd1) and 221 – 245 pm (Yb1). RE2 is ninefold coordinated in form of a tricapped trigonal prism by six oxygen atoms and three mixed positions O/F3, showing distances of 239 – 261 pm (Gd2) and 228 – 259 pm (Yb2). These characteristics are very similar to those of the coordination polyhedra of the trivalent rare-earth cations in Eu$_5$(SiO$_4$)$_3$F and Yb$_5$(SiO$_4$)$_3$S, which are coordinated in the same manner [243].

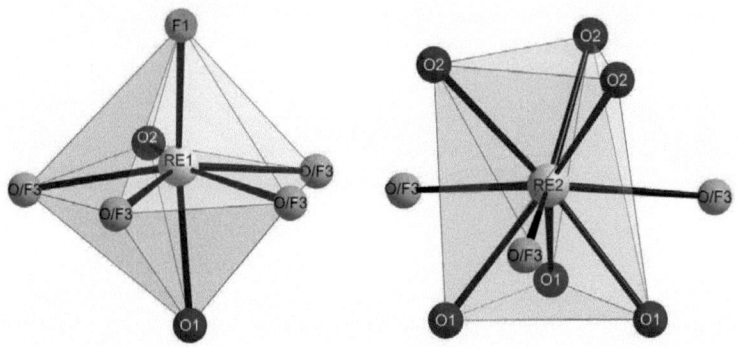

Figure 4.2-11
Coordination polyhedra of the rare-earth ions in "RE_5(BO$_{3.66}$F$_{0.34}$)$_3$F" (RE = Gd, Yb)

Another interesting aspect of the crystal structure is the fluoride ions located in channels along the *c*-axis. While borates, that contain only isolated fluoride anions, are generally termed "fluoride borates", "fluorido borates" contain fluorine atoms covalently bound to the boron atoms. To the best of our

4.2 Rare-Earth Fluoride and Fluorido Borates

current knowledge, the compounds "$RE_5(BO_{3.66}F_{0.34})_3F$" ($RE$ = Gd, Yb) are the first examples, in which both features occur.

The fluoride ions F1 in "$RE_5(BO_{3.66}F_{0.34})_3F$" ($RE$ = Gd, Yb) are located in channels, that run along the c-axis. In "$Gd_5(BO_{3.66}F_{0.34})_3F$", F1 is located at the position 2a (0, 0, ¼), which is typical for fluoroapatites. The structural refinement suggests the possibility to split this fluoride position into the positions (0, 0, 0.3327) and (0, 0, 0.1673). This is corroborated by Figure 4.2-12, where the results of difference Fourier syntheses after refinements without F1 are shown. In "$Gd_5(BO_{3.66}F_{0.34})_3F$", the electron density of the fluoride ion is smeared along the c-axis (Figure 4.2-12, left). This indicates that the fluoride ions are not fixed on a specific position, but rather disordered inside the channels. Similar displacements are observed in apatite-type oxide ion conductors and are considered to be essential for the conducting properties [244, 245]. This makes "$Gd_5(BO_{3.66}F_{0.34})_3F$" an exciting object for further investigations on its ionic conductivity.

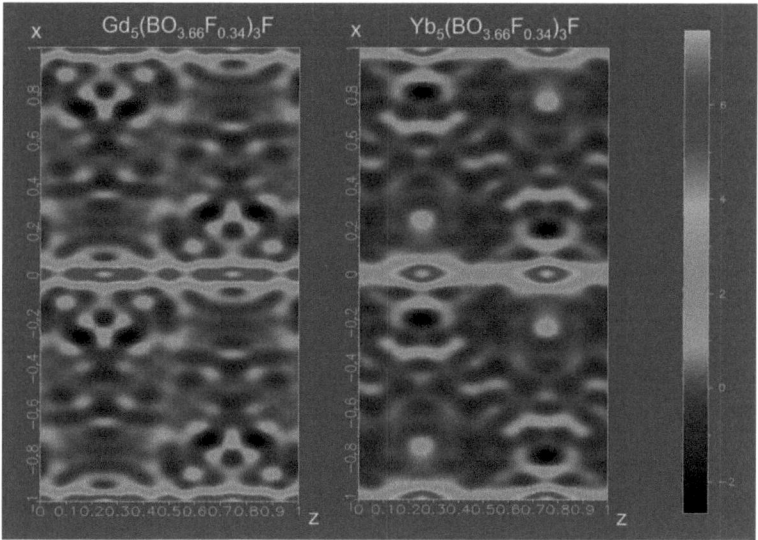

Figure 4.2-12

Difference Fourier maps of the fluoride channels running along the c-axes in "$Gd_5(BO_{3.66}F_{0.34})_3F$" (left) and "$Yb_5(BO_{3.66}F_{0.34})_3F$" (right)

In Yb$_5$(BO$_{3.66}$F$_{0.34}$)$_3$F, the fluoride ions are located off the (0, 0, ¼)-position to (0, 0, 0.198(7)). According to the symmetry of the structure, a mirror plane generates two F1 positions in a distance of 86.8 pm, that are occupied 50 % each. Figure 4.2-12 (right) shows the broadened electron density, evoked by the two possible F1 positions nearby. Compared to the Gd compound, the fluoride ions are more localized inside the channels.

Additionally, a bulk IR spectrum of "Yb$_5$(BO$_{3.66}$F$_{0.34}$)$_3$F" (Figure 4.2-13) was recorded between 400 and 4000 cm^{-1} on a Nicolet 5700 FT-IR spectrometer. Before measuring, the sample was thoroughly dried under high vacuum for several days. The absorption bands between 790 and 1150 cm^{-1} are those typical for the tetrahedral borate group BO$_4^{5-}$, as in π-GdBO$_3$ or in TaBO$_4$ [148, 246].

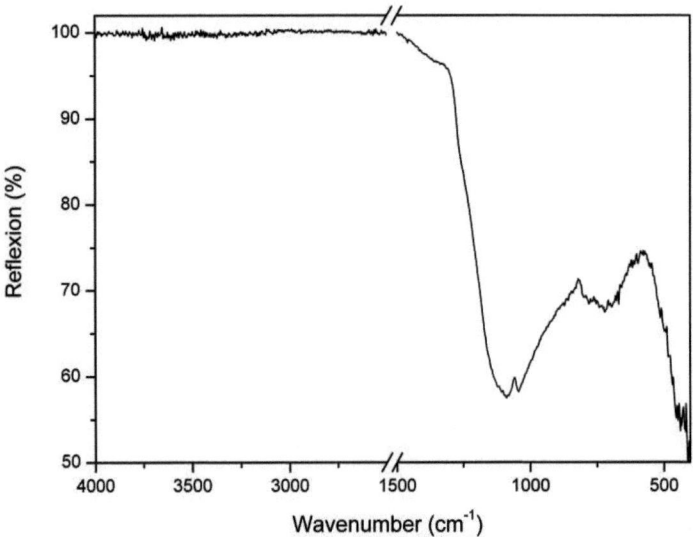

Figure 4.2-13
IR spectrum of "Yb$_5$(BO$_{3.66}$F$_{0.34}$)$_3$F"

In the region of 3000 to 3500 cm^{-1}, no absorption bands could be de-

4.2 Rare-Earth Fluoride and Fluorido Borates

tected. Bands at those wavelengths can be assigned to OH-groups and typically reveal water-containing borates. On the basis of this IR measurement, we can exclude the presence of OH-groups in "$Yb_5(BO_{3.66}F_{0.34})_3F$".

In order to complete vibrational spectroscopy, Raman spectra of single crystals of "$Yb_5(BO_{3.66}F_{0.34})_3F$" at two different excitation wavelengths were recorded (Figure 4.2-14). Because in both spectra bands were detected at the same wavenumbers, luminescence can be excluded. The strong bands between 800 and 1000 cm^{-1} can be assigned to symmetric stretching vibrations of the $BO_{3.66}F_{0.34}$-group in accordance with the vibrations of tetrahedral borate groups BO_4^{5-} [247-249]. A natural apatite from Moneta Island, Mayeguez, Puerto Rico, USA, with the chemical formula $Ca_5(PO_4)_3[(OH)_{0.67}F_{0.33}]$ [250] shows a simple spectrum with a few bands and one sharp, intense band, related to the PO_4-group vibration at 960 cm^{-1}. The spectrum of "$Yb_5(BO_{3.66}F_{0.34})_3F$" contains at least 12 bands; the most intense one is much

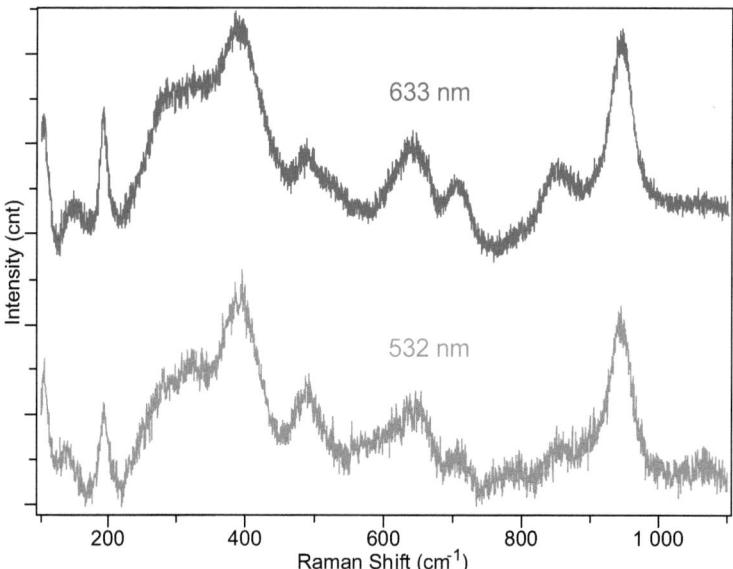

Figure 4.2-14
Raman spectra of "$Yb_5(BO_{3.66}F_{0.34})_3F$", excited at 532 and 633 nm

broader and shifted to 945 cm^{-1}; the bands at lower wavenumbers are much more intense than in the natural apatite. This reflects the structural differences, caused by the chemical substitutions.

The borates "$RE_5(BO_{3.66}F_{0.34})_3F$" ($RE$ = Gd, Yb) are the first apatite-type compounds, in which fourfold coordinated boron atoms fully replace PO_4-tetrahedra. In boron containing apatites (BAPs) with the general formula $Ca_wB_xP_yO_z$, only small amounts of boron are incorporated in the apatite structure ([251] and references therein). For further studies of BAPs, which are candidate materials for bioceramic applications, the syntheses of "$RE_5(BO_{3.66}F_{0.34})_3F$" ($RE$ = Gd, Yb) are a big step towards further diversity of these important compounds. Since we demonstrated that phosphorus could be fully replaced with boron in our experiments, high-pressure studies on the syntheses of BAPs are a promising field for the future.

4.2.9 $RE_3(BO_3)_2F_3$ (RE = Gd, Dy)

The first single crystals of the fluoride borates $Gd_3(BO_3)_2F_3$ and $Dy_3(BO_3)_2F_3$ were obtained in a Walker-type multianvil apparatus at a temperature of 900 °C and pressures of 3 and 5 GPa, respectively. Both compounds crystallize in the monoclinic space group $C2/c$ ($Z = 4$). While the synthesis of a powder sample of $Gd_3(BO_3)_2F_3$ was previously reported in the literature, the isotypic compound $Dy_3(BO_3)_2F_3$ is presented herein for the first time. Two different disorder models have been proposed previously for compounds with the composition $RE_3(BO_3)_2F_3$. The single crystal data now available, gives new insight into the disorder problem. The following results were still subject to final crystallographical checks so that it was not published yet at the end of this thesis.

In 1998, Corbel et al. synthesized the compounds $RE_3(BO_3)_2F_3$ (RE = Sm, Eu, Gd) and presented an *ab initio* structure determination for $Gd_3(BO_3)_2F_3$ based on X-ray powder diffraction data [218]. The structure solution revealed the positions of the heavy Gd^{3+} cations and the oxygen and fluorine positions were located by successive difference Fourier syntheses. Also the atomic positions of the boron atoms could be located in the final Fourier difference, but a refinement was impossible due to the weak scattering factor of boron. While no evidence of disorder was mentioned in this paper, Antic-Fidancev et al. later presented a disorder model based on broadened bands observed during luminescence measurements on $Er_3(BO_3)_2F_3$ [252]. A substitution model of $(BO_3)^{3-}$ versus 3 F⁻ was proposed, resulting in a differing sum formula $Eu_3(BO_3)_{2+x}F_{3-3x}$. In 2002, Müller-Bunz et al. tried to reproduce the compound $Gd_3(BO_3)_2F_3$ without success. Their attempts always resulted in the fluoride borate $Gd_2(BO_3)F_3$ [219], which is structurally closely related. For this compound, an anionic disorder evoked by two different orientations of the BO_3-groups in the structure was proposed. Due to the close relationship of the two crystal structures, Müller-Bunz et al. suggested the possibility of a similar model of disorder for $Gd_3(BO_3)_2F_3$. The Antic-Fidancev model was also put into doubt, because the exchange of $(BO_3)^{3-}$ versus 3 F⁻ would require a serious deformation of the crystal structure [219]. Without single crystal data, neither disorder model could be proved correct. Earlier high-pressure / high-temperature syntheses by our group yielded the compounds $RE_5(BO_3)_2F_9$ (RE = Er - Yb) (Sections 4.2.1 - 4.2.3), which are also structurally very closely related to the previously described fluoride borates. Here, large standard deviations of the structural parameters in the BO_3-groups of all three isotypic structures were observed. We simulated a Müller-Bunz disorder model for the case of $Yb_5(BO_3)_2F_9$, where a manual splitting of the boron position resulted in acceptable B–O distances, even though we did not find evidence for a second boron position in our crystals.

The prior syntheses of $RE_3(BO_3)_2F_3$ (RE = Sm, Eu, Gd) by Corbel et al. were performed by heating stoichiometric mixtures of RE_2O_3, B_2O_3, and REF_3 under ambient pressure, resulting in powder samples [218]. It was now possible to obtain single crystals of $RE_3(BO_3)_2F_3$ (RE = Gd, Dy) by high-pressure / high-temperature syntheses, whereof the dysprosium fluoride borate was synthesized for the first time. Both compounds show an explicit disorder in the anionic part (BO_3-groups) of the crystal structures. In the following, the syntheses and the structural details of $RE_3(BO_3)_2F_3$ (RE = Gd, Dy) are discussed, along with spectroscopic properties.

For the syntheses of $RE_3(BO_3)_2F_3$ (RE = Gd, Dy), reactions of the oxides RE_2O_3 and B_2O_3 with the corresponding rare-earth fluoride REF_3 were performed under high-pressure / high-temperature conditions. For $Gd_3(BO_3)_2F_3$, the synthesis was carried out at 3 GPa and 900 °C, while $Dy_3(BO_3)_2F_3$ was obtained at 5 GPa and 900 °C. Depending on the fluoride borate, mixtures of Gd_2O_3 (Strem Chemicals, 99.99 %) or Dy_2O_3 (Strem Chemicals, 99.9 %), B_2O_3 (Strem Chemicals, 99.9+ %), and GdF_3 (Strem Chemicals, 99.9 %) or DyF_3 (Strem Chemicals, 99.9 %) with a molar ratio of 1 : 3 : 1 were used. The samples were compressed up to 3 or 5 GPa in 1.5 or 2.25 h, respectively, then heated to 900 °C in 15 minutes and kept there for 20 minutes. Afterwards, both samples were cooled down to 700 °C in 20 minutes, and cooled down to room temperature by switching off the heating. The decompression of the assemblies required 4.5 and 6.75 h, respectively. Colorless, air- and water-resistant crystals of $Gd_3(BO_3)_2F_3$ and $Dy_3(BO_3)_2F_3$ could be obtained after the respective synthesis (Figure 4.2-15).

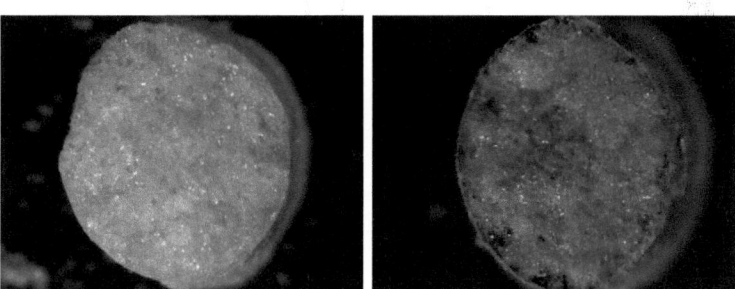

Figure 4.2-15
Crystalline samples of $Gd_3(BO_3)_2F_3$ (left) and $Dy_3(BO_3)_2F_3$ (right)

The sample characterizations were performed by powder X-ray diffraction, carried out in transmission geometry on flat samples of the reaction products. Both experimental powder patterns of $Gd_3(BO_3)_2F_3$ and $Dy_3(BO_3)_2F_3$ tally well with the corresponding theoretical patterns simulated from single-crystal data (Figure 4.2-16). In the case of $Gd_3(BO_3)_2F_3$, reflections of an unknown side phase could be detected. In the case of $Dy_3(BO_3)_2F_3$, reflections of the educt DyF_3 were detected. Both impurities are marked with asterisks in Figure 4.2-16.

Indexing the reflections of $Gd_3(BO_3)_2F_3$ resulted in the lattice parameters $a = 1248.4(5)$ pm, $b = 621.4(2)$ pm, and $c = 834.7(3)$ pm, with $\beta = 97.2(1)°$ and a unit-cell volume of 642.4(3) Å^3. For $Dy_3(BO_3)_2F_3$, values of $a = 1235.8(7)$ pm, $b = 614.5(3)$ pm, and $c = 827.6(3)$ pm, $\beta = 97.3(1)°$, and $V = 623.4(3)$ Å^3 were obtained. This confirmed the lattice parameters, received from the single-crystal X-ray diffraction study (Table 4.2-4) and is in good agreement with the parameters given for $Gd_3(BO_3)_2F_3$ by Corbel et al. ($a = 1253.4(1)$ pm, $b = 623.7(1)$ pm, $c = 836.0(1)$ pm, $\beta = 97.404(6)$, $V = 648.1(2)$ Å^3) [218]. It is often observed, that the application of high pressure during the synthesis of a compound leads to a slightly compressed unit cell compared to the synthesis under ambient pressure. For $Gd_3(BO_3)_2F_3$, this is in agreement with the largest deviation for the *a*-axis, along which the crystal structure should show the highest compressibility.

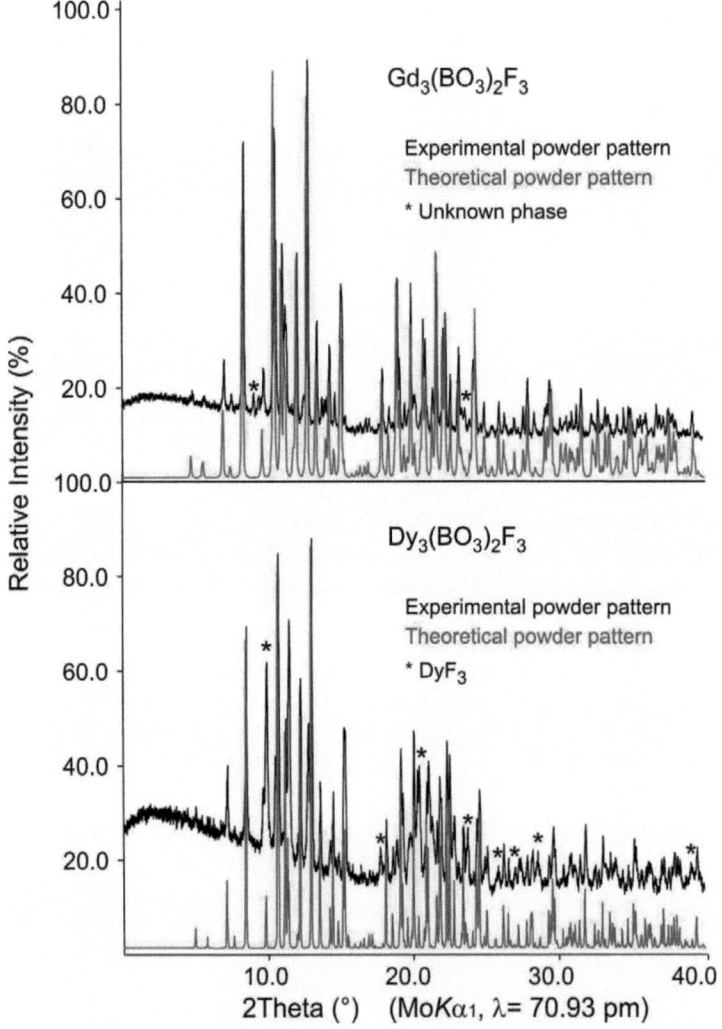

Figure 4.2-16

Comparison of the experimental to the theoretical powder patterns beneath of $Gd_3(BO_3)_2F_3$ (top) and $Dy_3(BO_3)_2F_3$ (bottom). Asterisks mark an unknown side phase for $Gd_3(BO_3)_2F_3$ and excess DyF_3 in the case of $Dy_3(BO_3)_2F_3$.

Table 4.2-4
Crystal data and structure refinement of $RE_3(BO_3)_2F_3$ (RE = Gd, Dy)

Empirical formula	$Gd_3(BO_3)_2F_3$	$Dy_3(BO_3)_2F_3$
Molar mass (g·mol^{-1})	646.37	662.12
Crystal system	monoclinic	
Space group	C2/c	
Lattice parameters from powder data		
Powder diffractometer	Stoe Stadi P	
Radiation	Mo$K_{\alpha 1}$ (λ = 70.93 pm)	
a (pm)	1248.4(5)	1235.8(7)
b (pm)	621.4(2)	614.5(3)
c (pm)	834.7(3)	827.6(3)
β (deg.)	97.2(1)	97.3(1)
Volume (Å^3)	642.4(3)	623.4(3)
Lattice parameters from single crystal data		
Single-crystal diffractometer	Nonius Kappa CCD	
Radiation	MoK_{α} (λ = 71.073 pm)	
a (pm)	1252.0(3)	1239.2(3)
b (pm)	623.2(2)	616.4(2)
c (pm)	836.9(2)	830.6(2)
β (deg.)	97.3(1)	97.3(1)
Volume (Å^3)	647.8(2)	629.3(2)
Formula units per cell	Z = 4	
Temperature (K)	293(2)	
Calculated density (g·cm^{-3})	6.628	6.988
Crystal size (mm^3)	0.02 × 0.03 × 0.04	0.03 × 0.03 × 0.03
Absorption coefficient (mm^{-1})	30.421	35.317
F (000), e	1108	1132
θ range (deg.)	3.28 ≤ θ ≤ 32.49	3.31 ≤ θ ≤ 32.49
Range in h k l	±18, ±9, ±12	±18, ±9, ±12
Total no. reflections	3763	3711
Independent reflections	1181	1146
	(R_{int} = 0.0484, R_{σ} = 0.0396)	(R_{int} = 0.0380, R_{σ} = 0.0303)
Reflections with $I > 2\sigma(I)$	1030	1066
Data / parameters	1181 / 76	1146 / 76
Absorption correction	multi-scan	multi-scan

Goodness-of-fit (F^2)	1.098	1.084
Final R indices ($I > 2\sigma(I)$)	R1 = 0.0278	R1 = 0.0181
	wR2 = 0.0601	wR2 = 0.0418
R indices (all data)	R1 = 0.0348	R1 = 0.0204
	wR2 = 0.0626	wR2 = 0.0426
Largest differ. peak, deepest hole (e/Å^{-3})	3.4 / −3.2	1.2 / −1.5

The intensity data of single crystals of $RE_3(BO_3)_2F_3$ (RE = Gd, Dy) were collected at room temperature by use of a Kappa CCD diffractometer. All relevant details of the data collection and evaluation are listed in Table 4.2-4.

According to the systematic extinctions, the monoclinic space groups $C2/c$ and Cc were derived. The structure solution in $C2/c$ (no. 15) succeeded for both compounds. A disordered site was detected in the structure, leading to the assumption that a fully ordered structure could be derived in a lower-symmetric space group, but attempts to solve the structure with a symmetry descent did not succeed. Nevertheless, additional crystallographic checks concerning the space groups Cc and $C2$ were undertaken in cooperation with Dr. R.-D. Hoffmann (University of Münster); results were not present at the end of this thesis.

The crystal structure of $RE_3(BO_3)_2F_3$ (RE = Gd, Dy) comprises of isolated BO_3-groups, rare-eath cations, and fluoride anions. The structure can be described via alternating layers of the formal compositions "$REBO_3$" and "REF_3", spreading into the bc-plane. Further information about the layer structure and the structural relationship of $RE_3(BO_3)_2F_3$ to the fluoride borates $Gd_2(BO_3)F_3$ and $RE_5(BO_3)_2F_9$ can be found in Sections 4.2.1 - 4.2.3.

The description of disordered crystal structures is contentious. It is argued, that for each disordered model a fully-ordered structure can be derived due to a symmetry descent or a superlattice. On the other hand, a disordered model is describing the same structural features without a commitment on the exact distribution in the crystal. The difference lies in the size of the domains inside of the crystal, depending whether they are smaller or larger than the

4.2 Rare-Earth Fluoride and Fluorido Borates

coherence wavelength. In our case, we were not able to determine the disorder problem with the analytical methods given and it was not possible to derive an ordered model from our single crystal data in the space groups Cc and $C2$ yet. We therefore stick to a disordered model, which has been previously described in the literature, since it satis-factoringly describes the structural features of $RE_3(BO_3)_2F_3$ (RE = Gd, Dy).

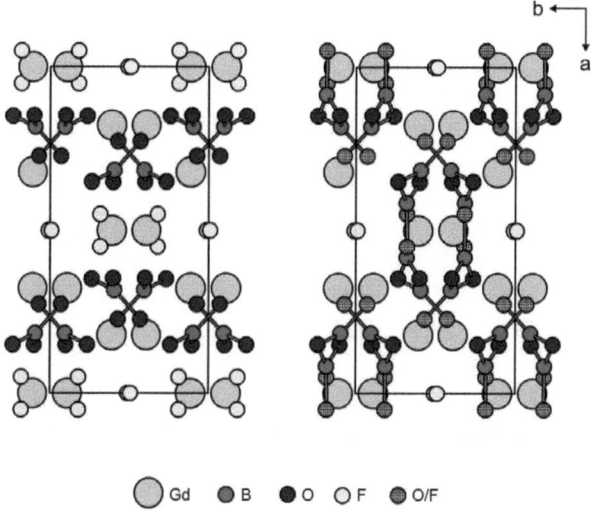

Figure 4.2-17
Fully ordered structure of $Gd_3(BO_3)_2F_3$ as proposed by Corbel et al. (left) and the disordered model from the single crystal structure determination (right), displaying two different orientations of the BO_3-group.

For $Gd_3(BO_3)_2F_3$, Corbel et al. refined a structure with a single crystallographically independent BO_3-group [218], as depicted in the left part of Figure 4.2-17. The single crystal data of the compounds $RE_3(BO_3)_2F_3$ (RE = Gd, Dy) now revealed a split position for the boron atom, leading to the atomic positions B1a and B1b (Figure 4.2-17 right and Table 4.2-5). Two positions for the boron atoms at the same time connote two possible different orientations of the BO_3-groups. In the fully ordered Corbel structure (Figure 4.2-17 left), a fluoride ion is present in close proximity to the BO_3-group. In the disordered structure derived from single crystal data, this fluoride ion is partially substituted by oxygen, while an oxygen anion of the BO_3-group is partially substituted by fluorine, leading to two mixed-occupied atom sites O3/F3 and F1/O4 (Figure 4.2-17 right and Figure 4.2-18). This results in two possible orientations of the BO_3-group, as depicted in Figure 4.2-18, which is in perfect agreement with the disorder detected in the fluoride borate $Gd_2(BO_3)F_3$ by Müller-Bunz et al. [219]. They were thus correct to propose a similar disorder model for $RE_3(BO_3)_2F_3$ on the basis of the relationship of the crystal structures.

Table 4.2-5

Atomic coordinates, isotropic equivalent displacement parameters (U_{eq} / Å^2), and side occupancy factors (s.o.f.) for $RE_3(BO_3)_2F_3$ (RE = Gd, Dy). U_{eq} is defined as one-third of the trace of the orthogonalized U_{ij} tensor.

Atom	Wyckoff site	x	y	z	U_{eq}	s.o.f.
$Gd_3(BO_3)_2F_3$						
Gd1	4e	0	0.88956(6)	1/4	0.0106(2)	
Gd2	8f	0.17832(2)	0.39016(4)	0.17765(3)	0.0086(2)	
B1a	8f	0.8164(8)	0.905(2)	0.480 (2)	0.012(2)	0.68(3)
B1b	8f	0.916(3)	0.816(5)	0.527(4)	0.05(2)	0.32(3)
O1	8f	0.6484(3)	0.6015(6)	0.3514(5)	0.0125(7)	
O2	8f	0.8502(3)	0.2364(6)	0.8677(5)	0.0110(7)	
O3/F3	8f	0.2727(4)	0.0713(7)	0.0819(5)	0.027(2)	0.68 / 0.32
F1/O4	8f	0.4499(3)	0.3178(7)	0.9630(5)	0.0245(9)	0.68 / 0.32
F2	4e	0	0.5097(8)	1/4	0.018(2)	
$Dy_3(BO_3)_2F_3$						
Dy1	4e	0	0.88591	1/4	0.00970(8)	
Dy2	8f	0.17871(2)	0.38921(3)	0.17749(2)	0.00754(7)	
B1a	8f	0.8162(6)	0.906(2)	0.4816(9)	0.018(2)	0.72(3)
B1b	8f	0.914(2)	0.813(3)	0.523(2)	0.020(5)	0.28(3)
O1	8f	0.6475(2)	0.6008(5)	0.3490(3)	0.0118(5)	
O2	8f	0.8509(2)	0.2385(4)	0.8671(3)	0.0093(5)	
O3/F3	8f	0.2742(3)	0.0713(5)	0.0828(4)	0.0229(7)	0.72 / 0.28
F1/O4	8f	0.4490(3)	0.3152(5)	0.9652(4)	0.0254(6)	0.72 / 0.28
F2	4e	0	0.5073(6)	1/4	0.0171(7)	

4.2 Rare-Earth Fluoride and Fluorido Borates

Neither Corbel et al. nor Müller-Bunz et al. were able to anisotropically refine disordered boron atoms in their crys-tal structures [218, 219]. The single crystal data of $RE_3(BO_3)_2F_3$ (RE = Gd, Dy) did not only allow the full refinement of both isotypic structures, additionally a statement on the occupancy of the split boron

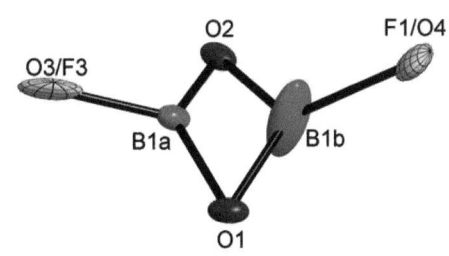

Figure 4.2-18
The two possible orientations of the BO_3-groups in $Gd_3(BO_3)_2F_3$. While position B1a is occupied to ~70 %, position B1b shows an occupancy of ~30 %.

atoms could be made. The free refinement of the boron positions resulted in a ration of B1a : B1b of 0.68(3) : 0.32(3) and 0.72(3) : 0.28(3) for $Gd_3(BO_3)_2F_3$ and $Dy_3(BO_3)_2F_3$, respectively (Table 4.2-5). The occupancy of the mixed-occupied atom sites O3/F3 and F1/O4 were then coupled to the refined occupancy factors of the boron positions, due to the fact that if B1a is occupied, O3 and F1 are present, while the occupancy of B1b leads to the presence of F3 and O4 (Figure 4.2-18). The single crystal structure refinement revealed that the atomic position of boron refined by Corbel et al. (which is B1a in our case) is the preferred orientation of the BO_3-groups (around 70 % occupancy).

High standard deviations can be observed for the isotropic and anisotropic displacement parameters of the boron atoms in both structures, due to the disorder. Especially for position B1b, larger ellipsoids are found, as depicted in Figure 4.2-18, as well as for the mixed-occupied atom sites O3/F3 and F1/O4. This also results in high standard deviations for the B–O distances (Table 4.2-6), and may be the average of two slightly shifted positions for the mixed-occupied atom sites. One can imagine an enlarged distance, when the O3/F3 and F1/O4 positions are occupied by fluorine, and a shorter distance when oxygen is present.

Table 4.2-6
Interatomic distances (pm) in $RE_3(BO_3)_2F_3$ (RE = Gd, Dy), calculated with the single-crystal lattice parameters.

$Gd_3(BO_3)_2F_3$					
Gd1–O1	234.9(4)	Gd2–O1	237.3(4)		
	2x	Gd2–O2	237.8(4)		
Gd1–O2	236.1(4)	Gd2–O1	257.5(4)		
	2x	Gd2–O2	263.0(4)		
Gd1–F2	236.8(5)	Gd2–F2	249.9(2)		
Gd1–F1/O4a	244.5(4)	Gd2–F1/O4	227.3(4)		
	2x	Gd2–O3/F3a	232.1(4)		
Gd1–F1/O4b	268.0(5)	Gd2–O3/F3b	234.3(4)		
	2x	Gd2–O3/F3c	249.5(4)		
	Ø = 244.9		Ø = 243.2		
B1a–O1	142(2)	B1b–O1	147(3)		
B1a–O2	139(2)	B1b–O2	151(2)		
B1a–O3	156(2)	B1b–O4	167(3)		
	Ø = 145.9		Ø = 155		
F1–Gd1a	244.5(4)	F2–Gd2	249.9(2)	F3–Gd2a	232.1(4)
F1–Gd1b	268.0(5)		2x	F3–Gd2b	234.3(4)
F1–Gd2	227.3(4)	F2–Gd1	236.8(5)	F3–Gd2c	249.5(4)
	Ø = 246.6		Ø = 245.6		Ø =238.6
$Dy_3(BO_3)_2F_3$					
Dy1–O1	232.2(3)	Dy2–O1	234.1(3)		
	2x	Dy2–O2	234.6(3)		
Dy1–O2	232.5(3)	Dy2–O1	255.1(3)		
	2x	Dy2–O2	260.8(3)		
Dy1–F2	233.3(4)	Dy2–F2	247.7(2)		
Dy1–F1/O4a	240.9(3)	Dy2–F1/O4	224.1(3)		
	2x	Dy2–O3/F3a	229.6(3)		
Dy1–F1/O4b	269.7(3)	Dy2–O3/F3b	232.2(3)		
	2x	Dy2–O3/F3c	246.9(3)		
	Ø = 242.7		Ø = 240.6		
B1a–O1	140.7(8)	B1b–O1	146(2)		
B1a–O2	142.3(8)	B1b–O2	149(2)		
B1a–O3	156.1(8)	B1b–O4	169(2)		
	Ø = 146.4		Ø = 155		

F1–Dy1a	240.9(3)	F2–Dy2	247.7(2)	F3–Dy2a	229.6(3)
F1–Dy1b	269.7(3)		2x	F3–Dy2b	232.2(3)
F1–Dy2	224.1(3)	F2–Dy1	233.3(4)	F3–Dy2c	246.9(3)
	Ø = 244.9		Ø = 242.9		Ø = 236.2

The average values for the B–O distances range between 145.9 and 155 pm (Table 4.2-6), which is extraordinarily large for interatomic distances in BO_3-groups. Typically, values around 137 pm are found, e.g. in borates with calcite structure ($AlBO_3$ (137.96(4) pm) [253], β-$YbBO_3$ (137.8(4) pm) [156], and $FeBO_3$ (137.9(2) pm) [254]). Corbel et al. reported B–O distances between 137 and 147 pm for the proposed position of the boron atom in $Gd_3(BO_3)_2F_3$.

The refined bond lengths in $RE_3(BO_3)_2F_3$ (RE = Gd, Dy) show high stan-

Table 4.2-7
Interatomic angles (deg.) in $RE_3(BO_3)_2F_3$ (RE = Gd, Dy), calculated with the single-crystal lattice parameters.

$Gd_3(BO_3)_2F_3$					
O1–B1a–O3	117.7(7)	O1–B1b–O4	126(3)		
O1–B1a–O2	124.7(7)	O1–B1b–O2	113(2)		
O3–B1a–O2	117.4(7)	O4–B1b–O2	119(2)		
	Ø = 119.9		Ø = 119.6		
Gd1a–F1–Gd1b	118.4(3)	Gd2a–F2–Gd2b	145.3(2)	Gd2a–F3–Gd2b	141.4(2)
Gd1a–F1–Gd2	133.6(2)	Gd2a–F2–Gd1	107.3(2)	Gd2a–F3–Gd2c	101.1(2)
Gd1b–F1–Gd2	104.6(2)	Gd2b–F2–Gd1	107.3(2)	Gd2b–F3–Gd2c	113.9(3)
	Ø = 118.9		Ø = 120.0		Ø = 118.8
$Dy_3(BO_3)_2F_3$					
O1–B1a–O3	118.6(5)	O1–B1b–O4	124(2)		
O1–B1a–O2	124.5(5)	O1–B1b–O2	117(2)		
O3–B1a–O2	116.7(5)	O4–B1b–O2	119(2)		
	Ø = 119.9		Ø = 119.3		
Dy1a–F1–Dy1b	118.2(2)	Dy2a–F2–Dy2b	145.8(2)	Dy2a–F3–Dy2b	142.4(2)
Dy1a–F1–Dy2	134.5(2)	Dy2a–F2–Dy1	107.1(2)	Dy2a–F3–Dy2c	101.0(2)
Dy1b–F1–Dy2	104.0(2)	Dy2b–F2–Dy1	107.1(2)	Dy2b–F3–Dy2c	113.6(2)
	Ø = 118.9		Ø = 120.0		Ø = 118.8

dard deviations, which holds also true for the O–B–O angles, especially for the BO_3-groups around B1b (Table 4.2-7). Large angles O1–B1a–O2 and O1–B1b–O4 between 124 and 126 ° underline the displacement of the boron atoms from the center of their BO_3-triangles. All these findings are evoked by the disorder in the BO_3-groups of the crystal structures. The large average B–O distances are mainly due to the large distances between the boron atoms and the mixed-occupied atom sites O3/F3 and F1/O4. This rather unfavorable stretching of the BO_3-groups might well be the reason for the disorder itself. The two possible BO_3-groups, which are formally sharing a common edge, are spanned by the coordination spheres of four surrounding rare-earth cations. This leads to a close packing of $RE^{3+}O/F_n$-polyhedra, which involves stretching of the BO_3-groups and thus enlarges the B–O-bond lengths. For $Gd_3(BO_3)_2F_3$, this is displayed in Figure 4.2-19. A similar enlargement of BO_3-bond lengths due to surrounding coordination spheres was previously observed in the fluoride borate $La_4B_4O_{11}F_2$ (Section 4.2.6).

In the crystal structure of $RE_3(BO_3)_2F_3$ (RE = Gd, Dy), there are two crystallographically independent RE^{3+} ions, which are ninefold coordinated by four oxygen atoms, one fluorine atom and four mixed-occupied positions (O/F) (Table 4.2-5 and Table 4.2-6). The average interatomic distances inside of the RE–O/F coordination polyhedra are 243.2 and

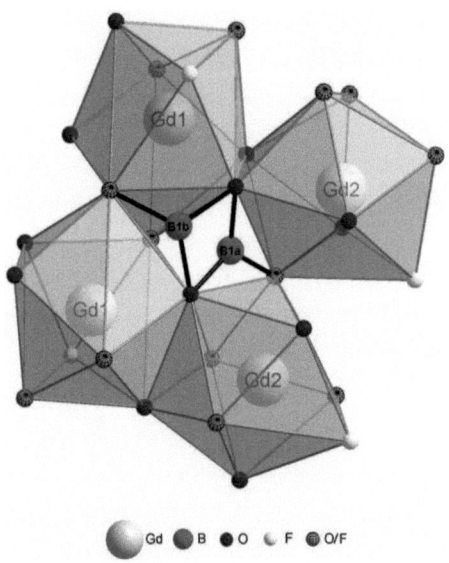

Figure 4.2-19

The two different orientations of the BO_3-groups in $Gd_3(BO_3)_2F_3$, which are spanned by the coordination spheres of the surrounding rare-earth cations.

4.2 Rare-Earth Fluoride and Fluorido Borates

244.9 pm for $Gd_3(BO_3)_2F_3$ and thus in the same range as the values given by Corbel et al. (242 and 246 pm) [218]. As expected from the smaller cationic radius, the average Dy–O/F distances are slightly shorter with 240.6 and 242.7 pm. The fluoride ions in $RE_3(BO_3)_2F_3$ (RE = Gd, Dy) are threefold coordinated by three RE^{3+} cations (Table 4.2-6).

Calculations of the charge distribution of the atoms in $RE_3(BO_3)_2F_3$ (RE = Gd, Dy) were performed for hypothetic, fully ordered compounds with position B1a occupied only. Bond valence sums (ΣV) with VaList (Bond Valence Calculation and Listing) [134] and with the CHARDI (Charge Distribution in Solids) concept (ΣQ) [135, 136] were obtained (Table 4.2-8). Here, especially the bond-length / bond strength calculations ΣV show remarkably low values of 2.41 and 2.37 for the boron atoms, which is a strong indicator for the described disorder. For the mixed-occupied positions O3/F3 and F1/O4 (O3 and F1 in Table 4.2-8, according to the preferred occupation) no remarkable deviations from a hypothetic fully ordered structure is found. The values for O3 in $Dy_3(BO_3)_2F_3$ are smaller than expected and thus in agreement with the disorder model, while the parameters for O3 in $Gd_3(BO_3)_2F_3$ and F1 in both structures are not significant.

Table 4.2-8
Bond valence sums in $RE_3(BO_3)_2F_3$ (RE = Gd, Dy), calculated with calculated with VaList (ΣV) and CHARDI (ΣQ).

$Gd_3(BO_3)_2F_3$								
	Gd1	Gd2	B1	O1	O2	O3	F1	F2
V	2.95	3.26	2.41	-2.02	-2.04	-1.88	-0.82	-0.78
Q	2.92	3.17	2.87	-1.65	-2.29	-2.14	-0.95	-0.94
$Dy_3(BO_3)_2F_3$								
	Dy1	Dy2	B1	O1	O2	O3	F1	F2
V	2.79	3.01	2.37	-1.91	-1.91	-1.75	-0.82	-0.78
Q	2.94	3.14	2.89	-2.19	-2.20	-1.71	-0.95	-0.94

The disorder model presented herein, was based on the findings of Müller-Bunz et al. [219]. It seems to be highly unvaforable to incorporate fluorine into the coordination sphere of the boron atoms instead of oxygen, when both elements are present. As a matter of fact, flurooxoborate anions were unknown, or not conclusively confirmed in inorganic solid state compounds, until Jansen et al. in 2009 presented the first and only unambiguously characterized fluoridoborate [215]. It is therefore unlikely to assume the presence of a BO_2F-group around B1b instead of two mixed-occupied O/F positions but it cannot be excluded on the basis of the data available. SS-NMR investigations are planned for the future to fully confirm the presented disorder model.

To assure the atom assignment in the structure, single crystals of our sample were subjected to elemental analysis via SEM/EDX experiments. The crystals showed average atomic Gd : B : O : F compositions (%) of 21 : 15 : 42 : 22 for $Gd_3(BO_3)_2F_3$ and average atomic Dy : B : O : F compositions (%) of 22 : 11 : 41 : 26 for $Dy_3(BO_3)_2F_3$. Due to the light weight of boron, measurements have to be taken with caution, but still, these results confirm the presence of all elements and the composition, obtained from the single crystal structure determination (calculated values (%) RE : B : O : F : 21.4 : 14.3 : 42.9 : 21.4).

Furthermore, we calculated the Madelung Part of Lattice Energy (MAPLE) [127-129] for $RE_3(BO_3)_2F_3$ (RE = Gd, Dy). The additive potential of the MAPLE values allows the calculation of hypothetical values for $RE_3(BO_3)_2F_3$ (RE = Gd, Dy), starting from binary oxides and fluorides. As result, we obtained a value of 41232 kJ/mol for $Gd_3(BO_3)_2F_3$ in comparison to 41315 kJ/mol (deviation: 0.2 %), starting from the binary components [Gd_2O_3 [257] (14987 kJ/mol) + B_2O_3–I [258] (20669 kJ/mol) + GdF_3 [259] (5659 kJ/mol)]. For $Dy_3(BO_3)_2F_3$, we derived a value of 41386 kJ/mol in comparison to 41688 kJ/mol (deviation: 0.7 %), starting from the binary components [Dy_2O_3 [257] (15291 kJ/mol) + B_2O_3–I [258] (20669 kJ/mol) + DyF_3 [259] (5728 kJ/mol)].

Raman FTIR spectra of single crystals of $RE_3(BO_3)_2F_3$ (RE = Gd, Dy) were recorded, as displayed in Figure 4.2-20. The absorption bands are in good agreement. The absorption patterns of the IR spectra are typical for borates

4.2 Rare-Earth Fluoride and Fluorido Borates

exhibiting triangular BO_3-groups [260, 261]. The strong and usually sharp absorptions derived from the out of plane bending (v_2) of the trigonal ion occur in the range of 600 –800 cm^{-1}. According to the crystalline environment, the absorption bands between 1000 - 1200 cm^{-1} can be classified as symmetric stretching vibrations (v_1). Asymmetric stretching vibrations (v_3) of the BO_3-groups are assigned in the areas of 1200 - 1600 cm^{-1}.

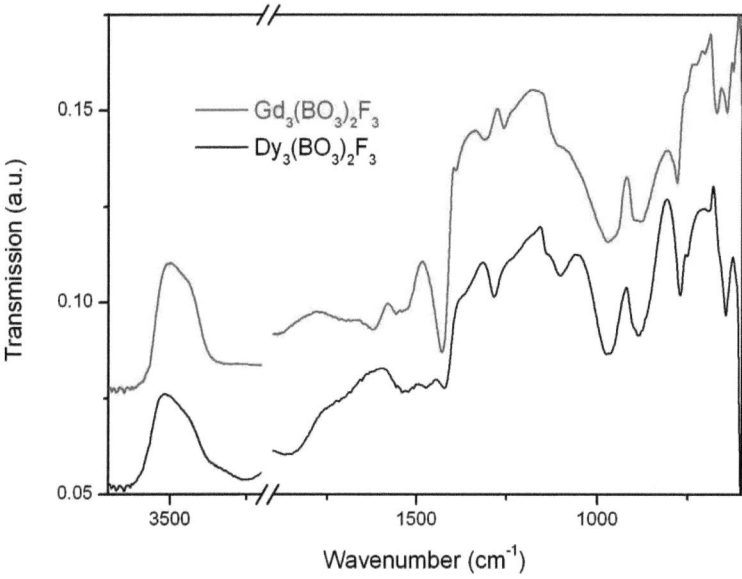

Figure 4.2-20
FTIR spectra of $Gd_3(BO_3)_2F_3$ (top) and $Dy_3(BO_3)_2F_3$ (bottom) in the range 600 – 4000 cm^{-1}.

Interestingly, two fairly strong bands between 3400 and 3600 cm^{-1} were detected in the single crystal spectra of $RE_3(BO_3)_2F_3$ (RE = Gd, Dy). Bands in this range usually indicate OH-groups in water-containing borates. Corbel *et al.* did not report the presence of hydroxyl groups in $RE_3(BO_3)_2F_3$ and it is nearly impossible to distinguish between fluoride ions and oxygen or hydroxyl ions by means of electron density or bond lengths. The substitution of fluoride with hydroxyl groups is a problem commonly known from fluoride borates [262, 263]. The single crystals, which were examined via IR spectroscopy, have been exposed to air and humidity. Thus, the exchange of fluoride with OH-groups occurs due to hygroscopic alteration. In apatites, the exchange of hydroxyl and fluoride ions was previously observed and studied [264, 265]. Nevertheless, from the elemental analyses, no deviation from the expected sum formula, indicating a loss of fluoride, could be detected.

Raman spectra of $RE_3(BO_3)_2F_3$ (RE = Gd, Dy) single crystals in the range 100 – 2000 cm^{-1} are displayed in Figure 4.2-21. In contrast to the FTIR spectra, only very small bands were observed between 3000 cm^{-1} and 4000 cm^{-1}, which is probably related to the comparatively low sensitivity of Raman spectroscopy for hydroxyl vibrational modes. The Raman spectra of the isostructural compounds $RE_3(BO_3)_2F_3$ (RE = Gd, Dy) are quite similar with the most intense bands < 500 cm^{-1} and the narrow, "isolated" band at ~950 cm^{-1} (Figure 4.2-21),

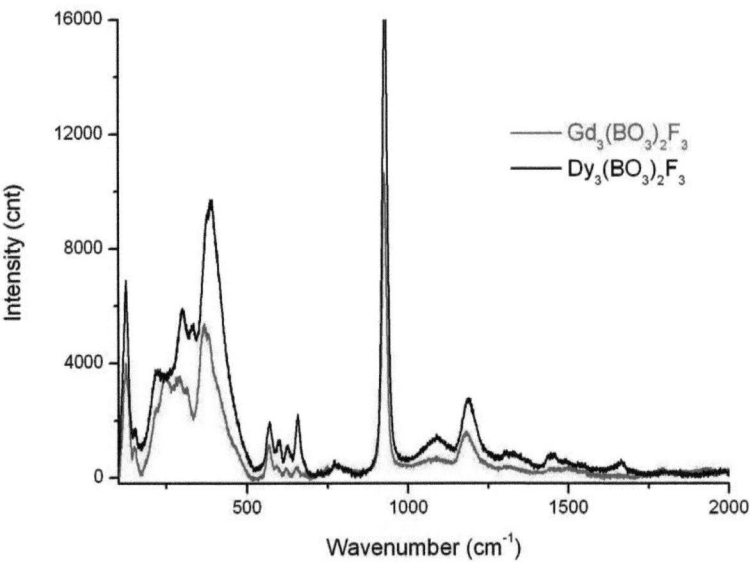

Figure 4.2-21
Raman spectrum of $Gd_3(BO_3)_2F_3$ (bottom) and $Dy_3(BO_3)_2F_3$ (top) in the range 100 – 2000 cm^{-1}.

and in good agreement with the literature [65, 147, 266-268]. Bands < 500 cm^{-1} can be assigned to the RE–O bond bending and stretching vibrations, as well as to the lattice vibrations. The most intense single band at ~950 cm^{-1} is assigned to symmetric stretching vibrations (v_1) of the isolated BO$_3$-groups and is also observed in other BO$_3$-group-containing phases, e.g. in $RE_5(BO_3)_2F_9$ (RE = Er - Yb) (Sections 4.2.1 - 4.2.3) and in λ-PrBO$_3$ (Section 4.1.3). Several small but sharp bands are observed between 550 and 700 cm^{-1} and correspond to pulse vibration modes (v_2) of the BO$_3$-groups. Above 1000 cm^{-1}, several broader bands are observed and assigned to the asymmetric stretching vibrations (v_3) of the BO$_3$-groups [147].

For the future, experiments on the optical and luminescent properties of single crystals of the compounds $RE_3(BO_3)_2F_3$ (RE = Gd, Dy) will be of great interest in respect to the disorder. Similar measurements were only performed on powder samples of Eu$_3$(BO$_3$)$_2$F$_3$ containing impurity phases [252] and should thus be further investigated. SS-NMR investigations are planned to fully confirm the presented disorder model and to exclude the presence of a BO$_2$F-group in the crystal structure.

4.2.10 $RE_{12}B_{11}O_{31}F_7$ (RE = La, Pr, Nd, Sm)

The compounds $RE_{12}B_{11}O_{31}F_7$ (RE = La, Pr, Nd, Sm) crystallize willingly in a wide pressure range: $La_{12}B_{11}O_{31}F_7$ was obtained at 6 GPa and 1300 °C, $Pr_{12}B_{11}O_{31}F_7$ and $Nd_{12}B_{11}O_{31}F_7$ at 3 GPa and 800 °C, and $Sm_{12}B_{11}O_{31}F_7$ at 5 GPa and 900 °C (Figure 4.2-22).

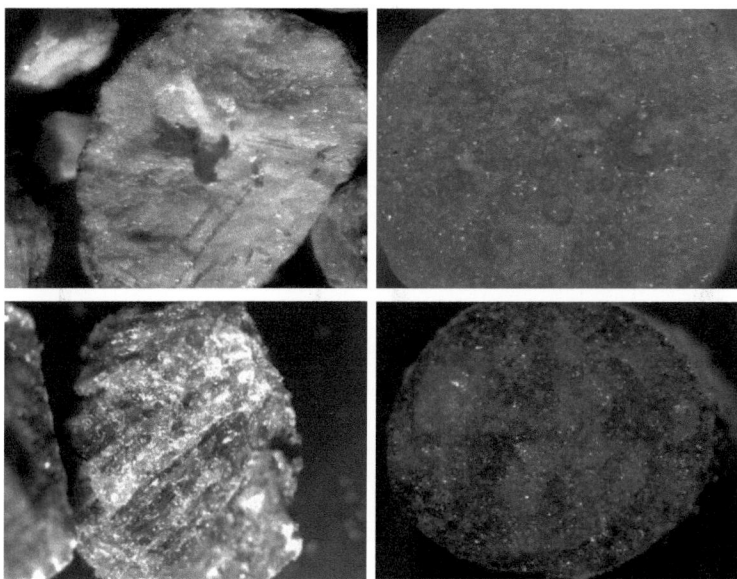

Figure 4.2-22
Crystalline samples of $La_{12}B_{11}O_{31}F_7$ (top left), $Pr_{12}B_{11}O_{31}F_7$ (top right), $Nd_{12}B_{11}O_{31}F_7$ (bottom left), and $Sm_{12}B_{11}O_{31}F_7$ (bottom right)

For all compounds, single crystals could be isolated and measured. The lattice parameters for all compounds are given in Table 4.2-9. Structural refinements in the triclinic space group $P\bar{1}$ were possible, leading to very good R-values. The obtained crystal structure is displayed in Figure 4.2-23. It is comprised of isolated BO_3-groups and the fundamental building block (FBB) 2 Δ☐ : Δ☐Δ which was reported for $La_4B_4O_{11}F_2$ (Section 4.2.6) for the first time.

Table 4.2-9

Lattice parameters for $RE_{12}B_{11}O_{31}F_7$ (RE = La, Pr, Nd, Sm)

	$La_{12}B_{11}O_{31}F_7$	$Pr_{12}B_{11}O_{31}F_7$	$Nd_{12}B_{11}O_{31}F_7$	$Sm_{12}B_{11}O_{31}F_7$
a (pm)	773.4(2)	763.5(1)	759.9(1)	751.3(1)
b (pm)	778.6(2)	767.1(1)	763.6(1)	755.2(1)
c (pm)	1386.9(3)	1371.8(1)	1366.3(1)	1351.7(1)
α (deg)	92.25(3)	92.62(1)	92.19(1)	92.23(1)
β (deg)	105.90(3)	105.67(1)	106.01(1)	105.79(1)
γ (deg)	113.72(3)	113.92(1)	114.21(1)	114.12(1)
V (Å^3)	724.6(3)	696.0(1)	684.6(1)	663.7(1)

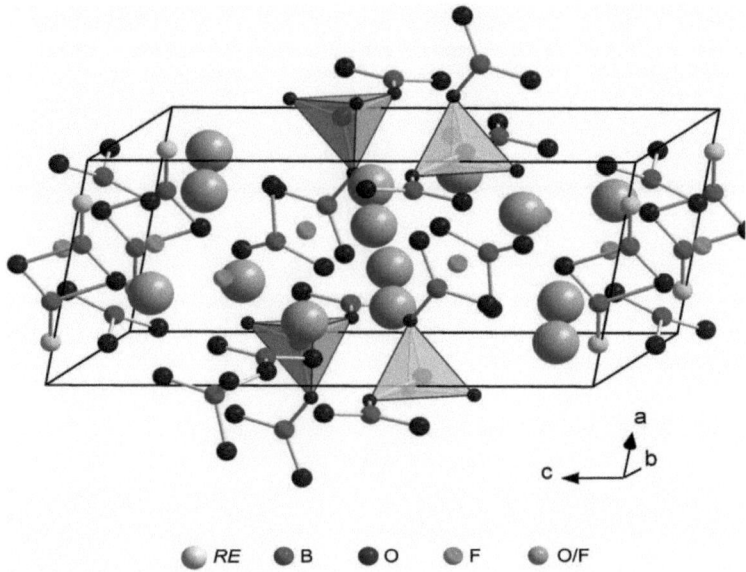

● RE ● B   ● O   ● F   ● O/F

Figure 4.2-23

Determined crystal structure of $RE_{12}B_{11}O_{31}F_7$ (RE = La, Pr, Nd, Sm)

4.2 Rare-Earth Fluoride and Fluorido Borates

Even though the crystal structures can be solved with very good R-values, several problems occur. All phases $RE_{12}B_{11}O_{31}F_7$ (RE = La, Pr, Nd, Sm) exhibit a site of disorder. Looking at the a-axis in Figure 4.2-23, there appear to be two edge-sharing BO_3-groups. This is not the case, because the two adjacent boron positions are only partially occupied. This means that two possible orientations of the BO_3-group can be found, as described earlier for the case of $Pr_4B_3O_{10}F$ (Section 0). A closer look at this disorder site is given in Figure 4.2-24. On the left side, the two possible orientations are displayed, leading to a full BO_3-group and a fluoride ion in close proximity to the BO_3-group. In the disordered averaged structure (Figure 4.2-24 right) derived from single crystal data, this fluoride ion is partially substituted by oxygen, while an oxygen anion of the BO_3-group is partially substituted by fluorine, leading to two mixed-occupied atom sites O/F.

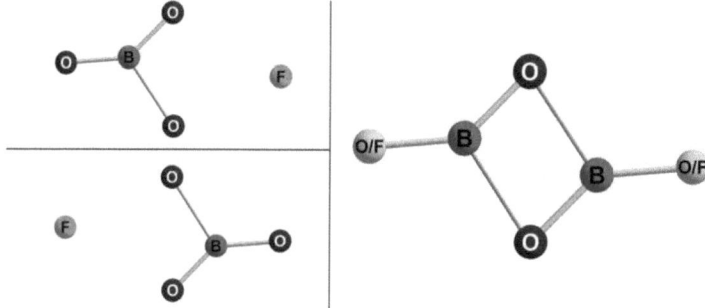

Figure 4.2-24
Disordered BO_3-group with the two possible orientations (left) and the averaged model (right).

The description of disordered crystal structures is contentious. It is argued that for each disordered model a fully-ordered structure can be derived due to a symmetry descent or a superlattice. Additionally, the anisotropic displacement parameters for $RE_{12}B_{11}O_{31}F_7$ (RE = La, Pr, Nd, Sm) were ill-behaving, indicating an intrinsic problem of the crystals. We therefore suspected an inversion twinning and tried to solve the structures in space group $P1$, resulting in even worse parameters. Up to now, it was not possible to obtain a satisfying crystal structure refinement of any of the isotypic phases. The presented disordered model is describing the structural features of $RE_{12}B_{11}O_{31}F_7$ (RE = La, Pr, Nd, Sm) without a commitment on the exact distribution in the crystal. Whether the inversion twinning or the disorder model is correct, depends on the size of the domains inside of the crystal. We were not able to determine this with the analytical methods given and it was not possible to derive a flawless model from our single crystal data in both the space groups $P\bar{1}$ and $P1$. Transmission electron microscopy and low-temperature X-ray measurements are planned on the crystals but the results are still outstanding.

Temperature-dependent magnetic susceptibility measurements of a polycrystalline $Nd_{12}B_{11}O_{31}F_7$ sample were performed in the group of Prof. Dr. R. Pöttgen. The magnetic measurements were carried out on a Quantum Design Physical Property Measurement System (PPMS) using the VSM option. For the VSM measurements 16.168 mg of $Nd_{12}B_{11}O_{31}F_7$ was packed in a polypropylene powder sample holder and attached to the sample holder rod for measuring the magnetic properties in the temperature range 3–300 K with magnetic flux densities up to 80 kOe. The temperature dependence of the magnetic and the inverse magnetic susceptibility of $Nd_{12}B_{11}O_{31}F_7$ are displayed in Figure 4.2-25. The compound shows Curie-Weiss behaviour above 75 K.

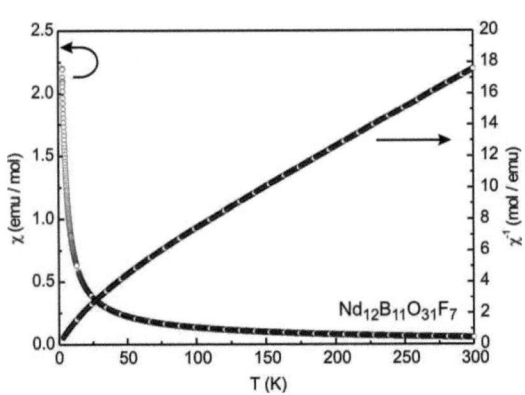

Figure 4.2-25

Temperature dependence (3–300 K data) of the magnetic susceptibility (χ and χ^{-1} data) of $Nd_{12}B_{11}O_{31}F_7$ measured at 10 kOe.

A fit of the 75–300 K data according to χ = C / (T − θ_p) revealed an effective magnetic moment of $\mu_{eff} = (8C)^{1/2} = 3.62(2)$ µ$_B$ / Nd atom and a paramagnetic Curie temperature of θ$_p$ = −46.5(5) K. The experimental magnetic moment is in perfect agreement with the free ion value of 3.62 µ$_B$ for Nd^{3+} [269]. Below 75 K we observe deviation from the Curie-Weiss law, most likely due to crystal field splitting of the $^4I_{9/2}$ ground term, similar to NdOF [270] or to Ba$_4$Nd$_7$[Si$_{12}$N$_{23}$O][BN$_3$] [271]. No magnetic ordering down to 3 K was observed for Nd$_{12}$B$_{11}$O$_{31}$F$_7$.

The magnetization isotherms measured at 5 and 50 K are presented in Figure 4.2-26. At 50 K we observe a linear increase of the magnetization with small absolute values as expected for a paramagnetic material. A field induced increase of the magnetization occurs at 5 K. The magnetization at 80 kOe and 5 K of 1.15(1) µ$_B$/Nd atom is much smaller than the theoretically possible ordered moment of $g \times J = 3.27$ µ$_B$ for Nd^{3+}.

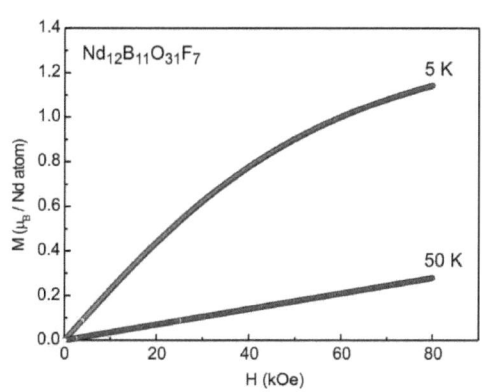

Figure 4.2-26
Magnetization isotherms of Nd$_{12}$B$_{11}$O$_{31}$F$_7$ at 5 and 50 K.

4.3 Rare-Earth Oxides

In high-pressure chemistry, the stability range of the obtained phases is usually quite small. Often side phases are observed as by-products of a desired reaction. In the case of fluoride borate syntheses, this can be rare-earth borates, oxyfluorides, or oxides. The latter can be present either due to an excess of the starting oxide or due to the evaporation of gaseous HF during the syntheses in the presence of humidity. Many materials show differing polymorphs under ambient and elevated pressure, which is of use for calibration purposes (Section 2.6). This is also the case for numerous rare-earth oxides, meaning that a pressure transformation of an educt can be observed besides a reaction. For the syntheses of praseodymium fluoride borates, Pr_6O_{11} was used as a starting reagent. In one of the samples, dark brown crystals of the high-pressure modification of PrO_2 (HP-PrO_2) could be found (Figure 4.3-1). While the isotypic phase HP-CeO_2 is well described in the literature, very few high-pressure studies were performed on the praseodymium compound and no single-crystal measurements and crystal structure parameters were reported so far. The following publication presents a detailed description of the crystal structure.

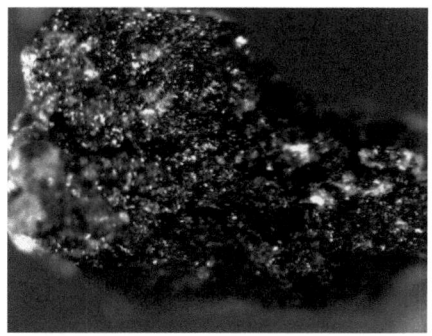

Figure 4.3-1
Crystalline sample of HP-PrO_2

ARTICLE

DOI: 10.1002/zaac.200900441

Orthorhombic HP-PrO₂ – High-Pressure Synthesis and Crystal Structure

Almut Haberer[a] and Hubert Huppertz*[a]

Dedicated to Professor Hans-Jörg Deiseroth on the Occasion of His 65th Birthday

Keywords: Praseodymium oxide; High-pressure; Solid-state structures; Rare Earths; Synthetic methods

Abstract. The high-pressure phase of PrO_2 was synthesized under conditions of 5.5 GPa and 1000 °C in a Walker-type multianvil apparatus. The crystal structure was determined on the basis of single-crystal X-ray diffraction data collected at room temperature. The title compound crystallizes in the orthorhombic α-$PbCl_2$-type structure with the lattice parameters $a = 637.6(3)$ pm, $b = 384.4(2)$ pm, and $c = 703.1(3)$ pm. The praseodymium ion is ninefold coordinated, the oxygen ions show either tetrahedral or square-pyramidal coordination spheres.

Introduction

The field of lanthanide dioxides LnO_2 with a cubic fluorite structure was extensively studied in the past. Among these components, which are potentially applicable as laser hosts and optical component materials, CeO_2 has a wide range of applications in the industry under the name of Ceria. Being of great interest for the catalytic industry, it is widely used as oxygen storage in automobile exhaust catalysts [1–3]. PrO_2 is an insulator isostructural with UO_2 and NpO_2 at room temperature and exhibits antiferromagnetic ordering below $T_N = 13.5$ K. It can be synthesized in high purity by electrocrystallization [4]. PrO_2 was found to have an anomalously small-ordered moment in the antiferromagnetic phase [5]. Further studies revealed a cooperative Jahn–Teller distortion at $T_D = 120 \pm 2$ K and a related distortion of the antiferromagnetic structure below T_N [6]. In order to determine the electronic ground state of PrO_2, many studies were performed, including core-level photoabsorption, photoemission spectra, and neutron scattering [7, 8]. The publication of an influential model [9], which describes an intermediate valence state in the anomalous cuprate $PrBa_2Cu_3O_{6+x}$ (PrBCO) similar to that proposed for the lanthanide dioxides, has increased interest in PrO_2, making it a key reference compound in high-energy spectroscopic studies of PrBCO and related materials.

Several studies were performed on the high-pressure modifications of CeO_2 and PrO_2 in the past years. Crystallizing into the cubic fluorite structure under ambient pressure conditions, a transition of these compounds into the orthorhombic α-$PbCl_2$-type (cotunite) structure under pressure was reported. In the case of CeO_2, many high-pressure studies were performed and a crystal structure, obtained from *in situ* X-ray powder diffraction data by *Dyclos* et al. is found in the database [10]. The phase transition in CeO_2 occurs around 31 GPa [11] and bulk moduli of 230 ± 10 GPa [10] and 220(9) GPa [12] were measured. The computational value of 357 GPa obtained by *Hill* et al., was significantly larger [13], whereas a study of *Gerward* et al. led to a lower calculated zero-pressure bulk modulus of 176.9 GPa [12]. Several computational studies concerning electronic, optical, and bonding properties were carried out [14–16]. The oxygen storage and surface properties were studied by *Skorodumova* et al. [17, 18], additionally the potential of CeO_2 as an UV absorber was investigated [19].

On PrO_2, very few high-pressure studies were performed. Whereas the transformation into the orthorhombic high-pressure phase was measured *in situ* in a diamond-anvil cell with laser heating [20, 21], no single-crystal measurements and crystal structure parameters were reported so far. In high-pressure experiments by *Liu*, the fluorite structure of PrO_2 was stable up to 20 GPa; a phase transition into a possible δ-Ni_2Si type was postulated [20]. A zero-pressure bulk modulus of 187.8 GPa for PrO_2 was investigated [12]; its computational values are 250 GPa and 176.8 GPa [22].

This work presents a synthesis, by which single-crystals of the high-pressure phase HP-PrO_2 were yielded for the first time. A detailed description of the crystal structure, obtained from single-crystal X-ray diffraction measurements, is given.

Experimental Section

Single-crystals of orthorhombic HP-PrO_2 were obtained as a by-product during a synthesis in the study of high-pressure rare-earth fluoride borates. Pr_6O_{11} (Strem Chemicals, 99.9 %), B_2O_3 (Strem Chemicals, 99.9+%), and PrF_3 (Strem Chemicals, 99.9 %) were mixed at a molar ratio of 1:2:1 and filled into a boron nitride crucible (Henze BNP GmbH, HeBoSint® S10, Kempten, Germany). This crucible was placed into the center of an 18/11-assembly, which was compressed by eight tungsten carbide cubes (TSM-20 Ceratizit, Reutte, Austria). The details of preparing the assembly can be found in ref. [23–27].

---

* Prof. Dr. H. Huppertz
  E-Mail: Hubert.Huppertz@uibk.ac.at
[a] Institut für Allgemeine, Anorganische und Theoretische Chemie
  Leopold-Franzens-Universität Innsbruck
  Innrain 52a
  6020 Innsbruck, Austria

4.3 Rare-Earth Oxides

ARTICLE

A. Haberer, H. Huppertz

Pressure was applied by a multianvil device, based on a Walker-type module, and a 1000-ton press (both devices from the company Voggenreiter, Mainleus, Germany). The sample was compressed up to 5.5 GPa for 2.5 h, afterwards heated to 1000 °C in 15 min and kept there for 20 min. Afterwards, the sample was cooled down to 600 °C in 20 min, followed by quenching to room temperature after switching off heating. Decompression required 7.5 h. The recovered experimental MgO-octahedron (pressure transmitting medium, Ceramic Substrates & Components Ltd., Newport, Isle of Wight, UK) was broken apart and the sample carefully separated from the surrounding boron nitride crucible, obtaining dark brown irregularly-shaped crystals of HP-PrO$_2$.

An experiment with pure Pr$_6$O$_{11}$ under the conditions mentioned above did not produce the orthorhombic HP-PrO$_2$ phase.

Crystal-Structure Analysis: The sample was characterized by powder X-ray diffraction, which was performed in transmission geometry on a flat sample of the reaction product with a STOE STADI P powder diffractometer with Mo-$K_{\alpha 1}$ radiation (Ge monochromator, λ = 71.073 pm). Reflections of orthorhombic HP-PrO$_2$ were identified, which tally well with the theoretical pattern simulated from single-crystal data. Additional reflections in the powder pattern of the sample show the presence of other yet unknown products. Because of this, a robust refinement with TOPAS was performed on the sample (Figure 1) [28]. The refined cell parameters for HP-PrO$_2$ from powder data resulted in a = 638.52(2) pm, b = 384.48(2) pm, and c = 702.70(2) pm, with a volume of 172.52(1) Å^3. This confirms the lattice parameters obtained from single-crystal X-ray diffraction (Table 1). The atomic position parameters were refined as well; they are in agreement with the parameters listed in Table 2. The whole robust refinement included 35 parameters.

Intensity data of a single-crystal of HP-PrO$_2$ were collected at room temperature with a Kappa CCD diffractometer (Bruker AXS/Nonius, Karlsruhe), equipped with a Miracol Fiber Optics Collimator and a Nonius FR590 generator (graphite-monochromatized Mo-$K_{\alpha 1}$ radiation, λ = 71.073 pm). An absorption correction, based on multi-scans [29], was applied to the data set. All relevant details of the data collection and evaluation are listed in Table 1.

The structure solution and the parameter refinement (full-matrix least-squares against F^2) were successfully performed, using the SHELX-97 software suite [30, 31] with anisotropic atomic displacement parameters for all atoms. According to the systematic extinctions, the orthorhombic space groups $Pnma$ and $Pn2_1a$ were derived. The structure solution was successfully performed in $Pnma$ (no. 62). The final difference Fourier syntheses did not reveal any significant residual peaks in all refinements. The positional parameters of the refinements, anisotropic displacement parameters, interatomic distances, and interatomic angles are listed in the Table 2, 3, 4. Further information of the crystal

Table 1. Crystal data and structure refinement of HP-PrO$_2$.

Empirical Formula	PrO$_2$
Molar mass /g·mol^{-1}	172.91
Crystal system	orthorhombic
Space group	$Pnma$ (No. 62)
Lattice parameters from powder data	
Powder diffractometer	Stoe Stadi P
Radiation	Mo-$K_{\alpha 1}$ (λ = 71.073 pm)
a /pm	638.52(2)
b /pm	384.48(2)
c /pm	702.70(2)
Volume /Å^3	172.52(1)
Single-crystal data	
Single-crystal diffractometer	Bruker AXS/Nonius Kappa CCD
Radiation	Mo-$K_{\alpha 1}$ (λ = 71.073 pm)
a /pm	637.1(2)
b /pm	383.90(8)
c /pm	702.6(2)
Volume /Å^3	171.84(6)
Formula units per cell	4
Temperature /K	293(2)
Calculated density /g·cm^{-3}	6.684
Crystal size /mm	0.04 × 0.03 × 0.02
Absorption coefficient /mm^{-1}	27.85
$F(000)$	300
θ range /deg	4.3 ≤ θ ≤ 32.4
Range in $h\,k\,l$	±9, ±5, ±10
Total no. reflections	2222
Independent reflections	349 (R_{int} = 0.0304)
Reflections with $I > 2\sigma(I)$	341 (R_σ = 0.0181)
Data/parameters	349/20
Absorption correction	Multi-scan [29]
Goodness-of-fit [F^2]	1.137
Final R indices [$I > 2\sigma(I)$]	R_1 = 0.0144, wR_2 = 0.0380
R indices (all data)	R_1 = 0.0150, wR_2 = 0.0384
Largest differ. peak, deepest hole /e·Å^{-3}	1.33/−1.18

Orthorhombic HP-PrO₂ – High-Pressure Synthesis and Crystal Structure

Figure 1. Robust refinement of the experimental powder pattern (top) of the reaction sample, identifying orthorhombic HP-PrO₂ (top) as a product. The resulting difference curve (bottom) shows the reflection profile of the yet unidentified side products.

Table 2. Atomic coordinates and isotropic equivalent displacement parameters (U_{eq} /Å^2) for HP-PrO₂ (space group: *Pnma*). U_{eq} is defined as one-third of the trace of the orthogonalized U_{ij} tensor.

Atom	Wyckoff site	x	y	z	U_{eq}
Pr	4c	0.27334(4)	1/4	0.39881(4)	0.0091(2)
O1	4c	0.1462(6)	1/4	0.0799(5)	0.0103(6)
O2	4c	0.0240(5)	1/4	0.6651(5)	0.0115(6)

structure is available from the Fachinformationszentrum Karlsruhe (E-Mail: crysdata@fiz-karlsruhe.de), 76344 Eggenstein-Leopoldshafen, Germany, by quoting the Registry No. CSD-380398.

Table 3. Anisotropic displacement parameters (U_{ij} /Å^2) for HP-PrO₂ (space group: *Pnma*).

Atom	U_{11}	U_{22}	U_{33}
Pr	0.0092(2)	0.0085(2)	0.0098(2)
O1	0.010(2)	0.010(2)	0.012(2)
O2	0.012(2)	0.015(2)	0.008(2)

	U_{12}	U_{13}	U_{23}
Pr	0	–0.00078(8)	0
O1	0	–0.002(2)	0
O2	0	0.005(2)	0

Results and Discussion

Several studies were performed on the high-pressure modifications of CeO₂ and PrO₂ in the past years. Whereas these compounds crystallize in the cubic fluorite structure under ambient pressure conditions, a transition into the orthorhombic α-PbCl₂-type structure was reported under high-pressure conditions. It was also postulated that large metal-ion dioxides like CeO₂ and PrO₂ might be transformed into a δ-Ni₂Si-type struc-

Table 4. Interatomic distances /pm and angles /deg in HP-PrO₂, calculated with the single-crystal lattice parameters.

Interatomic distances /pm		
Pr–O1a (2 ×) 235.9(2)	Pr–O1b 238.0(4)	Pr–O1c 238.3(3)
Pr–O2a 245.4(3)	Pr–O2b (2 ×) 273.4(3)	Pr–O2c (2 ×) 283.7(3)

Interatomic angles /deg		
Pr–O1–Pr (2 ×) 104.5(2)	Pr–O2–Pr	85.2(2)
Pr–O1–Pr 106.3(2)	Pr–O2–Pr (2 ×)	86.29(3)
Pr–O1–Pr (2 ×) 108.86(9)	Pr–O2–Pr	89.2(2)
Pr–O1–Pr 108.9(2)	Pr–O2–Pr (2 ×)	98.4(2)
Pr–O1–Pr (2 ×) 115.67(9)	Pr–O2–Pr (2 ×)	108.86(9)
	Pr–O2–Pr (2 ×)	152.3(2)
Ø = 109.2		Ø = 106.6

ture at high pressures [20]. According to *Jeitschko*, axial ratios can be used to distinguish between these very similar structure types, which differ only slightly in their atomic parameters [32]. For HP-CeO₂, the axial ratio led to a classification as an α-PbCl₂-type structure [10].

Single-crystal data for HP-PrO₂ were now obtained for the first time, leading to lattice parameters of a = 637.1(2) pm, b = 383.90(8) pm, and c = 702.6(2) pm (Table 1). The lattice parameters reported for the corresponding cerium phase are about 10 % smaller (a = 545.7 pm, b = 342.7 pm, c = 652.1 pm) [10]. This can be easily explained, because the *in situ* X-ray measurement of the lattice parameters was performed on a compressed structure at 70 GPa, whereas our metastable crystals were measured at ambient pressure conditions.

Looking at the axial ratios of HP-PrO₂, we obtain values of c/b = 1.83 and a/b = 1.66. These results tally well with the values known for HP-CeO₂ (c/b = 1.9 and a/b = 1.59) [10]. For the differentiation between the δ-Ni₂Si-type and the α-PbCl₂-type structure, the parameter a/b is important. For the δ-Ni₂Si type, axial ratios between 1.3 and 1.4 are reported, whereas the a/b ratios for the α-PbCl₂ type are in the range of 1.5 to 1.75 [10, 33]. Hence, HP-PrO₂ can be classified as an α-PbCl₂-type structure. The α-PbCl₂ structure type contains cations in an almost hexagonal closest-packing, whereas the anions are coordinated tetrahedrally or square-pyramidally, but both distorted [34]. Figure 2 gives an overall view of the structure, showing the coordination of the praseodymium cations.

As listed in Table 2, one praseodymium cation and two oxygen anions can be identified in the structure. The atomic parameters correspond well with the parameters reported for the high-pressure phase of CeO₂ [10]. In Figure 3, the coordination spheres of the oxygen anions are displayed; the dark grey polyhedra indicate tetrahedra, whereas the light grey polyhedra stand for square pyramids.

The praseodymium ion is ninefold coordinated with Pr–O distances ranging from 235.9(2) to 283.7(3) pm (Table 4). These distances are larger than the Pr–O distances of 233.5 pm measured for the ambient-pressure phase of PrO₂, as expected because of the increase of the coordination number from eight

ARTICLE

A. Haberer, H. Huppertz

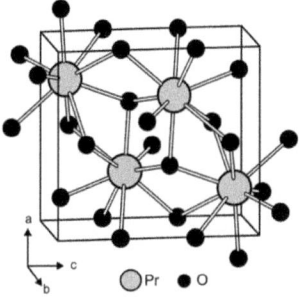

Figure 2. HP-PrO$_2$ in the α-PbCl$_2$ structure type.

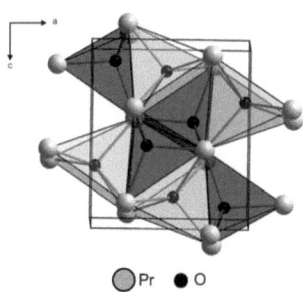

Figure 3. View of the HP-PrO$_2$ crystal structure along [0$\bar{1}$0]. The oxygen ions are either tetrahedrally (dark grey polyhedra) or square-pyramidally coordinated (light grey polyhedra).

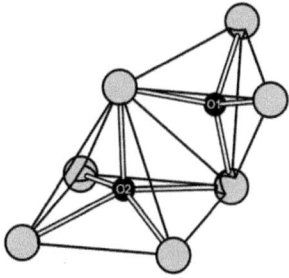

Figure 4. Tetrahedral and square-pyramidal coordination of the oxygen ions in HP-PrO$_2$.

distances between the praseodymium ions spanning the basal planes are 381.04(6) and 383.90(8) pm and the angles are 90°, thus the basal plane of the pyramid forms an almost perfect square. The Pr–O bond length between the oxygen ion and the praseodymium ion, which forms the tip of the pyramid, is 245.4(3) pm. If the central atom is located in the center of the pyramid, all Pr–O distances should be equal. This would result in angles of around 85° and 145° between the praseodymium ions of the basal plane [35]. In HP-PrO$_2$ we find values of 85.2(2), 86.29(3), and 89.2(2)° for the small angle and of 152.3(2)° for the large angle because of the off-center position of the central atom. The angles between the basal plane atoms and the pyramid tip atom should be 108°; in HP-PrO$_2$, angles of 98.4(2) and 108.86(9)° are reached. According to the shorter Ce–O bond lengths in HP-CeO$_2$, the coordination polyhedra are smaller than in the HP-PrO$_2$ phase, but their characteristics are similar.

to nine. The Ce–O distances, reported for the orthorhombic HP-CeO$_2$, vary between 205.4 and 266.9 pm, being in the same range [10]. The shorter bond lengths in HP-CeO$_2$ agree with the smaller lattice parameters relative to HP-PrO$_2$, because of the *in situ* measurement at 70 GPa.

The two oxygen ions are either fourfold or fivefold coordinated, as displayed in Figure 4. The interatomic angles of the coordination polyhedra are listed in Table 4. For the tetrahedrally coordinated O1 atom, the angles show an average value of 109.2°, very close to the ideal tetrahedral angle. The Pr–O distances lay all between 235.9(2) and 238.3(3) pm, so no distortion of the tetrahedron is found. The coordination sphere of O2 is square-pyramidal. The oxygen atom is located 66.6 pm from the basal plane and slightly shifted towards one side of the pyramid. This results in Pr–O distances of 273.4(3) pm to two neighboring praseodymium ions of the basal plane and of 283.7(3) pm to the other two praseodymium ions. The

Whereas in the literature no phase transformation of PrO$_2$ was reported up to 20 GPa at room temperature [20], we obtained the single-crystals of the high-pressure phase at only 5.5 GPa and 1000 °C, starting from a reaction mixture of Pr$_6$O$_{11}$, B$_2$O$_3$, and PrF$_3$. A similar experiment with only Pr$_6$O$_{11}$ as a reactant failed in the orthorhombic high-pressure phase. Therefore, probably, the other starting materials took an active part in the phase transition, because of the fact that in Pr$_6$O$_{11}$ not every praseodymium cation is in the oxidation state (IV) and thus the reaction to a byproduct influenced the formation of HP-PrO$_2$. An experiment with cubic PrO$_2$ as starting material will therefore be of great interest.

The calculations of the MAPLE value (madelung part of lattice energy according to *Hoppe* [36–38]) resulted in a value of 11225 kJ·mol^{-1} for HP-PrO$_2$, which is similar to the value of 11987 kJ·mol^{-1} of the ambient-pressure phase. The same calculations for the corresponding cerium phases led to 12609 kJ·mol^{-1} for HP-CeO$_2$ and 11948 kJ·mol^{-1} for CeO$_2$.

Conclusions

Under high-pressure/high temperature conditions, crystals of orthorhombic HP-PrO$_2$ were obtained. A study of the axial ratios of HP-PrO$_2$ clarified that the high-pressure form of PrO$_2$ crystallizes in an α-PbCl$_2$-type structure. Whereas phase transformations of the ambient pressure phase previously were not observed up to 20 GPa, in this case the formation of HP-PrO$_2$ took place at 5.5 GPa and 1000 °C. Starting from a reaction mixture of Pr$_6$O$_{11}$, B$_2$O$_3$, and PrF$_3$, single-crystals were obtained, allowing the first single-crystal structure determination of HP-PrO$_2$.

Acknowledgement

We thank *Dr. Gunter Heymann* for collecting the single-crystal data. Special thanks go to *Dipl.-Chem. Marcus Tegel* (LMU Munich) for the robust refinement of the powder data.

References

[1] A. Trovarelli, *Catal. Rev. Sci. Eng.* **1996**, *38*, 439–520.
[2] A. Trovarelli, *Catalysis by Ceria and Related Materials*, Imperial College Press, **2002**.
[3] T. Yamamoto, H. Momida, T. Hamada, T. Uda, T. Ohno, *Thin Solid Film.* **2005**, *486*, 136–140.
[4] M. Malchus, M. Jansen, *Solid State Sci.* **2000**, *2*, 65–70.
[5] S. Kern, C.-K. Loong, J. Faber Jr., G. H. Lander, *Solid State Commun.* **1984**, *49*, 295–298.
[6] C. H. Gardiner, A. T. Boothroyd, P. Pattison, M. J. McKelvy, G. J. McIntyre, S. J. S. Lister, *Phys. Rev. B: Condens. Matter* **2004**, *70*, 024415.
[7] A. Kotani, T. Jo, J. C. Parlebas, *Adv. Phys.* **1988**, *37*, 37–85.
[8] A. T. Boothroyd, C. H. Gardiner, S. J. S. Lister, P. Santini, B. D. Rainford, L. D. Noailles, D. B. Currie, R. S. Eccleston, R. I. Bewley, *Phys. Rev. Lett.* **2001**, *86*, 2082–2085.
[9] R. Fehrenbacher, T. M. Rice, *Phys. Rev. Lett.* **1993**, *70*, 3471–3474.
[10] S. J. Duclos, Y. K. Vohra, A. L. Ruoff, A. Jayaraman, G. P. Espinosa, *Phys. Rev. B: Condens. Matter* **1998**, *38*, 7755–7758.
[11] A. Kourouklis, A. Jayaraman, G. P. Espinosa, *Phys. Rev. B: Condens. Matter* **1988**, *37*, 4250–4253.
[12] L. Gerward, J. Staun Olsen, L. Petit, G. Vaitheeswaran, V. Kanchana, A. Svane, *J. Alloys Compd.* **2005**, *400*, 56–61.
[13] S. E. Hill, C. R. A. Catlow, *J. Phys. Chem. Solids* **1993**, *54*, 411–419.
[14] D. D. Koelling, A. M. Boring, J. H. Wood, *Solid State Commun.* **1983**, *47*, 227–232.
[15] G. A. Landrum, R. Dronskowski, R. Niewa, F. J. DiSalvo, *Chem. Eur. J.* **1999**, *5*, 515–522.
[16] N. V. Skorodumova, R. Ahuja, S. I. Simak, I. A. Abrikosov, B. Johansson, B. I. Lundqvist, *Phys. Rev. B: Condens. Matter* **2001**, *64*, 115108.
[17] N. V. Skorodumova, S. I. Simak, B. I. Lundqvist, I. A. Abrikosov, B. Johansson, *Phys. Rev. Lett.* **2002**, *89*, 166601.
[18] N. V. Skorodumova, M. Baudin, K. Hermansson, *Phys. Rev. B: Condens. Matter* **2004**, *69*, 075401.
[19] F. Goubin, X. Racquefelte, M.-H. Whangbo, Y. Montardi, R. Brec, S. Jobic, *Chem. Mater.* **2004**, *16*, 662–672.
[20] L.-G. Liu, *Earth Planet. Sci. Lett.* **1980**, *49*, 166–172.
[21] Z. Hu, S. Bertram, G. Kaindl, *Phys. Rev. B: Condens. Matter* **1994**, *49*, 39–43.
[22] J. Dabrowski, V. Zavodinsky, A. Fleszar, *Microelectron. Reliab.* **2001**, *41*, 1093–1096.
[23] D. Walker, M. A. Carpenter, C. M. Hitch, *Am. Mineral.* **1990**, *75*, 1020–1029.
[24] D. Walker, *Am. Mineral.* **1991**, *76*, 1092–1100.
[25] H. Huppertz, *Z. Kristallogr.* **2004**, *219*, 330–338.
[26] D. C. Rubie, *Phase Transitions* **1999**, *68*, 431–451.
[27] N. Kawai, S. Endo, *Rev. Sci. Instrum.* **1970**, *41*, 1178–1181.
[28] K. H. Stone, S. H. Lapidus, P. W. Stephens, *J. Appl. Crystallogr.* **2009**, *42*, 385–391.
[29] Z. Otwinowski, W. Minor, *Methods Enzymol.* **1997**, *276*, 307–326.
[30] G. M. Sheldrick, *SHELXS-97 and SHELXL-97, Program Suite for the Solution and Refinement of Crystal Structures*, University of Göttingen, Göttingen, Germany, **1997**.
[31] G. M. Sheldrick, *Acta Crystallogr. Sect. A* **2008**, *64*, 112–122.
[32] W. Jeitschko, *Acta Crystallogr. Sect. B* **1968**, *24*, 930–934.
[33] R. W. G. Wyckoff in *Crystal Structures*, Interscience, New York, **1964**.
[34] H.-J. Meyer in *Moderne Anorganische Chemie* (Ed.: E. Riedel), de Gruyter, Berlin, **2003**.
[35] J. S. Knyrim, J. Friedrichs, S. Neumair, F. Roessner, Y. Floredo, S. Jakob, D. Johrendt, R. Glaum, H. Huppertz, *Solid State Sci.* **2008**, *10*, 168–176.
[36] R. Hoppe, *Angew. Chem.* **1966**, *78*, 52–63; *Angew. Chem. Int. Ed. Engl.* **1966**, *5*, 95–106.
[37] R. Hoppe, *Angew. Chem.* **1970**, *82*, 7–16; *Angew. Chem. Int. Ed. Engl.* **1970**, *9*, 25–34.
[38] R. Hübenthal, *MAPLE (Version 4), Program for the Calculation of MAPLE Values*, University of Gießen, Gießen, Germany, **1993**.

Received: September 13, 2009
Published Online: October 30, 2009

5 Prospects

With this thesis, the family of rare-earth oxoborates obtained under high-pressure conditions could be successfully enlarged. The first single crystal structures of λ-PrBO$_3$ and π-ErBO$_3$ were obtained. The latter is a very important contribution to the ongoing discussion concerning the crystal structure of these orthoborates. Additionally, spectroscopic properties of the new holmium borate Ho$_{31}$O$_{27}$(BO$_3$)$_3$(BO$_4$)$_6$, synthesized by S. Hering, could be obtained, which represents the rare-earth richest rare-earth oxoborate known up to now.

Besides these contributions to the well-studied field of rare-earth oxoborates, the main goal of this thesis was to establish the high-pressure / high-temperature route for rare-earth fluoride and fluorido borates. Here, only the rare-earth fluoride borates RE_3(BO$_3$)$_2$F$_3$ (RE = Sm, Eu, Gd) and Gd$_2$(BO$_3$)F$_3$, synthesized under ambient pressure, were known at the beginning of our investigations. Systematic research on this field led to numerous new compounds with interesting crystal structures.

The isotypic phases RE_5(BO$_3$)$_2$F$_9$ (RE = Er - Yb) show a structural relationship to the above-mentioned ambient-pressure phases. A separation into alternating layers with the formal compositions "REBO$_3$" and "REF$_3$" resulted in a descriptive model for these structures. High-pressure investigations also led to isotypic phases of RE_3(BO$_3$)$_2$F$_3$ (RE = Sm, Eu, Gd) and Gd$_2$(BO$_3$)F$_3$. These characterizations are still in progress. During this thesis, it was also possible to obtain the compound Er$_4$(BO$_3$)F$_9$. Unfortunately, only one single crystal could be measured and the crystal structure could not be fully refined. Nevertheless, the provisional result indicated that this compound also belongs to this structural family, showing the same unit cell as Gd$_2$(BO$_3$)F$_3$ with a different layer sequence. Further attempts on this structural family of rare-earth fluoride borates are needed in order to clarify the formation and stability range.

For the new composition $RE_4B_4O_{11}F_2$, two polymorphs with completely different crystal structures were discovered. $RE_4B_4O_{11}F_2$ (RE = Gd, Eu, Dy) shows an interesting intermediate state between a BO_3-group and a BO_4-tetrahedron. Another point of interest in $RE_4B_4O_{11}F_2$ (RE = Eu, Gd, Dy) are the BO_3-groups of two neighboring building blocks, facing each other at a relatively short B–B-distance. The application of higher pressures might transform these neighboring BO_3-groups into edge-sharing BO_4-tetrahedra, which are an extremely rare structural feature of high-pressure oxoborates.

The other polymorph, $La_4B_4O_{11}F_2$, shows an interesting wave-like modulation along one very large axis of the unit cell. Both polymorphs exhibit fundamental building blocks, which were unknown in classical borate chemistry. Additional research on the formation and stability range for these polymorphs should be performed. It will be of great interest to see, which crystal structure is favored by the rare-earth ions of intermediate size and whether it will be possible to obtain both structure types as polymorphs for one and the same cation.

The triclinic high-pressure phase $Pr_4B_3O_{10}F$ represents the first praseodymium fluoride borate with a new crystal structure as well as a new composition. One of the boron atoms is located on a split position, leading to a BO_3-group, which can be orientated in the crystal structure in two different ways. This can be interpreted as another special case of an intermediate state between a BO_3-group and a BO_4-tetrahedron. This gives further insight in the pressure transformation of borate units. Temperature programmed X-ray powder diffraction investigations indicate that $Pr_4B_3O_{10}F$ is a thermodynamically stable phase, which might also be obtained under ambient pressure conditions. Here, the application of pressure might induce the crystallization, as known from oxoborates. The synthesis of a metastable high-pressure praseodymium fluoride borate, in which the split position results in a tetrahedrally coordinated boron position, can be part of further investigations.

The apatite-structured compounds "$RE_5(BO_{3.66}F_{0.34})_3F$" (RE = Gd, Yb) seem to be unique in two different aspects. They may be the first examples, in

4.3 Rare-Earth Oxides

which both isolated fluoride anions and fluorine atoms covalently bound to the boron atoms occur. Unfortunately, this could not be fully clarified up to now. What can be said is that the structural diversity of apatites is extended towards an element of the 13th group as a full substitute of the original phosphorus atom. For further studies of boron-containing apatites (BAPs), which are candidates for bioceramic applications, the syntheses of "$RE_5(BO_{3.66}F_{0.34})_3F$" ($RE$ = Gd, Yb) are a big step towards the further diversi-ty of these important compounds.

The compounds $RE_{12}B_{11}O_{31}F_7$ (RE = La, Pr, Nd, Sm) were synthesized but their crystal structure could not be fully refined. Future examinations on these phases might reveal whether the proposed disorder model for the structure will be verified or whether an ordered structure can be derived.

The field of rare-earth fluoride and fluorido borates has just been superficially touched by this thesis. Still, six new compounds in four different structure-types (three of them were unknown previously) could be added during this work. Due to their high glass formation tendency, not many crystalline fluorido and fluoride borates were known previously. The structural characterization of glasses is not possible *via* X-ray diffraction experiments. The above mentioned high-pressure/high-temperature syntheses all led to coarse-crystalline powder samples, which are perfectly suitable for X-ray diffraction analyses. Single crystals could be isolated from all compounds. Therewith, the pressure induced crystallization is the key factor for a successful structural characterization of new fluorido and fluoride borates.

As expected, the transformation of BO_3-groups into BO_4-tetrahedra succeeded. A new structural motif, not yet reported from borate chemistry, was found and the motif of edge-sharing tetrahedra, which has been a unique feature of high-pressure oxoborates, might be obtained for fluorido and fluoride borates as well. Analogous to the oxoborates, we expect that the building units in the structures can be connected to chains, sheets, and networks opening up higher condensed structures with outstanding material properties.

6 Summary

6.1 High-Pressure / High-Temperature Synthesis

The high-pressure route has proven to be a versatile tool for the syntheses of new borates under extreme reaction conditions. This applies to the oxoborates, which have been studied for years, as well as to the fluoride and fluorido borates, which were investigated under high-pressure / high-temperature conditions for the first time during this thesis. The increase of the boron coordination number from three to four, the densification of structures by interconnection of existing building blocks implies a tremendous number of possible, higher condensed structures for both borate types, in analogy to silicates. Furthermore, the tendency of glass-formation in the class of borates can be suppressed *via* pressure induced crystallization, giving access to structural details of the compounds.

The well studied oxoborates still reveal new crystal structures under high-pressure / high-temperature conditions. The field of high-pressure fluorido and fluoride borates has just been opened during this thesis, implying an enormous potential for further high-pressure investigations.

The summarized results reflect the variety in oxoborate, fluorido, and fluoride borate chemistry, leaving room for many other fascinating structures and properties still to be discovered.

6.2 Rare-Earth Oxoborates

$Ho_{31}O_{27}(BO_3)_3(BO_4)_6$ *(Section 4.1.2, page 59)*

The new rare-earth oxoborate $Ho_{31}O_{27}(BO_3)_3(BO_4)_6$ was synthesized in a Walker-type module under high-pressure / high-temperature conditions of 7.5-

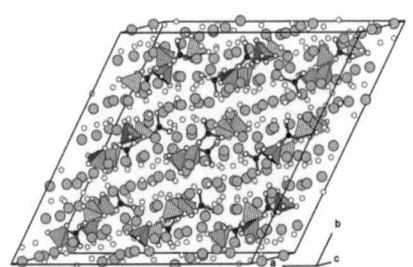

Figure 6.2-1
Crystal structure of $Ho_{31}O_{27}(BO_3)_3(BO_4)_6$

11.5 GPa and 1200 °C. The compound crystallizes in the space group $R\bar{3}$ (no. 148) with the lattice parameters $a = 2657.9(4)$ pm and $c = 1146.9(2)$ pm ($Z = 6$). The structure exhibits 11 different Ho cations next to three and six independent BO_3- and BO_4-groups, respectively. With an elemental ratio of Ho : B = 3.44 (31 : 9), the compound $Ho_{31}O_{27}(BO_3)_3(BO_4)_6$ represents the rare-earth richest rare-earth oxoborate known to this day. Figure 6.2-1 shows the structure of the holmium oxoborate $Ho_{31}O_{27}(BO_3)_3(BO_4)_6$, displaying isolated BO_4-tetrahedra as dark hatched polyhedra and BO_3-units in form of ball-stick models, embedded in a complex holmium oxide network.

λ-PrBO₃   (Section 4.1.3, page 71)

The praseodymium orthoborate λ-$PrBO_3$ was synthesized from Pr_6O_{11}, B_2O_3, and PrF_3 under high-pressure / high-temperature conditions of 3 GPa and 800 °C in a Walker-type multianvil apparatus. The crystal structure was determined on the basis of single-crystal X-ray diffraction data, collected at room temperature. The title compound crystallizes in the orthorhombic aragonite-type structure, space group *Pnma*, with the lattice parameters $a = 577.1(2)$, $b = 506.7(2)$, $c = 813.3(2)$ pm, and $V = 0.2378(2)$ nm³, with $R_1 =$

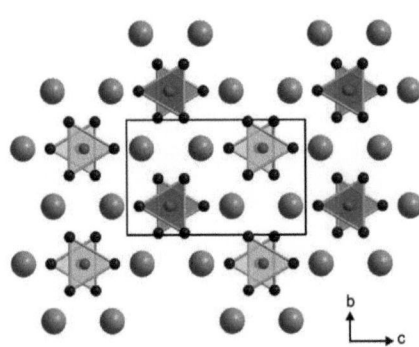

Figure 6.2-2
Crystal structure of λ-$PrBO_3$

6.2 Rare-Earth Oxoborates

0.0400 and wR_2 = 0.0495 (all data). The crystal structure is composed of stacked trigonal planar BO_3-groups (Figure 6.2-2), the average B–O distance is 137.2 pm. The praseodymium atoms are ninefold coordinated by oxygen atoms. For λ-$PrBO_3$, only cell parameters from powder data were given in the literature. High-pressure / high-temperature synthesis now made the first crystal structure determination of λ-$PrBO_3$ from single-crystals possible.

π-ErBO$_3$ (Section 4.1.4, page 79)

Single crystals of the orthoborate π-ErBO$_3$ were synthesized from Er$_2$O$_3$ and B$_2$O$_3$ under high-pressure / high-temperature conditions of 3 GPa and 800 °C in a Walker-type multianvil apparatus. The crystal structure was determined on the basis of single-crystal X-ray diffraction data, collected at room temperature. The title compound crystallizes in the monoclinic pseudowollastonite-type structure, space group $C2/c$, with the lattice a = 1128.4(2) pm, b = 652.6(2) pm, c = 954.0(2) pm, and β = 112.81(3) °, with R_1 = 0.0124 and wR_2 = 0.0404 (all data). Over the past 50 years, five different structural models for the orthoborates π-REBO$_3$ were proposed, revised, doubted, and supported. Remarkably, only one single-crystal measurement on π-YBO$_3$ was reported, while all other models rely on powder measurements.

High-pressure/high-temperature analysis now resulted in single crystals of π-ErBO$_3$, which made the first satisfying single-crystal structure determination possible.

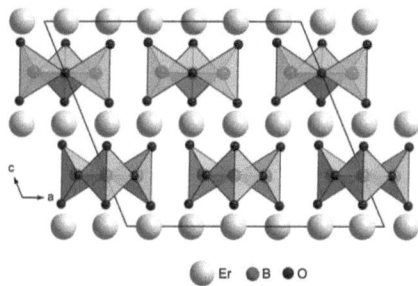

Figure 6.2-3
Crystal structure of π-ErBO$_3$

6.3 Rare-Earth Fluoride and Fluorido Borates

$RE_5(BO_3)_2F_9$ (RE = Er - Yb)   *(Sections 4.2.1 - 4.2.3, page 89)*

$Yb_5(BO_3)_2F_9$ was synthesized under high-pressure / high-temperature conditions in a Walker-type multianvil apparatus at 7.5 GPa and 1100 °C, representing the first known ytterbium fluoride borate. The isotypic fluoride borates $Tm_5(BO_3)_2F_9$ and $Er_5(BO_3)_2F_9$ were synthesized at 5 GPa and 900 °C or 3 GPa and 800 °C, respectively. The compounds exhibit isolated BO_3-groups next to rare-earth cations and fluoride anions (Figure 6.3-1), showing a structure closely related to the other known rare-earth fluoride borates $RE_3(BO_3)_2F_3$ (RE = Sm, Eu, Gd) and $Gd_2(BO_3)F_3$.

$RE_5(BO_3)_2F_9$ (RE = Er - Yb) crystallizes in the monoclinic space group $C2/c$. Three different rare-earth cations can be identified in the crystal structures, each coordinated by nine fluoride and oxygen anions. None of the five crystallographically independent fluoride ions is coordinated by boron atoms, solely by trigonally-planar arranged rare-earth cations. In close proximity to the above mentioned compounds $RE_3(BO_3)_2F_3$ (RE = Sm, Eu, Gd) and $Gd_2(BO_3)F_3$, $RE_5(BO_3)_2F_9$ can be described *via* alternating layers with the formal compositions "$REBO_3$" and "REF_3" in the bc-plane.

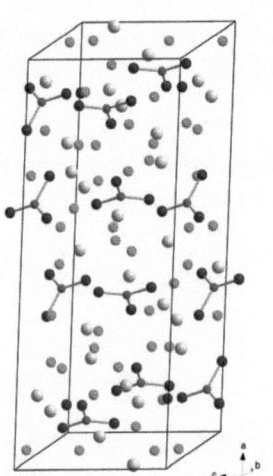

Figure 6.3-1
Crystal structure of $RE_5(BO_3)_2F_9$ (RE = Er - Yb)

6.3 Rare-Earth Fluoride and Fluorido Borates

$RE_4B_4O_{11}F_2$ (RE = Eu, Gd, Dy)   (Sections 4.2.4 and 4.2.5, page 115)

A new gadolinium fluoride borate $Gd_4B_4O_{11}F_2$ was yielded in a Walker-type multianvil apparatus at 7.5 GPa and 1100 °C. Soon after, the isotypic compounds $RE_4B_4O_{11}F_2$ (RE = Eu, Dy) were synthesized at 5 GPa / 900 °C, and 8 GPa / 1000 °C, respectively. $RE_4B_4O_{11}F_2$ (RE = Eu, Gd, Dy) crystallize in the space group $C2/c$. They differ by all means from the polymorph $La_4B_4O_{11}F_2$ (space group $P2_1/c$).

The crystal structure exhibits a structural motif not yet reported from borate chemistry: two BO_4-tetrahedra (□) and two BO_3-groups (Δ) are connected via common corners, leading to the fundamental building block 2Δ2□: ΔΩΩΔ. The crystal structure with these building blocks is depicted in Figure 6.3-2. In the two crystallographically identical BO_4-tetrahedra, a distortion resulting in a very long B–O bond is found. A closer look reveals that the coordination sphere of the boron atom can be described as an intermediate state between a BO_3-group and a BO_4-tetrahedron.

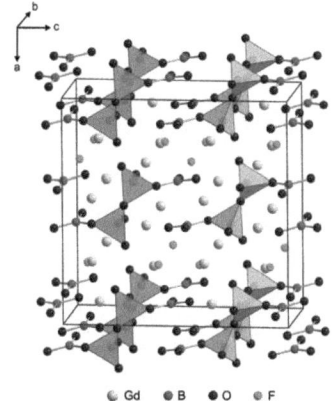

Figure 6.3-2
Crystal structure of $RE_4B_4O_{11}F_2$ (RE = Eu, Gd, Dy)

$La_4B_4O_{11}F_2$ (Section 4.2.6, page 133)

The first lanthanum fluoride borate $La_4B_4O_{11}F_2$ was yielded in a Walker-type multianvil apparatus at 6 GPa and 1300 °C. $La_4B_4O_{11}F_2$ crystallizes monoclinically in the space group $P2_1/c$ with the lattice parameters $a = 779.6(6)$ pm, $b = 3573.4(17)$ pm, $c = 765.8(5)$ pm, and $\beta = 113.94(7)°$ (Z = 8) and is thus a new polymorph of $RE_4B_4O_{11}F_2$. The crystal structure is built up from BO_4-tetrahedra and BO_3-groups, which are either isolated or connected via common corners, forming the building blocks ΔΔ and ΔΩΔ. The latter is also a novelty in borate chemistry. The crystal structure of $La_4B_4O_{11}F_2$ shows a wave-like modulation along the b-axis (Figure 6.3-3).

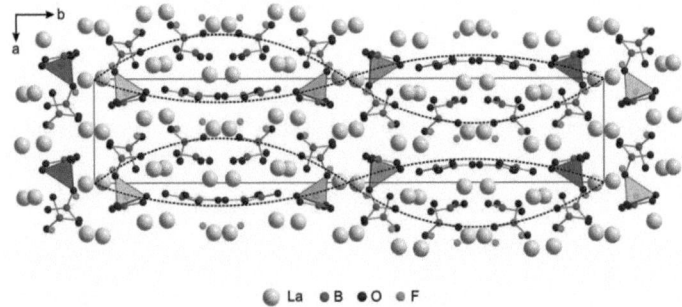

Figure 6.3-3
Crystal structure of $La_4B_4O_{11}F_2$

$Pr_4B_3O_{10}F$ (Section 4.2.7, page 145)

The first praseodymium fluoride borate $Pr_4B_3O_{10}F$ was synthesized under high-pressure/high-temperature conditions in a Walker-type multianvil apparatus at 3 GPa and 800 °C, starting from stoichiometric mixtures of Pr_6O_{11}, B_2O_3, and PrF_3. $Pr_4B_3O_{10}F$ crystallizes in space group $P\bar{1}$ with $Z = 2$ and lattice parameters $a = 662.8(5)$ pm, $b = 882.5(4)$ pm, and $c = 894.1(5)$ pm, with $\alpha = 106.77(4)°$, $\beta = 108.67(3)°$, $\gamma = 104.92(4)°$ ($R1 = 0.0235$ and $wR2 = 0.0537$ (all data)). The crystal structure is built up exclusively from isolated BO_3-groups (Figure 6.3-4). One of the boron atoms is located on a split position, leading to a BO_3-group, which can be orientated in the crystal structure

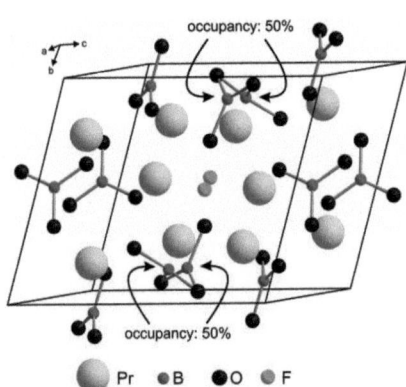

Figure 6.3-4
Crystal structure of $Pr_4B_3O_{10}F$

in two different ways. This can be interpreted as a special case of an intermediate state between a BO_3-group and a BO_4-tetrahedron. The fluoride position is occupied to a small percentage by OH^- ions.

"$RE_5(BO_{3.66}F_{0.34})_3F$" ($RE$ = Gd, Yb)   (Section 4.2.8, page 153)

The borate apatites $RE_5(BO_{3.66}F_{0.34})_3F$ (RE = Gd, Yb) were synthesized in high-pressure / high-temperature reactions at 10 GPa and 1000 °C, starting from RE_2O_3, REF_3, and B_2O_3. With "$RE_5(BO_{3.66}F_{0.34})_3F$" ($RE$ = Gd, Yb), I synthesized the first two apatite-type compounds, with fourfold coordinated boron atoms instead of the original phosphorus atom in apatite. The structural diversity of apatites is thus extended towards an element of the 13th group.

The isotypic crystal structures of "$RE_5(BO_{3.66}F_{0.34})_3F$" ($RE$ = Gd, Yb) were solved and refined on the basis of single-crystal data (space group: $P6_3/m$). The exact distribution of oxygen and fluorine in the structure could not yet be verified and are further examined, thus the sum formulas are still written in parentheses. Figure 6.3-5 shows the crystal structure of compounds, which (in the current model) contain isolated "$BO_{3.66}F_{0.34}$"-tetrahedra and fluoride ions located in channels along the c-axis. While borates, that contain only isolated fluoride anions, are generally termed "fluoride borates", "fluorido borates" contain fluorine atoms covalently bound to the boron atoms. To the best of our knowledge, the compounds $RE_5(BO_{3.66}F_{0.34})_3F$ (RE = Gd, Yb) are the first examples, in which both features occur.

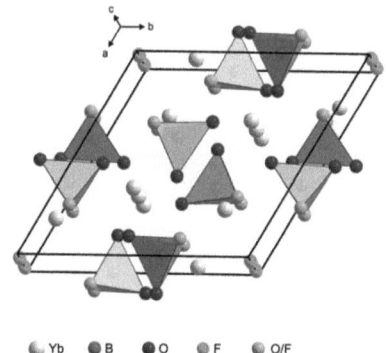

Figure 6.3-5
Crystal structure of $RE_5(BO_{3.66}F_{0.34})_3F$ (RE = Gd, Yb)

$RE_3(BO_3)_2F_3$ (RE = Gd, Dy)   (Section 4.2.9 page 163)

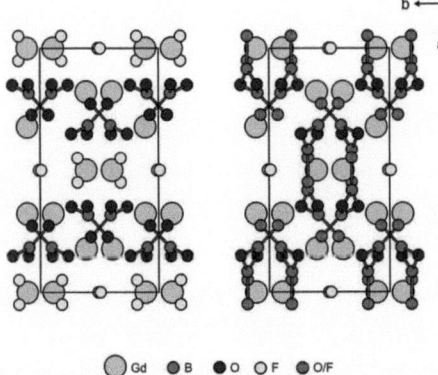

Figure 6.3-6
Ordered and disordered crystal structure of $Gd_3(BO_3)_2F_3$

The first single crystals of the fluoride borates $Gd_3(BO_3)_2F_3$ and $Dy_3(BO_3)_2F_3$ were obtained in a Walker-type multianvil apparatus at a temperature of 900 °C and pressures of 3 and 5 GPa, respectively. Both compounds crystallize in the monoclinic space group $C2/c$ ($Z = 4$). While the synthesis of a powder sample of $Gd_3(BO_3)_2F_3$ was previously reported in the literature, the isotypic compound $Dy_3(BO_3)_2F_3$ is presented herein for the first time. Two different disorder models have been proposed previously for compounds with the composition $RE_3(BO_3)_2F_3$. The single crystal data now available, gives new insight into the disorder problem. The following results were still subject to final crystallographical checks so that it was not published until the end of this thesis.

$RE_{12}B_{11}O_{31}F_7$ (RE = La, Pr, Nd, Sm)   (Section 4.2.10, page 181)

The compounds $RE_{12}B_{11}O_{31}F_7$ (RE = La, Pr, Nd, Sm) crystallize willingly in a wide pressure range: $La_{12}B_{11}O_{31}F_7$ was obtained at 6 GPa and 1300 °C, the phases $Pr_{12}B_{11}O_{31}F_7$ and $Nd_{12}B_{11}O_{31}F_7$ at 3 GPa and 800 °C, and $Sm_{12}B_{11}O_{31}F_7$ at 5 GPa and 900 °C. Figure 6.3-7 shows the crystal structure which comprises of isolated BO_3-groups and the fundamental building block $2\Delta\square{:}\Delta\square\Delta$. Even though the crystal structures can be solved with very good R-values, several problems occur. All phases $RE_{12}B_{11}O_{31}F_7$ (RE = La, Pr, Nd, Sm) exhibit a site of disorder. Additionally, the anisotropic displacement parameters were ill-behaving, indicating an intrinsic problem of the crystals. Up to now, it was not possible to obtain a satisfying crystal structure refinement of any of the isotypic phases in both possible space groups $P\overline{1}$ and $P1$. The compounds are objects to further investigations.

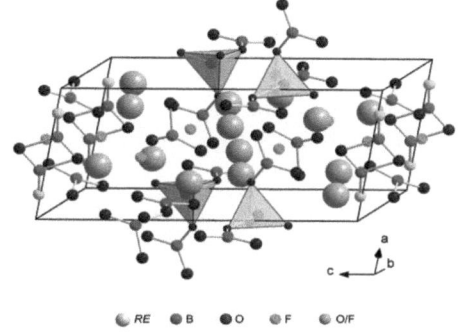

Figure 6.3-7
Determined crystal structure of $RE_{12}B_{11}O_{31}F_7$ (RE = La, Pr, Nd, Sm)

6.4   Rare-earth oxides

HP-PrO$_2$ (Section 4.3, page 187)

The high-pressure phase of PrO_2 was synthesized under conditions of 5.5 GPa and 1000 °C in a Walker-type multianvil apparatus. The crystal structure was determined on the basis of single-crystal X-ray diffraction data, collected at room temperature. The praseodymium oxide crystallizes in the orthorhombic α-$PbCl_2$-type structure with the lattice parameters a = 637.6(3) pm, b = 384.4(2) pm, and c = 703.1(3) pm. The praseodymium ion is ninefold coordinated, the oxygen ions show either tetrahedral or square-pyramidal coordination spheres.

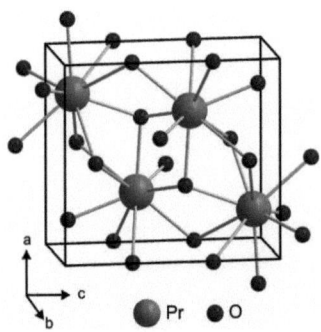

Figure 6.4-1
Crystal structure of HP-PrO$_2$

7 Appendix

7.1 List of Abbreviations and Special Characters

Å	Ångstrøm	IR	Infrared
Ø	Average	JCPDS	Joint Committee on Powder Diffraction Standards
δ	Chemical shift parameter		
°C	Degree Celsius	K	Kelvin
θ	Diffraction angle	KBBF	KBe_2BO_3F
ρ	Density	kbar	Kilobar
Σ	Sum	kJ	Kilojoule
λ	Wavelength	LED	Light-Emitting Diode
ATR	Attenuated Total Reflection	LMU	Ludwig-Maximilians-University
BAPs	Boron-containing Apatites	LPP	Low-Pressure Phase
BIBO	$\alpha\text{-}BiB_3O_6$	LPSSiC	liquid phase sintered SiC
ca., ~	circa	mg	Milligramme
c-BN	Cubic boron nitride	min	Minute
CCD	Charged Coupled Device	MIR	Middle Infra Red
cm	Centimeter	mm	Millimetre
cm^{-1}	Wavenumber	MPa	Megapascal
C.N.	Coordination Number	mW	Milliwatt
CSD	Cambridge Structural Database	N	Newton
d	Distance	N.A.	Numerical Aperture
DAC	Diamond Anvil Cell	NIR	Near Infra Red
ECoN	Effective Coordination Number	NLO	Non-linear optical
EDX	Energy Dispersive X-Ray Analysis	nm	Nanometer
e.g.	exempli gratia	OEL	octahedral edge length
EPMA	Electronprobe Microanalysis	Pa	Pascal
et al.	et alii	PET	Polyethylen terephtalate
F	Structure factor	PLC	Programmable Logic Controller
FBB	Fundamental Building Block	pm	Picometer
FT	Fourier Transform	PSD	Position Sensitive Detector
FTIR	Fourier Transformed Infrared	PTFE	Polytetrafluoroethylene
g	Gramme	*RE*	Rare-earth
GOOF	Goodness of Fit	SEM	Scanning Electron Microscope
GPa	Gigapascal	SS-NMR	Solid State NMR
h	Hour	t	Ton
h-BN	Hexagonal boron nitride	TCO	Tin-based composite oxide
HP	High Pressure	TEL	truncation edge length
HPP	High-Pressure Phase	UV	Ultraviolet
HPSiC	hot pressed SiC	UV-Vis	Ultraviolet-visible
I	Intensity	V	Volume
ICSD	Inorganic Crystal Structure Database	YAG	Yttrium aluminium garnet
		Z	Formula Units

7.2 Awards

Lupe Award 2009 for outstanding efforts in the dissemination of science at the „ Lange Nacht der Forschung" in Austria

Römer-Award 2008 (Dr. Klaus Römer-Foundation) for outstanding scientific research during the Master's thesis.

7.3 Publications

7.3.1 Conference Contributions

(5) Almut Haberer
„Fluoro- und Fluoridoborate der Lanthanoide unter Hochdruck" (Talk)
6. Workshop Anorganische Chemie in Österreich, Linz (Austria)
29. – 31.03.2010

(4) Almut Haberer, Hubert Huppertz
"Rare-earth fluoride borates – a structural comparison" (Poster)
13. Österreichische Chemietage, Vienna, Austria, 24. – 27.08.2009

(3) Almut Haberer
"Synthesis and structure of the first ytterbium fluoride borate $Yb_5(BO_3)_2F_9$" (Talk)
International High-Pressure Conference JOINT AIRAPT-22 & HPCJ-50, Tokyo, Japan, 26. – 31.07.2009

(2) Almut Haberer
„Hochdrucksynthese und Kristallstruktur des ersten Ytterbiumfluoridborates $Yb_5(BO_3)_2F_9$" (Talk)
Festkörperseminar Hirschegg, Hirschegg, Austria, 11. – 13.06.2009

7.3 Publications

(1) Almut Haberer, Hubert Huppertz

„$Ti_5B_{12}O_{26}$: Ein gemischt-valentes Hochdruck-Titanborat mit NaTl-verwandter Struktur" (Poster)

14. Vortragstagung der Fachgruppe Festkörperchemie und Materialforschung, Bayreuth, Germany, 24. – 26.09.2008

7.3.2 Papers

(11) "$RE_4B_4O_{11}F_2$ (RE = Eu, Dy): New Phases Isotypic to the Fluoride Borate $Gd_4B_4O_{11}F_2$"

Almut Pitscheider, Michael Enders, and Hubert Huppertz

Z. Naturforsch. 65b (2010), 1439.

(10) "The Crystal Structure of π-$REBO_3$: New Single Crystal Data for an Old Problem"

Almut Pitscheider, Reinhard Kaindl, Oliver Oeckler, and Hubert Huppertz

J. Solid State Chem. 184 (2011) 149.

(9) "High-pressure synthesis and crystal structure of the new holmium oxoborate $Ho_{31}O_{27}(BO_3)_3(BO_4)_6$"

Stefanie A. Hering, Almut Haberer, Reinhard Kaindl, and Hubert Huppertz

Solid State Sci. 12 (2010), 1993.

(8) "High-pressure Synthesis and Characterization of the Fluoride Borate $Tm_5(BO_3)_2F_9$"

Almut Haberer, Michael Enders, Reinhard Kaindl, and Hubert Huppertz

Z. Naturforsch. 65b (2010), 1213.

(7) "Synthesis and Crystal Structure of the Praseodymium Orthoborate λ-$PrBO_3$"

Almut Haberer, Reinhard Kaindl, and Hubert Huppertz

Z. Naturforsch. 65b (2010), 1206.

(6) *"A new polymorph of $RE_4B_4O_{11}F_2$: High-pressure synthesis and crystal structure of $La_4B_4O_{11}F_2$"*
Almut Haberer, Reinhard Kaindl, Oliver Oeckler, and Hubert Huppertz
J. Solid State Chem. 183 (2010) 1970.

(5) *"$Gd_4B_4O_{11}F_2$: Synthesis and crystal structure of a rare-earth fluoride borate, exhibiting a new "fundamental building block" in borate chemistry"*
Almut Haberer, Reinhard Kaindl, and Hubert Huppertz
J. Solid State Chem. 183 (2010) 471.

(4) *"Orthorhombic HP-PrO_2 – High-Pressure Synthesis and Crystal Structure"*
Almut Haberer and Hubert Huppertz
Z. Anorg. Allg. Chem. 636 (2010) 363.

(3) *"The first praesodymium fluoride borate $Pr_4B_3O_{10}F$ – High-pressure synthesis and characterization"*
Almut Haberer, Reinhard Kaindl, and Hubert Huppertz
Solid State Sci. 12 (2010) 515.

(2) *"Synthesis of $Er_5(BO_3)_2F_9$ and Properties of $RE_5(BO_3)_2F_9$ (RE = Er, Yb)"*
Almut Haberer, Reinhard Kaindl, Jürgen Konzett, Robert Glaum, and Hubert Huppertz
Z. Anorg. Allg. Chem. 636 (2010) 1326.

(1) *"High-pressure synthesis, crystal structure, and structural relationship of the first ytterbium fluoride borate $Yb_5(BO_3)_2F_9$"*
Almut Haberer and Hubert Huppertz
J. Solid State Chem. 182 (2009) 888.

7.4 CSD Numbers

The CIF (Crystallographic Information File) of the following compounds were deposited at the Fachinformationszentrum Karlsruhe (crysdata@fiz-karlsruhe.de), D-76344 Eggenstein-Leopoldshafen, Germany.

$Ho_{31}O_{27}(BO_3)_3(BO_4)_6$	CSD-421761
λ-$PrBO_3$	CSD-421745
π-$ErBO_3$	CSD-422094
$Yb_5(BO_3)_2F_9$	CSD-420182
$Er_5(BO_3)_2F_9$	CSD-421469
$Tm_5(BO_3)_2F_9$	CSD-421889
$Gd_4B_4O_{11}F_2$	CSD-420809
$Eu_4B_4O_{11}F_2$	CSD-422159
$Dy_4B_4O_{11}F_2$	CSD-422160
$La_4B_4O_{11}F_2$	CSD-421688
$Pr_4B_3O_{10}F$	CSD-421268
$Gd_5(BO_{3.66}F_{0.34})_3F$	CSD-421304
$Yb_5(BO_{3.66}F_{0.34})_3F$	CSD-421303
HP-PrO_2	CSD-380398

7.5 References

[1] J. R. Holloway, B. J. Wood, Simulating the Earth, Harper Collins Academic, London, 1988.
[2] D. Bröll, C. Kaul, A. Krämer, P. Krammer, T. Richter, M. Jung, H. Vogel, P. Zehner, Angew. Chem. 111 (1999) 3180; Angew. Chem. Int. Ed. 38 (1999) 2998.
[3] F. Benito-López, R. J. M. Egberink, D. N. Reinhoudt, W. Verboom, Tetrahedron 64 (2008) 10023.
[4] F. R. Boyd, J. L. England, J. Geophys. Res. 65 (1960) 741.
[5] H. Huppertz, Habilitationsschrift, Ludwig-Maximilians-Universität München (2003).
[6] H. T. Hall, Rev. Sci. Instrum. 31 (1960) 125.
[7] Y. Kanke, M. Akaishi, S. Yamaoka, T. Taniguchi, Rev. Sci. Instrum. 73 (2002) 3268.
[8] P. W. Bridgman, Proc. Roy. Soc. (London) A 203 (1950) 1.
[9] H. T. Hall, Rev. Sci. Instrum. 29 (1958) 267.
[10] M. Contre, High Temp.-High Pressures 1 (1969) 339.
[11] T. Yage, S. Akimoto, J. Appl. Phys. 47 (1976) 3350.
[12] N. Kawai, Proc. Jpn Acad. 42 (1966) 385.
[13] D. Walker, M. A. Carpenter, C. M. Hitch, Am. Mineral. 75 (1990) 1020.
[14] A. Onodera, High Temp.-High Pressures 19 (1987) 579.
[15] E. Ito, A. Kubo, T. Katsura, M. Akaogi, T. Fujita, Geophys. Res. Let. 25 (1998) 821.
[16] J. A. Xu, H. K. Mao, P. M. Bell, Science, 232 (1986) 1404.
[17] A. Zerr, G. Serghiou, R. Boehler, Handbook of Ceramics and Hard Materials, Vol. 1, ed. R. Riedel, Wiley VCH, Weinheim, 2000, p.41.
[18] S. T. Weir, A. C. Mitchell, W. J. Nellis, Phys. Rev. Lett. 76 (1996) 1860.
[19] A. Neuhaus, Chimia 18 (1964) 93.
[20] U. Griesbach, D. Zollinger, H. Pütter, C. Comninellis, J. Appl. Electrochem. 35 (2005) 1265.

[21] E. A. Ekimov, V. A. Sidorov, E. D. Bauer, N. N. Mel'nik, N. J. Curro, J. D. Thompson, S. M. Stishov, Nature 428 (2004) 542.
[22] http://de.wikipedia.org/wiki/Siliciumcarbid
[23] A. F. Holleman, E. Wiberg, N. Wiberg, Lehrbuch der Anorganischen Chemie, 102. Auflage, Walter de Gruyter, Berlin (2007) 1113 f.
[24] V. L. Solozhenko, D. Andrault, G. Fiquet, M. Mezouar, D. C. Rubie, Appl. Phys. Lett. 78 (2001) 1385.
[25] S. Chen, X. G. Gong, S.-H. Wie, Phys. Rev. Lett. 98 (2007) 015502.
[26] W. Schnick, Angew. Chem. 111 (1999) 3511; Angew. Chem. Int. Ed. 38 (1999) 3309.
[27] P. Kroll, J. Appen, Phys. Stat. Sol. B 1 (2001) R6.
[28] A. Liu, M. Cohen, Science 245 (1985) 841.
[29] C. Niu, Y. Z. Lu, C. M. Lieber, Science 261 (1993) 334.
[30] J. H. Nguyen, R. Jeanloz, Mater. Sci. Eng. A 209 (1996) 23.
[31] L. W. Yin, M. S. Li, Y. X. Liu, J. L. Sui, J. M. Wang, J. Phys. Condens. Matter 15 (2003) 309.
[32] G. Demazeau, Eur. J. Solid State Inorg. Chem. 34 (1997) 759.
[33] G. Demazeau, Chem. Scrip. 28 (1988) 21.
[34] M. C. Johnson, D. Walker, S. M. Clark, R. L. Jones, Am. Mineral. 86 (2001) 1367.
[35] R. G. Delaplane, U. Dahlborg, B. Granéli, P. Fischer, T. Lundström, J. Non-Cryst. Solids 104 (1988) 249.
[36] U.S. Geological Survey, Mineral Commodity Summaries (2006).
[37] S. Schlag, H. Mori, SRI Consulting, CEH Report on Boron Minerals and Chemicals (2008).
[38] Inorganic Crystal Structure Database (version 1.2.0.), Fachinformationszentrum Karlsruhe (2010).
[39] J. Liebertz, S. Stähr, Z. Kristallogr. 165 (1983) 91.
[40] C. Chen, B. Wu, A. Jiang, G. You, Sci. Sin. B 28 (1985) 235.
[41] D. Xue, S. J. Zhang, Acta Crystallogr. B 54 (1998) 652.
[42] C. Mazzetti, F. D. Carli, Chim. Ital. 56 (1926) 23.

7.5 References

[43] B. S. R. Sastry, F. A. Hummel, J. Am. Ceram. Soc. 41 (1958) 11.
[44] H. König, R. Hoppe, Z. Anorg. Allg. Chem. 439 (1978) 71.
[45] C. Chen, Y. Wu, A. Jiang, B. Wu, G. You, R. Li, S. Lin, J. Opt. Soc. Am. B 6 (1989) 616.
[46] J. Liebertz, Z. Kristallogr. 158 (1982) 319.
[47] R. Fröhlich, L. Bohatý, J. Liebertz, Acta Crystallogr. C 40 (1984) 343.
[48] P. Becker, J. Liebertz, L. Bohatý, J. Cryst. Growth 203 (1999) 149.
[49] Y. Mori, I. Kuroda, S. Nakajima, T. Sasaki, S. Nakai, Appl. Phys. Lett. 67 (1995) 1818.
[50] G. Ryu, C. S. Yoon, T. P. J. Han, H. G. Gallagher, J. Cryst. Growth 191 (1998) 492.
[51] Y. Mori, Y. K. Yap, T. Kamimura, M. Yoshimura, T. Sasaki, Opt. Mater. 19 (2002) 1.
[52] T. Sasaki, Y. Mori, M. Yoshimura, J. Nonlinear Opt. Phys. Mater. 2 (2001) 249.
[53] G. E. Gurr, P. W. Montgomery, C. D. Knutson, B. T. Gorres, Acta Crystallogr. B 26 (1970) 906.
[54] C. T. Prewitt, R. D. Shannon, Acta Crystallogr. B 24 (1968) 869.
[55] T. Nikelski, Th. Schleid, Z. Anorg. Allg. Chem. 629 (2003) 1017.
[56] H. Emme, T. Nikelski, Th. Schleid, R. Pöttgen, M. H. Möller, H. Huppertz, Z. Naturforsch. 59b, (2004) 202.
[57] H. Emme, G. Heymann, A. Haberer, H. Huppertz, Z. Naturforsch 62b (2007) 765.
[58] H. Emme, C. Despotopoulou, H. Huppertz, Z. Anorg. Allg. Chem. 630 (2004) 2450.
[59] G. Heymann, T. Soltner, H. Huppertz, Solid State Sci. 8 (2006) 821.
[60] A. Haberer, G. Heymann, H. Huppertz, Z. Naturforsch. 62b (2007) 759.
[61] H. Huppertz, B. von der Eltz, J. Am. Chem. Soc. 124 (2002) 9376.
[62] H. Huppertz, Z. Naturforsch. B 58 (2003) 278.
[63] H. Emme, H. Huppertz, Chem. Eur. J. 9 (2003) 3623.
[64] H. Emme, H. Huppertz, Z. Anorg. Allg. Chem. 628 (2002) 2165.

[65] H. Huppertz, J. Solid State Chem. 177 (2004) 3700.
[66] J. S. Knyrim, F. Roeßner, S. Jakob, D. Johrendt, I. Kinski, R. Glaum, H. Huppertz, Angew. Chem., Int. Ed. Engl. 46 (2007) 9097.
[67] S. C. Neumair, R. Glaum, H. Huppertz, Z. Naturforsch. 64b (2009) 883.
[68] S. C. Neumair, H. Huppertz, Z. Naturforsch. 65b (2010), in preparation.
[69] Y. Wu, J.-Y. Yao, J.-X. Zhang, P.-Z. Fu and Y.-C. Wu, Acta Crystallogr. Sect. E: Struct. Rep. Online E66 (2010) i45.
[70] S. Jin, G. Cai, W. Wang, M. He, S. Wang and X. Chen, Angew. Chem., Int. Ed. 49 (2010) 4967.
[71] H. Emme, M. Valldor, R. Pöttgen, H. Huppertz, Chem. Mater. 17 (2005) 2707.
[72] J. S. Knyrim, H. Huppertz, Z. Naturforsch. B 63 (2008) 707.
[73] J. S. Knyrim, H. Huppertz, J. Solid State Chem. 180 (2007) 742.
[74] Y. Idota, T. Kubota, A.Matsufuji, Y. Maekawa, T. Miyasaka, Science 276 (1997) 1395.
[75] J. S. Knyrim, F. M. Schappacher, R. Pöttgen, J. Schmedt auf der Günne, D. Johrendt, H. Huppertz, Chem. Mater. 19 (2007) 254.
[76] D. J. Frost, Personal Message.
[77] W. Wünschheim, H. Huppertz, PRESSCONTROL - A Program for Communication, Calibration and Surveillance of a Hydraulic Press with Heating Device *via* RS232C Interfaces, Ludwig-Maximilians-Universität München (1999-2003).
[78] D. C. Rubie, Phase Trans. 68 (1999) 431.
[79] I. C. Getting, G. L. Chen, J. A. Brown, Pure Appl. Geophys. 141 (1993) 545.
[80] E. Takahashi, H. Yamada, E. Ito, Geophys. Res. Lett. 9 (1982) 805.
[81] P. W. Bridgman, Phys. Rev. 48 (1935) 825.
[82] V. E. Bean, S. Akimoto, P. M. Bell, S. Block, W. B. Holzapfel, M. H. Manghnani, M. F. Nicol, S. M. Stishov, Physica 139 &140B (1986) 52.
[83] F. P. Bundy, Natl. Bur. Stand. Sp. Publ. 326 (1971) 263.
[84] P. W. Bridgman, Phys. Rev. 48 (1935) 893.

7.5 References

[85] A. A. Giardini, G. A. Samara, J. Phys. Chem. Solids 26 (1965) 1523.
[86] M. J. Duggin, J. Phys. Chem. Solids 33 (1972) 1267.
[87] A. Yoneda, S. Endo, J. Appl. Phys. 51 (1980) 3216.
[88] J. H. Chen, H. Iwasaki, T. Kikegawa, High Press. Res. 15 (1996) 143.
[89] E. C. Lloyd, C. W. Becket, F. R. Boyd Jr., Accurate Characterization of the High-Pressure Environment ed. E. C. Lloyd (Washington, DC: Natl. Bur. Stand. (US) Sp. Publ. 326 (1971) 1.
[90] H. G. Drickamer, Rev. Sci. Instr. 41 (1970) 1667.
[91] C. G. Homan, J. Phys. Chem. Solids 36 (1975) 1249.
[92] M. I. McMahon, O. Degtyareva, R. J. Nelmes, Phys. Rev. Let. 85 (2000) 4896.
[93] K. Kusaba, L. Galoisy, Y. Wang, M. T. Vaughan, D. J. Weidner, Pure Appl. Geophys. 141 (1993) 643.
[94] J. Camacho, I. Loa, A. Cantarero, K. Syassen, J. Phys.: Condens. Matter 14 (2002) 739.
[95] A. San Miguel, A. Polian, M. Gautier, J. P. Itie, Phys. Rev. B 48 (1993) 8683.
[96] R. J. Nelmes, M. I. McMahon, N. G. Wright, D. R. Allan, J. Phys. Chem. Solids 56 (1995) 545.
[97] R. J. Nelmes, M. I. McMahon, N. G. Wright, D. R. Allan, Phys. Rev. Lett. 73 (1994) 1805.
[98] M. J. Walter, Y. Thibault, K. Wei, R. W. Luth, Can. J. Phys. 73 (1995) 273.
[99] W. Kleber, Einführung in die Kristallographie, Verlag Technik Berlin, 17th ed (1990).
[100] J. Karle, Angew. Chem. 98 (1986) 611; Angew. Chem. Int. Ed. 25 (1986) 614.
[101] H. Hauptmann, Angew. Chem. 98 (1986) 600; Angew. Chem. Int. Ed. 25 (1986) 603.
[102] B. D. Cullity, Elements of X-Ray Diffraction, Addison-Wesley Pub. Co., 2nd ed. (1978).
[103] Stoe WinXpow, Vers. 1.2, Stoe & Cie, Darmstadt, Germany (2001).

[104] P. – E. Werner, TREOR90, Universität Stockholm, (1990).
[105] P. – E. Werner, Z. Kristallogr. 120 (1964) 375.
[106] P. – E. Werner, L. Errikson, M. Westdahl, J. Appl. Crystallogr. 18 (1985) 367.
[107] J. W. Visser, J. Appl. Crystallogr. 2 (1969) 89.
[108] A. Boultif, D. Louër, J. Appl. Crystallogr. 24 (1991) 987.
[109] STOE WinXPOW THEO, Vers. 1.18, STOE & Cie, Darmstadt, Germany (2000).
[110] WinXPOW Search, Vers. 1.22, (STOE & Cie, Darmstadt, Germany 1999).
[111] JCPDS, International Center for Diffraktion Data, Swathmore, USA (1992).
[112] Z. Otwinowski, W. Minor, Methods Enzymol. 276 (1997) 307.
[113] STOE X-RED, v1.19, STOE Data Reduction Program, STOE & Cie, Darmstadt, Germany (1999).
[114] STOE X-SHAPE, v1.06, Crystal Optimization for Numerical Absorption Correction, STOE & Cie, Darmstadt, Germany (1999).
[115] W. Herrendorf, H. Bärnighausen, HABITUS – Program for Numerical Absorption Correction, Universities of Karlsruhe/Giessen, Germany (1993/1997).
[116] G. M. Sheldrick, X-PREP, Data Preparation and Reciprocal Space Exploration, v6.12, Siemens Analytical X-Ray Instruments (1996).
[117] G. M. Sheldrick, Acta Crystallogr. A64, (2008) 112.
[118] G. M. Sheldrick, SHELXS-97, Program for the Solution of Crystal Structures, Universität Göttingen, (1997).
[119] G. M. Sheldrick, SHELXL-97, Program for Crystal Structure Refinement, Universität Göttingen, (1997).
[120] L. J. Farrugia, J. Appl. Cryst. 32 (1999) 837.
[121] A. L. Spek PLATON, A Multipurpose Crystallographic Tool, Utrecht University, Utrecht, The Netherlands (2010).
[122] K. Brandenburg, Diamond, Program for X-Ray Structure Analysis, v.3.2e, Crystal ImpactBonn (2010).

7.5 References

[123] ORIGIN 8, OriginLab Corporation, Northampton, USA (2007).

[124] F.M. Mirabella Jr. (Ed.), Principles, Theory, and Practice of Internal Reflection Spectroscopy in Internal Reflection Spectroscopy, Theory and Applications, Marcel Dekker Inc., New York (1993) 17–53.

[125] E. Krausz, AOS News 12 (1998) 21.

[126] E. Krausz, Aust. J. Chem. 46 (1993) 1041.

[127] R. Hoppe, Angew. Chem. 78 (1966) 52; Angew. Chem. Int. Ed. 5 (1966) 95.

[128] R. Hoppe, Angew. Chem. 82 (1970) 7; Angew. Chem. Int. Ed. 9 (1970) 25.

[129] R. Hübenthal, M. Serafin, R. Hoppe, MAPLE (version 4.0), Program for the Calculation of Distances, Angles, Effective Coordination Numbers, Coordination Spheres, and Lattice Energies, University of Gießen, Gießen (Germany), (1993).

[130] L. Pauling, J. Am. Chem. Soc. 69 (1947) 542.

[131] I. D. Brown, D. Altermatt, Acta Crystallogr. B41 (1985) 244.

[132] N. E. Brese, M. O`Keeffe, Acta Crystallogr. B47 (1991) 192.

[133] A. Trzesowska, R. Kruszynski, T. J. Bartczak, Acta Crystallog. B 60 (2004) 174.

[134] A.S. Wills, VALIST, v3.0.13, University College London, UK (1998-2008).

[135] R. Hoppe, Z. Kristallogr. 150 (1979) 23.

[136] R. Hoppe, S. Voigt, H. Glaum, J. Kissel, H. P. Müller, K. Bernet, J. Less-Common Met. 156 (1989) 105.

[137] E. M. Levin, R. S. Roth, J. B. Martin, Am. Mineral. 46 (1961) 1030.

[138] S. Hosokawa, Y. Tanaka, S. Iwamoto, M. Inoue, J. Mater. Sci. 43 (2008) 2276.

[139] J. H. Lin, S. Sheptyakov, Y. Wang, P. Allenspach, Chem. Mater. 16 (2004) 2418.

[140] T. A. Bither, H. S. Young, J. Solid State Chem. 6 (1973) 502.

[141] F. Bartram, E. J. Felten, Rare Earth Res. Conf. (1961) 329.

[142] W. F. Bradley, D. L. Graf, R. S. Roth, Acta Crystallogr. 20 (1966) 283.

[143] J.-Y. Henry, Mat. Res. Bull. 11 (1976) 577.
[144] R. E. Newnham, M. J. Redman, R. P. Santoro, J. Am. Ceram. Soc. 46 (1963) 253.
[145] R. S. Roth, J. L. Waring, E. M. Levin, Proc. 3rd Conf. Rare Earth Res., Clearwater, Fla., 153 (1963).
[146] P. E. D. Morgan, P. J. Carroll, F. F. Lange, Mat. Res. Bull. 12 (1977) 251.
[147] G. Chadeyron, M. El-Ghozzi, R. Mahiou, A. Arbus, J. C. Cousseins, J. Solid State Chem. 128 (1997) 261.
[148] M. Ren, J. H. Lin, Y. Dong, L. Q. Yang, M. Z. Su, L. P. You, Chem. Mater. 11 (1999) 1576.
[149] K. K. Palkina, V. G. Kuznetsov, L. A. Butman, B. F. Dzhurinskii, Acad. Sci. USSR 2 (1976) 286.
[150] H. J. Meyer, Naturwissenschaften 56 (1969) 458.
[151] H. J. Meyer, A. Skokan, Naturwissenschaften 58 (1971) 566.
[152] H. J. Meyer, Naturwissenschaften 59 (1972) 215.
[153] M. Th. Cohen-Adad, O. Aloui-Lebbou, C. Goutaudier, G. Panczer, C. Dujardin, C. Pedrini, P. Florian, D. Massiot, F. Gerard, Ch. Kappenstein, J. Solid State Chem. 154 (2000) 204.
[154] D. A. Keszler, H. Sun, Acta Crystallogr. C44 (1988) 1505.
[155] S. C. Abrahams, J. L. Bernstein, E. T. Keve, J. Appl. Crystallogr. 4 (1971) 284.
[156] H. Huppertz, Z. Naturforsch. 56b (2001) 697.
[157] G. Corbel, M. Leblanc, E. Antic-Fidancev, M. Lemaître-Blaise, J. C. Krupa, J. Alloys Comp. 287 (1999) 71.
[158] R. Böhlhoff, H. U. Bambauer, W. Hoffmann, Z. Kristallogr. Krist. 133 (1971) 386.
[159] R. Böhlhoff, H. U. Bambauer, W. Hoffmann, Naturwissenschaften 57 (1970) 129.
[160] S. Lemanceau, G. Bertrand-Chadeyron, R. Mahiou, M. El-Ghozzi, J. C. Cousseins, P. Conflant, R. N. Vannier, J. Solid State Chem. 148 (1999) 229.

[161] J. Weidelt, H. U. Bambauer, Naturwissenschaften 55 (1968) 342.
[162] G. K. Abdullaev, Kh. S. Mamedov, G. G. Dzhafarov, Azerbaidzhanskii Khimicheskii Zhurnal (1976) 117.
[163] S. Noirault, O. Joubert, M. T. Caldes, Y. Piffard, Acta Crystallogr. E 62 (2006) i228.
[164] H. Emme, H. Huppertz, Acta Crystallogr. C 60 (2004) i117.
[165] F. Goubin, Y. Montardi, P. Deniard, X. Rocquefelte, R. Brec, S. Jobic, J. Solid State Chem. 177 (2004) 89.
[166] H. Müller-Bunz, T. Nikelski, Th. Schleid, Z. Naturforsch. 58b (2003) 375.
[167] H. Huppertz, B. von der Eltz, R.-D. Hoffmann, H. Piotrowski, J. Solid State Chem. 166 (2002) 203.
[168] R.-D. Hoffmann, H. Huppertz, unpublished results.
[169] H. U. Bambauer, J. Weidelt, J.-St. Ysker, Z. Kristallogr. 130 (1969) 207.
[170] H. U. Bambauer, J. Weidelt, J.-St. Ysker, Naturwissenschaften 55 (1968) 81.
[171] J.-St. Ysker, W. Hoffmann, Naturwissenschaften 57 (1970) 129.
[172] G. K. Abdullaev, Kh. S. Mamedov, G. G. Dzhafarov, Sov. Phys. Crystallogr. 26 (1981) 473.
[173] A. Goriounova, P. Held, P. Becker, L. Bohatý, Acta Crystallogr. E 60 (2004) i131.
[174] A. Goriounova, P. Held, P. Becker, L. Bohatý, Acta Crystallogr. E 59 (2003) i83.
[175] G. Canneri, Gazz. Chim. Ital. 56 (1926) 450.
[176] P. Becker, R. Fröhlich, Cryst. Res. Technol. 43 (2008) 1240.
[177] J. Weidelt, Z. Anorg. Allg. Chem. 374 (1970) 26.
[178] I. V. Tananaev, B. F. Dzhurinskii, B. F. Chistova, Inorg. Mater. 11 (1975) 69.
[179] I. V. Tananaev, B. F. Dzhurinskii, I. M. Belyakov, Izv. Akad. Nauk SSSR Neorgan. Materialy 2 (1966) 1791.
[180] I. V. Tananaev, B. F. Dzhurinskii, B. F. Chistova, Izv. Akad. Nauk SSSR Neorgan. Materialy 11 (1975) 165.

[181] G. K. Abdullaev, Kh. S. Mamedov, G. G. Dzhafarov, Sov. Phys. Crystallogr. 20 (1975) 161.
[182] C. Sieke, T. Nikelski, Th. Schleid, Z. Anorg. Allg. Chem. 628 (2002) 819.
[183] H. Emme, C. Despotopoulou, H. Huppertz, Z. Anorg. Allg. Chem. 630 (2004) 1717.
[184] V. I. Pakhomov, G. B. Sil'nitskaya, A. V. Medvedev, B. F. Dzhurinskii, Inorg. Mater. 8 (1972) 1107.
[185] T. Nikelski, M. C. Schäfer, H. Huppertz, Th. Schleid, Z. Kristallogr. NCS 223 (2008) 177.
[186] Gmelin, *Handbook of Inorganic and Organometallic Chemistry* C11b, 8th edition, Springer Verlag, Berlin, (1991).
[187] M. Leskelä, L. Niinistö, Handbook on the Physics and Chemistry of Rare-Earth, K. A. Gschneider, Jr., L. Eyring, Ed. 9, Elsevier Science, Amsterdam (1986) p. 203.
[188] K. I. Machida, G. Y. Adachi, H. Hata, J. Shiokawa, Bull. Chem. Soc. Jpn. 54 (1981) 1052.
[189] J. H. Lin, M. Z. Su, K. Wurst, E. Schweda, J. Solid State Chem. 126 (1996) 287.
[190] S. Noirault, S. Celerier, O. Joubert, M. T. Caldes, Y. Piffard, Inorg. Chem. 46 (2007) 9961.
[191] J. H. Lin, S. Zhou, L. Q. Yang, G. Q. Yao, M. Z. Su, L. P. You, J. Solid State Chem. 134 (1997) 158.
[192] J. H. Lin, L. P. You, G. X. Lu, L. Q. Yang, M. Z. Su, J. Mater. Chem. 8 (1998) 1051.
[193] B. F. Dzhurinskii, A. B. Ilyukhin, Kristallografiya 47 (2002) 442.
[194] L. Li, P. Lu, Y. Wang, X. Jin, G. Li, Y. Wang, L. You, J. Lin, Chem. Mater. 14 (2002) 4963.
[195] L. Li, X. Jin, G. Li, Y. Wang, F. Liao, G. Yao, J. Lin, Chem. Mater. 15 (2003) 2253.
[196] T. Nikelski, M. C. Schäfer, Th. Schleid, Z. Anorg. Allg. Chem. 634 (2008) 49.

[197] H. Emme, H. Huppertz, Acta Crystallogr. C61 (2005) i29.
[198] H. Huppertz, S. Altmannshofer, G. Heymann, J. Solid State Chem. 170 (2003) 320.
[199] H. Emme, H. Huppertz, Acta Crystallogr. C61 (2005) i23.
[200] S. C. Neumair, H. Huppertz, Z. Naturforsch. 64b (2009) 1339.
[201] A. Haberer, G. Heymann, H. Huppertz, J. Solid State Chem. 180 (2007) 1595.
[202] A. Haberer, G. Heymann, H. Huppertz, Z. Anorg. Allg. Chem. 632 (2006) 2079.
[203] C. A. Gressler, J. E. Shelby, J. Appl. Phys. 64 (1988) 4450.
[204] E. V. Grishchuk, N. P. Efryushina, V. P. Dotsenko, E. R. Gubanova, Inorg. Mater. 34 (1998) 520.
[205] J. E. Shelby, L. D. Baker, Phys. Chem. Glasses 39 (1998) 23.
[206] L. R. P. Kassab, L. C. Courrol, A. S. Morais, S. H. Tatumi, N. U. Wetter, L. Gomes, J. Opt. Soc. Am. B: Opt. Phys. 19 (2002) 2921.
[207] C. K. Jayasankar, V. Venkatramu, P. Babu, Th. Troster, W. Sievers, G. Wortmann, W. B. Holzapfel, J. Appl. Phys. 97 (2005) 093523/1.
[208] W. A. Pisarski, J. Pisarska, M. Maczka,W. Ryba-Romanowski, J. Mol. Struct. 792-793 (2006) 207.
[209] A. V. Ravi Kumar, B. Apparao, N. Veeraiah, Bull. Mater. Sci. 21 (1998) 341.
[210] G. Su, H. Toratani, Jpn. Kokai Tokkyo Koho (1997) 6.
[211] T. Suzuki, M. Hirano, H. Hosono, J. Appl. Phys. 91 (2002) 4149.
[212] C. Chen, Z. Xu, D. Deng, J. Zhang, G. K. L. Wong, B. Wu, N. Ye, D. Tang, Appl. Phys. Lett. 68 (1996) 2930.
[213] T. A. Bither, H. S. Young, J. Solid State Chem. 10 (1974) 302.
[214] T. Alekel, D. A. Keszler, J. Solid State Chem. 106 (1993) 310.
[215] G. Cakmak, J. Nuss, M. Jansen, Z. Anorg. Allg. Chem. 635 (2009) 631.
[216] D. M. Chackraburtty, Acta Crystallogr. 10 (1957) 199.
[217] L. Li, G. Li, Y.Wangand F. Liao, J. Lin, Chem. Mater. 17 (2005) 4174.
[218] G. Corbel, R. Retoux, M. Leblanc, J. Solid State Chem. 139 (1998) 52.

[219] H. Müller-Bunz, Th. Schleid, Z. Anorg. Allg. Chem. 628 (2002) 2750.
[220] I. A. Baidina, V. V. Bakakin, N. V. Podberezskaya, V. I. Alekseev, L. R. Batsanova, V. S. Pavlyuchenko, Zh. Strukt. Khim. 19 (1978) 125.
[221] A. A. Borovkin, L. V. Nikishova, Kristallografiya 20 (1975) 740.
[222] L. V. Nikishova, A. A. Brovkin, E. A. Kuz'min, S. L. Pyatkin, Zh. Strukt. Khim. 12 (1971) 183.
[223] A. dal Negro, C. Tadini, Acta Crystallogr. 3 (1950) 208.
[224] J. G. Fletcher, F. P. Glasser, R. A. Howie, Acta Crystallogr. C 47 (1991) 12.
[225] A. A. Brovkin, L. V. Nikishova, Sov. Phys. Crystallogr. 20 (1975) 252.
[226] T. Alekel, D. A. Keszler, Inorg. Chem. 32 (1993) 101.
[227] K. Kazmierczak, H. A. Höppe, Eur. J. Inorg. Chem. (2010) 2678.
[228] C. Rodellas, S. Garcia Blanco, A. Vegas, Z. Kristallogr. 165 (1983) 255.
[229] J. F. Sawyer, G. J. Schrobilgen, Acta Crystallogr. B 38 (1982) 1561.
[230] H. Park, J. Barbier, J. Solid State Chem. 155 (2000) 354.
[231] L. Mei, Y. Wang, C. Chen, Mater. Res. Bull. 29 (1994) 81.
[232] L. Mei, X. Huang, Y. Wang, Q. Wu, B. Wu, C. Chen, Z. Kristallogr. 210 (1995) 93.
[233] I. A. Baidina, V. V. Bakakin, L. P. Bacanova, N. A. Pal'chik, N. V. Podberezskaya, L. P. Solov'eva, Zh. Strukt. Khim. 16 (1975) 1050.
[234] P. C. Burns, J. D. Grice, F. C. Hawthorne, Can. Mineral. 33 (1995) 1131.
[235] F. C. Hawthorne, P. C. Burns, J. D. Grice, The crystal chemistry of boron. In: Boron, Mineralogy, Petrology and Geochemistry. (Eds. E. S. Grew, L. M. Anovitz), Reviews in Mineralogy 33, Mineralogical Society of America, Washington (1996).
[236] J. D. Grice, P. C. Burns, F. C. Hawthorne, Can. Mineral. 37 (1999) 731.
[237] P.C. Burns, J. D. Grice, F. C. Hawthorne, Can. Mineral. 33 (1995) 1131.
[238] T. J. White, D. ZhiLi, Acta Crystallogr. B59 (2003) 1.
[239] C. Wickleder, I. Hartenbach, P. Lauxmann, Th. Schleid, Z. Anorg. Allg. Chem. 628 (2002) 1602.
[240] E. Zobetz, Z. Kristallogr. 191 (1990) 45.

7.5 References

[241] D. Yu. Pushcharovsky, S. Merlino, O. Ferro, S. A. Vinogradova, O. V. Dimitrova, J. Alloys Compd. 306 (2000) 163.
[242] J. S. Knyrim, J. Friedrichs, S. Neumair, F. Roessner, Y. Floredo, S. Jakob, D. Johrendt, R. Glaum, H. Huppertz, Solid State Sci. 10 (2008) 168.
[243] R. Ternane, G. Panczer, M. Th. Cohen-Adad, C. Goutaudier, G. Boulon, N. Kbir-Ariguib, M. Trabelsi-Ayedi, Opt. Mater. 16 (2001) 291.
[244] E. Kendrick, A. Orera, P. R. Slater, J. Mater. Chem. 19 (2009) 7955.
[245] E. Béchade, O. Masson, T. Iwata, I. Julien, K. Fukuda, P. Thomas, E. Champion, Chem. Mater. 21 (2009) 2508.
[246] G. Blasse, G. P. M. van den Heuvel, Phy. Status Solidi 19 (1973) 111.
[247] J. S. Knyrim, H. Huppertz, J. Solid State Chem. 181 (2008) 2092.
[248] S. C. Neumair, J. S. Knyrim, R. Glaum, H. Huppertz, Z. Anorg. Allg. Chem. 635 (2009) 2002.
[249] J. P. Attfield, A. M. T. Bell, L. M. Rodriguez-Martinez, J. M. Greneche, R. Retoux, M. Leblanc, R. J. Cernik, J. F. Clarke, D. A. Perkins, J. Mater. Chem. 9 (1999) 205.
[250] R. T. Downs, The RRUFF Project: an integrated study of the chemistry, crystallography, Raman and infrared spectroscopy of minerals. Program and Abstracts of the 19th General Meeting of the International Mineralogical Association in Kobe, Japan (2006); O03-13. Kobe, Japan.
[251] A. Baykal, G. Gürbüz, M. Kizilyalli, R. Kniep, Key Eng. Mater. 264 (2004) 2017.
[252] E. Antic-Fidancev, G. Corbel, N. Mercier, M. Leblanc, J. Solid State Chem. 153 (2000) 270.
[253] A. Vegas, Acta Crystallogr. B33 (1977) 3607.
[254] R. Diehl, Solid State Commun. 17 (1975) 743.
[255] N. E. Brese, M. O'Keeffe, Acta Crystallogr. B47 (1991) 192.
[256] R. Hoppe, Angew. Chem. 82 (1970) 7; Angew. Chem. Int. Ed. 9 (1970) 25.
[257] Z. K. Heiba, L. Arda, Y. S. Hascicek, J.Appl.Crystallogr. 38 (2005) 306.
[258] S. V. Berger, Acta Crystallogr. 5 (1952) 389.

[259] L. S. Garashina, B. P. Sobolev, V. B. Aleksandrov, Y. S. Vishnyakov, Sov. Phys. Crystallogr. (Engl. Transl.) 25 (1980) 171.
[260] G. Heymann, K. Beyer, H. Huppertz, Z. Naturforsch. B59 (2004) 1200.
[261] J. P. Laperches, P. Tarte, Spectrochim. Acta. 22 (1966) 1201.
[262] A. Haberer, R. Kaindl, H. Huppertz, J. Solid State Chem. 183 (2010) 471.
[263] A. Haberer, R. Kaindl, H. Huppertz, Solid State Sci. 12 (2010) 515.
[264] A. Knappwost, Naturwiss. 46 (1959) 555.
[265] V. P. Orlovskii, S. P. Ionov, T. V. Belyaevskaya, S. M. Barinov, Inorg. Mater. 38 (2002) 182.
[266] L. Jun, X. Shuping, G. Shiyang, Spectrochim. Acta A51 (1995) 519.
[267] C. Zhang, Y. H. Wang, X. Guo, J. Lumin. 122–123 (2007) 980.
[268] G. Padmaja, P. Kistaiah, J. Phys. Chem. A113 (2009) 2397.
[269] S. Blundell, Magnetism in Condensed Matter, Oxford master Series in Condensed Matter Physics, Oxford University Press, New York, (2001).
[270] L. Beaury, G. Calvarin, J. Derouet, J. Hölsä, E. Säilynoja, J. Alloys Compd. 275-277 (1998) 646.
[271] M. Orth, R.-D. Hoffmann, R. Pöttgen, W. Schnick, Chem. Eur. J. 7 (2001) 2791.

Die VDM Verlagsservicegesellschaft sucht für wissenschaftliche Verlage abgeschlossene und herausragende

Dissertationen, Habilitationen, Diplomarbeiten, Master Theses, Magisterarbeiten usw.

für die kostenlose Publikation als Fachbuch.

Sie verfügen über eine Arbeit, die hohen inhaltlichen und formalen Ansprüchen genügt, und haben Interesse an einer honorarvergüteten Publikation?

Dann senden Sie bitte erste Informationen über sich und Ihre Arbeit per Email an *info@vdm-vsg.de*.

Sie erhalten kurzfristig unser Feedback!

VDM Verlagsservicegesellschaft mbH
Dudweiler Landstr. 99   Telefon +49 681 3720 174
D - 66123 Saarbrücken   Fax   +49 681 3720 1749
www.vdm-vsg.de

Die VDM Verlagsservicegesellschaft mbH vertritt

Printed by Books on Demand GmbH, Norderstedt / Germany